Quantum Machine Learning: An Applied Approach

The Theory and Application of Quantum Machine Learning in Science and Industry

Santanu Ganguly

Apress®

Quantum Machine Learning: An Applied Approach: The Theory and Application of Quantum Machine Learning in Science and Industry

Santanu Ganguly
Ashford, UK

ISBN-13 (pbk): 978-1-4842-7097-4   ISBN-13 (electronic): 978-1-4842-7098-1
https://doi.org/10.1007/978-1-4842-7098-1

Managing Director, Apress Media LLC: Welmoed Spahr
Acquisitions Editor: Aditee Mirashi
Development Editor: Matthew Moodie
Coordinating Editor: Aditee Mirashi

Cover designed by eStudioCalamar

Cover image designed by Freepik (www.freepik.com)

Distributed to the book trade worldwide by Springer Science+Business Media New York, 1 New York Plaza, Suite 4600, New York, NY 10004-1562, USA. Phone 1-800-SPRINGER, fax (201) 348-4505, e-mail orders-ny@ springer-sbm.com, or visit www.springeronline.com. Apress Media, LLC is a California LLC and the sole member (owner) is Springer Science + Business Media Finance Inc (SSBM Finance Inc). SSBM Finance Inc is a **Delaware** corporation.

For information on translations, please e-mail booktranslations@springernature.com; for reprint, paperback, or audio rights, please e-mail bookpermissions@springernature.com.

Apress titles may be purchased in bulk for academic, corporate, or promotional use. eBook versions and licenses are also available for most titles. For more information, reference our Print and eBook Bulk Sales web page at http://www.apress.com/bulk-sales.

Any source code or other supplementary material referenced by the author in this book is available to readers on GitHub via the book's product page, located at www.apress.com/978-1-4842-7097-4. For more detailed information, please visit http://www.apress.com/source-code.

Printed on acid-free paper

I dedicate this book to

Deben and Rishika, for being the lights of my life

Sigrid, for being an unwavering source of support for the last 17 years

My late mother, Subhra Ganguly, my now late father, Samir Kumar Ganguly, my late mother-in-law, Jeanne Van de Velde, and my father-in-law, Jean Van Lierde

Table of Contents

About the Author

Santanu Ganguly has worked in quantum technologies, cloud computing, data networking, and security (research, design, and delivery) for more than 22 years. He has worked in Switzerland for ISPs and is currently based in the United Kingdom, where he has held senior-level positions at various Silicon Valley vendors. He has two postgraduate degrees (one in mathematics and another in observational astrophysics) and research experience and publications in quantum technologies, silicon photonics and laser spectroscopy. He is currently leading global projects related to quantum communication and machine learning, among other technologies.

About the Technical Reviewer

Tina Sebastian is the co-founder of Quacoon, a quantum AI–based platform that provides collaborative supply chain intelligence for building and managing resilient and sustainable supply chains, and solving product shortages, congestion, and other disruptions due to unexpected events. With a computer engineering degree from CUSAT and an MBA from the University of Massachusetts, Amherst, Tina has 12 years of software engineering and consulting experience. She is certified in quantum computing and cybersecurity from the Massachusetts Institute of Technology (MIT) and is a Certified Scrum Master (CSM).

Acknowledgments

I have studied, lived, and worked in four different countries on three different continents. I have been fortunate enough to see much of the wonders of the world that I had aspired to see growing up wild in the Himalayas as a boy in a middle-class Indian family: from sperm whales along the US Pacific Coast to the Kinkaku-ji Temple in Kyoto, Japan; from the aurora borealis in Iceland to the magnificent Nile, pyramids, and temples of Egypt; from rhinos and tigers in Kaziranga, India, to following lions and the wildebeest migration line in Serengeti, Tanzania. In a life that has experienced the joy of diversity as much as mine, it is not unusual to have come across many amazing people whose profound influence was instrumental in getting me to plunge into the madness of writing a book.

This was my first (book) authoring experience, which, albeit proving to be a challenge far beyond my expectations, was a joy thanks to the opportunity very kindly given to me by **Aditee Mirashi** of Apress. Without her help, patience, guidance, and constant shepherding, this project would not have been possible. I would also like to express my gratitude to **Matthew Moodie**, development editor, for his stellar support.

I extend my very deep gratitude to **Tina Anne Sebastian**, co-founder of quantum computing start-up Quacoon and the technical reviewer of this book, who selflessly devoted her valuable time when she was also setting up her start-up company. Her conscientious due diligence and constructive criticism have been invaluable in keeping me honest.

I thank my wife, **Sigrid**, for always supporting me through the nine odd months of frequent periods of solitude that this project demanded at some very trying times during the pandemic-induced serial lockdowns in the United Kingdom. This extends to my eight-year-old son, **Deben**, for allowing his daddy some peaceful periods to work on the book between incessant demands to play cricket. Deben remains convinced that no one will ever want to read "that boring book" written by his daddy!

I thank my dear father, **Samir Kumar Ganguly**, who, as I write this, remains in a critical health condition in Calcutta, without me being able to be at his side due to Covid-related travel restrictions. He has been an inspiration all my life. He always wished for me to see the world and infused an addiction to adventures in my DNA very early in my life. His kind, open-minded guidance gave me the courage to get rid of traditional baggage and opened my mind to the joy of pursuing the unknown.

ACKNOWLEDGMENTS

I thank my childhood friend **Saumyadip Dasgupta, MD**, of Carmel, Indiana (and his brilliant family) for his lifelong friendship and unselfish support. My appreciation knows no bounds.

In addition, I would like to express my very special thanks of gratitude to the following people.

- **Dr. Markus Hoffmann** of Google Research for his kind, unconditional help and guidance about using Google's open source libraries in this work

- **D-Wave** for reviewing the chapters related to their technology and for permitting me to use their libraries and some material

- **Dr. Maria Schuld** of Xanadu Quantum Technologies Inc. and the University of KwaZulu-Natal for being kind enough to respond to my queries—her pioneering work in this field has been abundantly referred to in this book.

- **Rafal Janik** and **Sepehr Taghavi** of Xanadu Quantum Technologies Inc. for being kind enough to spare some time and guidance on Xanadu and PennyLane

- **Dr. Louis "Sam" Samuel**, Director of Emerging Technologies and Incubation (ET&I) at Cisco Systems, for his leadership and guidance in everything related to "quantum tech" and research

- **Jeff Apcar**, Distinguished Engineer, Customer Experience (CX) at Cisco Systems, for his leadership in quantum-related publications and for allowing me to use his very creative diagrams in Chapter 1

- **Dr. Joel Gottlieb** of Quantum Computing Inc. (formerly of D-Wave) for his kind offer to review the book and for his patience with my outlandish serial requests since that fateful day when I met him at a D-Wave training

Introduction

I shall be telling this with a sigh

Somewhere ages and ages hence:

Two roads diverged in a wood, and I —

I took the one less traveled by,

And that has made all the difference.

—Robert Frost, 1916

Quantum machine learning (QML) is a cross-disciplinary subject made up of two of the most exciting research areas: quantum computing and classical machine learning. Quantum computing and machine learning are arguably two of the "hottest" topics in science and industry. Quantum computation and quantum information have appeared in various areas of computer science, such as information theory, cryptography, emotion representation, and image processing because tasks that appear inefficient on classical computers can be achieved by exploiting the power of quantum computation. Novel methods for speeding up certain tasks and interdisciplinary research between computer science and several other scientific fields have attracted deep scientific interest to this emerging field.

The growing confidence in the promises harbored by quantum computing and machine learning fields has given rise to many start-ups boosted by the quantum promise. Besides the well-known outfits, such as Google, IBM, Rigetti, Xanadu, and D-Wave, to name a few, there are quite a few young, exciting, and enthusiastic endeavors taking place around the world.

This book is the result of my deep interest and involvement in quantum computing and machine learning. There are two notable, ground-breaking books on the subject: Dr. Peter Wittek's *Quantum Machine Learning: What Quantum Computing Means to Data Mining* (Academic Press, 2014) and *Supervised Learning with Quantum Computers*, a monumental volume, by Dr. Maria Schuld and Prof. Francesco Petruccione (Springer, 2018). In parallel to these two great works, there has been an explosion of research publications in various areas of quantum science. Besides diverse research groups in academia, great industrial research outfits from Google, IBM, Rigetti, Xanadu, and D-Wave, along with national laboratories, such as Brookhaven and Los Alamos, have made profound impacts on the field of quantum computing and machine learning.

In my humble journey into the world of quantum machine learning, I noticed a few rudimentary facts:

- The great works that I mentioned are largely meant for an expert-level audience or an audience with insights into quantum physics and machine learning.

- These works do not address the coding aspect of theories. In other words, they appear to assume that readers are familiar with the context of the material and the coding (if any) that was used.

- The efficiencies of several existing quantum machine learning algorithms are dependent on the type of hardware they are run on; in other words, what works great and gives amazing results on a quantum annealing platform may not do so on a gate-based quantum system and vice versa.

These three realizations prompted me to put together this book. I attempt to address the most important aspects of quantum computing and machine learning algorithms for newcomers to this subject. This involves an effort to bridge the gap between the theoretical and coding side of things, utilizing libraries such as Google's Cirq and TensorFlow Quantum, D-Wave's dimod and qbsolve, and Xanadu's PennyLane, Qiskit, pyQuil, and QVM. I hope that this approach of complementing the theory with code helps readers with a more applied mindset find a quicker and less laborious path to advance further in the field.

The first chapter of this book addresses the basics of quantum mechanics. The next two chapters cover aspects and techniques of classical machine learning. The subsequent chapters follow a systematic, structured deep dive into various quantum machine learning algorithms, quantum optimization, applications of advanced QML algorithms, inference, Hamiltonian simulation, and so forth, finishing with advanced and up-to-date research areas, such as quantum walks, QBoost, and more. These subjects complement hands-on exercises from open source libraries regularly used today in industry and research, such as Qiskit, Rigetti's Forest, D-Wave's dOcean, Google's Cirq and TensorFlow Quantum, and Xanadu's PennyLane, accompanied by guided implementation instructions. Wherever applicable, the book also states various options of accessing quantum computing and machine learning ecosystems as may be relevant to specific algorithms.

The most important mathematical fundamentals have been addressed in Appendix A. The book offers a hands-on approach to QML using updated libraries and algorithms in this emerging field accompanied by a constantly updated website for the book, which provides all the code examples used. My deepest wish is to make a difference to the readers who are either quantum physicists without much insight into machine learning or are machine learning professionals who aspire to get into the "quantum" side of things.

CHAPTER 1

Rise of the Quantum Machines: Fundamentals

The small wisdom is like water in a glass: clear, transparent, pure. The great wisdom is like the water in the sea: dark, mysterious, impenetrable.

—Rabindranath Tagore

Los Alamos Laboratory was celebrating its fortieth anniversary on April 14, 1983. Physics Laureate Richard P. Feynman stated in a lecture titled "Tiny Computers Obeying Quantum-Mechanical Laws" that computing based on classical logic could not easily and efficiently process calculations describing quantum phenomena. He offered his vision of computing that could operate in a quantum manner. Prior to making these historical comments, Feynman had always been a champion of thoughts leading toward computers leveraging laws of quantum mechanics when he stated, "There's plenty of room at the bottom" and, "Nature isn't classical. If you want to make a simulation of nature, you'd better make it quantum mechanical," which means to properly simulate quantum systems, you must use a quantum computer.

Quantum computing grew in the context of quantum simulation and efforts to build computing devices that follow quantum mechanical laws. Caltech released Prof. John Preskill's lecture notes on quantum computing for public access in 1999 [1]. Subsequently, interest grew in this exciting field which promises to be the singularity where the laws of physics meet the practices of applied technology.

There are various definitions in literature for a *quantum computer*. For the general purpose of this book, a quantum computer is a computing device whose computational processes can be explicitly described with the laws of quantum theory. Why do we expect a quantum computer to be more advantageous over a classical one? This is because of the inherent nature of the way a classical computer is built. Doubling the

1

© Santanu Ganguly 2021
S. Ganguly, *Quantum Machine Learning: An Applied Approach*, https://doi.org/10.1007/978-1-4842-7098-1_1

power of a classical computer requires about double the number of electrical circuitry and associated gates working on a problem. In contrast, the power of a quantum computer can be approximately doubled every time only one qubit is added.

How This Book Is Organized

The first three chapters offer insights into the combination of the science of quantum mechanics and the techniques of classical machine learning, which is poised to become the singularity where powers of classical information technology meet the power of physics. The subsequent chapters take a systematic, structured look at various quantum machine learning algorithms, quantum optimization, and applications of advanced QML algorithms (e.g., HHL, quantum annealing, quantum phase estimation, optimization, quantum neural networks). The book ends with discussions on advanced research areas, such as quantum walks, Hamiltonian simulation, and QBoost.

Everything is complemented with hands-on exercises from open source libraries regularly used today in industry and research, such as Rigetti's Forest and QVM, Google's TensorFlow Quantum (released in March 2020), and Xanadu's PennyLane, accompanied by implementation instructions. Wherever applicable, the various options for accessing quantum computing and machine learning ecosystems relevant to specific algorithms are discussed. This book is accompanied by all the code examples used in it for the exercises. The codes will be updated regularly to keep up with progress in the field and can be downloaded from the following site: `https://www.apress.com/de/book/9781484270974`.

The Essentials of Quantum Computing

Modern-day computers are physical devices built based on electronic circuits which process information. Algorithms, which are the computer programs, or software, are our conduit to manipulate these circuits to execute our desired computations and obtain their outputs. The physical processes of these computing devices involve microscopic particles such as electrons, atoms, and molecules, which can be described with a macroscopic, classical theory of the electric properties of the circuits.

If very small-scale systems such as photons, biological DNA, electrons, and atoms are directly used to process the same information, that would involve using a set of specific mathematical structures, different from the ones in our existing classical computing case. This mathematical framework, called *quantum theory*, is needed to capture the fact that nature behaves very differently from what our intuition teaches us. Quantum theory, since its development at the beginning of the twentieth century, has generally been the most comprehensive description of small-scale physics.

When considering the foundation of digital computation and communication, one often thinks of a binary digit or a bit. A bit can be a 0 or 1 at any given time, the same way that a coin on our desk could either be showing heads or tails. Quantum computers use *quantum bits*, or *qubits*, which can be either a 0 or a 1 or in both states (called a *superposition*) at any given time. It is like flipping a coin; while the coin is spinning in the air, it is definitely not showing heads or tails; we can think of it as being in a *superposition* (elaborated later in this chapter and further in Chapter 4).

This allows certain operations to run in parallel rather than sequentially, meaning they exponentially reduce the number of operations required in certain algorithms. So, they're not universally faster (i.e., each operation itself still takes the same amount of time to complete), which means browsing the Internet, writing a Word document, or streaming a video wouldn't necessarily be faster on a quantum computer (QC). It is why a quantum computer cannot replace standard classical computing devices for day-to-day use.

What exactly is the advantage of a quantum computer? One example is useful in the world of molecular simulations. To simulate a penicillin molecule with 42 atoms, the exponentially large parameter space of electron configurations would require 10^{86} states, which are more states than the number of atoms in the Universe; the quantum systems can do this with only 286 qubits. Simulation of a caffeine molecule would require 10^{48} bits in a classical system, which is about 5 % − 10% of atoms on planet Earth. A quantum system could execute the same task with only 160 qubits.

As another example, let's say we wanted to do something "simple" on a 64-bit computer: keep adding the number 1 until the 64-bit register overflows in a "fast" classical computer. You can reasonably estimate that a modern-day "fast" computer can execute 2 billion instances of +1 additions per second; but, at this rate, to achieve the task of overflowing the register, the computer is crunching away for about 400 years, since it is adding bit by bit. However, a 64-qubit quantum computer can have all those numbers 0 to 2^{64} *all at the same time*! This is shown in Figure 1-1. Every classical computer ever built can be described fully by principles of quantum physics, but the reverse is not true.

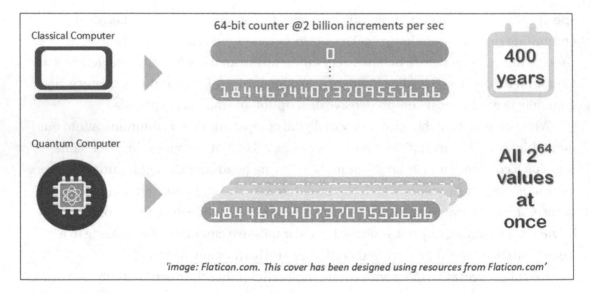

Figure 1-1. *Classical vs. quantum computing*

QC has been adopted in many companies and federal outfits. Google, IBM, Xanadu, Rigetti, D-Wave, and many others compete in a quest for "quantum supremacy." Airbus famously opened a competition on quantum computing to address long-standing computational challenges in the aerospace industry. D-Wave's 2000Q-based quantum annealing-based computers were installed at NASA,[1] Google, and USRA (Universities Space Research Association).

The era of QC is upon us, perhaps a lot quicker than many people envisaged. With the dawn of this era, as governments around the world continue to invest in this enticing futuristic technology, the broader industrial bodies have also realized a rising challenge: that of a qualified and trained workforce. Fundamentals of QC are cross-functional to the extreme, requiring physics, mathematics, and coding. Growth in this area requires investment in personnel development and commitments from Universities and academic bodies to open more and more relevant course works for interested students.

This book draws inspiration, excitement, algorithms, and frameworks from existing machine learning algorithms and those currently being worked on by researchers. The appendices cover some mathematical background as a reference, such as tensor products and Fourier transforms.

---

[1] www.nas.nasa.gov/projects/quantum.html

The Qubit

A quantum bit or *qubit* is a two-level quantum mechanical system and is represented by quantum states. Any quantum particle that can be measured in two discrete states could be used as a qubit; for example, a trapped ion, such as a single calcium ion (confined to an optical cavity using electromagnetic fields), polarized photons, and electron spin.

A qubit is analogous to a classical bit in a classical computing system. The basic difference between the two is that in a classical system, a bit can take up values of either 0 or 1 as opposed to a qubit, which can take up a whole set of values between $|0\rangle$ and $|1\rangle$ representing the *superposition* of states as depicted in Figure 1-4. In quantum mechanics, it is a general convention to denote an element ψ of an abstract complex vector spaces as a *ket* $|\psi\rangle$ using vertical bars and angular brackets and refer to them as *kets* rather than as vectors.

State of a Qubit

The state of a classical bit is a number (00 or 11), the state of a qubit is a vector in a two-dimensional vector space. This vector space is also known as *state space*. The *state* of a quantum system is given by a vector $|\psi(t)\rangle$ that contains all possible information about the system at any given time. The vector $|\psi(t)\rangle$ is a member of the Hilbert Space $\mathcal{H}$ (described later) and can be a time variable (i.e., may change with time). In quantum mechanics, it is typical to normalize the states (i.e., to find a way to set the inner product $\langle\psi|\psi\rangle = 1$). Figure 1-2 shows a two-dimensional representation of the quantum state of a qubit.

The two top diagrams in Figure 1-2 show the position of the vector if the basis state is $|0\rangle$ and $|1\rangle$. Please note that the top part of the vector is the position on the conventional X axis, or $|0\rangle$ axis, and the bottom is the position on the conventional Y axis, or $|1\rangle$. The top two diagrams show the state $|0\rangle$ with a state vector of $|0\rangle$ and $|1\rangle$. There is a probability of 1 (or 100%) that the qubit reads an output of $|0\rangle$ and vice versa for the $|1\rangle$.

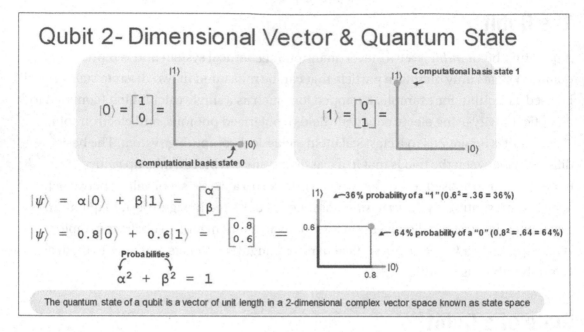

Figure 1-2. *Two-dimensional quantum state for quantum computing*

If we suppose that a qubit state is given by $|\psi\rangle$, then the *squares* of α and β are the individual probabilities that $|\psi\rangle$ may be found in state $|0\rangle$ and $|1\rangle$ respectively and the sums of the squares of α and β is 1, where α and β are known as the *amplitudes* of the states and generally speaking can be complex numbers. In other words, there is a 100% chance of an outcome that $|\psi\rangle$ is either a $|0\rangle$ or a $|1\rangle$ and the sum of the probabilities of the outcome is 1. The bottom diagram of Figure 1-2 shows an example of manipulating the quantum state by 0.8 for $|0\rangle$ and 0.6 for $|1\rangle$ which translates into 36% and 64% probability, respectively.

The Bloch Sphere

The Bloch sphere is a three-dimensional geometric representation of qubit state space as points on the surface of an imaginary unit sphere. This is one of two ways of representing a qubit. The other way represents the qubit in Dirac notation. Simply put, the Bloch representation takes the two-dimensional (2-D) graph representation and depicts it in a 3-D representation with the state of a qubit represented by a point on the sphere as depicted in Figure 1-3.

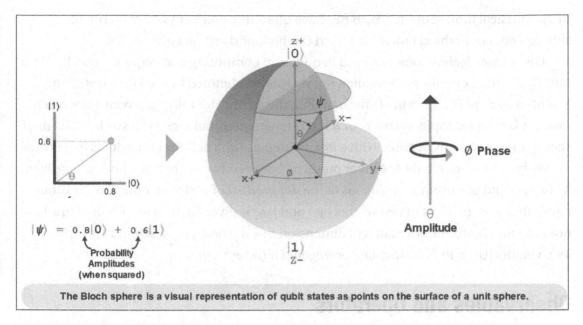

The Bloch sphere is a visual representation of qubit states as points on the surface of a unit sphere.

Figure 1-3. *The Bloch sphere*

The angle ϕ of the Bloch[2] sphere in Figure 1-3 is called the *azimuthal* angle and measured from the positive X axis to the projection of state $|\psi\rangle$ onto the $x - y$ plane. Angle θ is called the *polar* angle and is measured from the positive Z axis to the Bloch vector representing the state $|\psi\rangle$.

The Bloch sphere allows for negative and complex numbers in the probabilities. Operations on single qubits that are commonly used in quantum information processing can be completely described within the Bloch sphere description. The Bloch sphere is particularly useful to explain quantum gates. The probabilities used (α and β) in our previous examples and relevant discussions can be changed into representing amplitudes or latitudes on the sphere where the state is positioned. States in a quantum computation can be represented as a vector that starts at the origin and ends on the surface of the unit Bloch sphere. By applying unitary operations (described later) on the state vector $|\psi\rangle$, it can be rotated and moved around the surface of the sphere in Figure 1-3. As per convention, the two poles of the sphere are taken as $|0\rangle$ on top of the sphere at z_+ (or at the north pole) and $|1\rangle$ at the bottom of the sphere at z_- (or at the south pole).

---

[2]A great Bloch sphere simulation can be found on the University of St. Andrews' site at www. st-andrews.ac.uk/physics/quvis/simulations_html5/sims/blochsphere/blochsphere.html

In the classical limit, a qubit, which can have quantum states anywhere on the Bloch sphere, reduces to the classical bit, which can be found only at either poles.

The *linear algebra* notations used in quantum computing may require a quick introduction. In quantum mechanics, the vectors are denoted by the Dirac notation, invented by Paul Dirac. In the Dirac notation, the symbol identifying a vector is written inside a *ket*, for example, vector $\vec{a}$ or $\boldsymbol{a}$ is written in quantum mechanics as $|a\rangle$. The dual vector for a vector a is denoted with a *bra*, written as $\langle a|$. Their inner products are written as *bra-kets*. In other words, the inner products between two vectors $|\psi_1\rangle$ and $|\psi_2\rangle$ is given by $\langle \psi_1|\psi_2\rangle$ and the result is analogous to the *dot product* of vector algebra. On the other hand, the outer product of two vectors $|\psi_1\rangle$ and $|\psi_2\rangle$ is given $|\psi_1\rangle\langle\psi_2|$ in the Dirac bra-ket notation and it produces a matrix of dimension $m \times n$ where $|\psi_1\rangle \in \mathbb{R}^m$ and $|\psi_2\rangle \in \mathbb{R}^n$. Two vectors $|\psi_1\rangle$ and $|\psi_2\rangle$ are called *orthogonal* if $\langle \psi_1|\psi_2\rangle = 0$.

Observables and Operators

In quantum computing, for any variable x that changes dynamically and can be measured physically as a quantity, there is a corresponding operator $\mathcal{O}$. The operator $\mathcal{O}$ is Hermitian in nature. It is composed of a basis of orthogonal eigenvectors in the vector space. The eigenvector definitions are covered in the appendices.

The Hilbert Space

From a mathematical point of view, the vector $|\psi\rangle$ is a complete vector on which an inner product is defined and is an element of a class of vector spaces called *Hilbert spaces* denoted by $\mathcal{H}$. The vector spaces referred to in this book include real and complex number spaces and are finite-dimensional, which simplifies the mathematics needed for the treatment of QC. Since $\mathcal{H}$ is finite-dimensional, a basis can be chosen to represent vectors in this basis as finite column vectors and represent operators with finite matrices. The Hilbert spaces of interest for quantum computing typically has dimension 2^n for some positive integer n. The mathematical representation of the state of a *qubit* lives in a Hilbert space of dimension 2. In principle, a state vector of a quantum system could be an element of a Hilbert space of arbitrary dimensions. In this book, let us consider *finite-dimensional* Hilbert spaces.

While performing computations in quantum systems, it is often convenient to fix a basis and refer to it as the *computational basis*. On this basis, we label the 2^n basis vectors in the Dirac notation using the binary strings of length n as follows.

$$|00...00\rangle_n, |00...01\rangle_n, ..., |11...10\rangle_n, |11...11\rangle_n \qquad 1.1$$

In equation 1.1, the n refers to the length of the binary strings.

The Hilbert space $\mathcal{H}$ consists of a set of vectors ψ, ϕ, φ... and a set of scalars such as $a, b, c, ..$, which exhibit and follow the four properties.

    a. $\mathcal{H}$ **is a linear space.** A linear vector space is made up of two sets of elements and two algebraic rules.

        I.   A set of vectors ψ, ϕ, φ... and a set of scalars such as $a, b, c, ..$

        II.   A rule for vector addition and a rule for scalar multiplication.

    b.   **The scalar product in $\mathcal{H}$ is strictly positive.**

    c. $\mathcal{H}$ **is separable.**

    d. $\mathcal{H}$ **is complete**

If you are interested in further discussion of Hilbert space, try *Principles of Quantum Mechanics* by R. Shankar (Plenum, 2010) [2].

Two of the most important fundamentals of quantum computing and quantum machine learning are *superposition* and *entanglement*. Let's take a quick look at the fundamentals of both and then expand on that as we go through various examples in the book where each of the properties are leveraged.

Generally, an operator is said to be linear if it satisfies the following relation.

$$\hat{P}\left(\alpha f_1 + \beta f_2\right) = \alpha\left(\hat{P}f_1\right) + \beta\left(\hat{P}f_2\right) \qquad 1.2$$

Measurements

Measurement in quantum mechanics is challenging in comparison with classical physics. In the classical world, measurement is simple and assumed to not affect the item whose parameter or characteristic is being measured; for example, we are free to measure the weight of a ball and proceed to measure the circumference of the ball, confident that another measurement of the weight of the ball produces the same result.

In the quantum world, any measurement of a given quantum state causes a collapse of the corresponding wave function. As per postulates of quantum mechanics, a physical variable must have real *expectation values* and *eigenvalues*. This implies that the operators representing physical variables have special properties. It can be shown, by computing the complex conjugate of the expectation value of a physical variable, that physical operators are their own Hermitian conjugate $\mathbb{H}^\dagger = \mathbb{H}$. Operators that are their own Hermitian conjugates are called *Hermitian operators*.

As per postulates of quantum mechanics, the results that are possible outcomes of a measurement of a dynamical variable x are the eigenvalues a_n of the linear Hermitian operator A applied on the state ψ. In this case, it is the Hermitian operator which is the observable, such as, for example, an observable position vector of a particle, $\vec{x}$. The probability of obtaining either a $|0\rangle$ or a $|1\rangle$ is related to the projection of the qubit onto the measurement basis. Hermitian conjugates can be used as a mathematical tool to calculate the projection of one state onto another.

One of the properties of the Hermitian complex number is as follows: the adjoint or conjugate $v^\dagger$ of a number, v is the complex conjugate $v^\star$ of this number: $v^\dagger = v^\star$. Another property of the Hermitian operators states that their eigenvalues are real numbers, which helps measure certain physical properties. The outcome of a measurement is guaranteed to be an eigenvalue, say, λ of the observed system. This measurement outcome is described by the *projection operator P. P* is Hermitian in character, hence $P = P^\dagger$, as per the definition of Hermitian operators, and it is equal to its own square: $P^2 = P$.

After a quantum system is measured, as per postulates of quantum mechanics, the wave function describing the state collapses. Therefore, whereas the state of the system *before* the measurement may have been a superposition of basis. But, *after* the measurement is performed, the system collapses to the basis state that corresponds to the result of the measurement. Mathematically, if the original state before the measurement was $|\psi\rangle$, the state of the system after measurement $|\psi'\rangle$ is given by

$$|\psi'\rangle = \frac{P|\psi\rangle}{\sqrt{\langle\psi|P|\psi\rangle}}$$

1.3

Projective measurements can also be described in terms of an *observable,* which is a Hermitian operator A acting on the state space of the system. As A is Hermitian, its *spectral decomposition* is given by

$$A = \sum a_i P_i$$

1.4

where a_i is the eigenvalue of A and P_i is the orthogonal projector on the eigenspace of A. In this context, performing a projective measurement is the same as measuring the observable with respect to decomposition $I = \sum P_i$, where eigenvalue a_i corresponds to the i-th measurement result.

The *spectral decomposition* theorem (SDT) tells us that for every normal operator T acting on a finite-dimensional Hilbert space $\mathcal{H}$, there is an orthonormal basis of $\mathcal{H}$ consisting of eigenvectors $|T_j\rangle$ of T. In other words, an operator T belonging to some vector space that has a diagonal matrix representation with respect to some basis of that vector space. This result is known as the spectral decomposition theorem. Suppose that operator A satisfies the spectral decomposition theorem for some basis. As per the SDT, we can always diagonalize normal operators in finite dimensions. We know from linear algebra that the diagonalization can be accomplished by a change of basis to the basis consisting of eigenvectors. The change of basis is accomplished by conjugating the operator T with a unitary operator. Suppose that an operator T with eigenvalues T_j satisfies the spectral decomposition theorem for some basis $|v_j\rangle$. Then, the operator T, by the SDT, is given by

$$T = \sum_{j=1}^{n} T_j |v_j\rangle\langle v_j|$$

Superposition

The principle of superposition is fundamental to quantum physics. The principle states that states of a quantum system may be superimposed to form a combination of several states, such as waves in classical physics to form a coherent quantum state that is a separate and distinct state from its component states. Hence, whereas a qubit can exist in the state $|0\rangle$ or state $|1\rangle$, it can also exist in a state that is a linear combination of the states $|0\rangle$ and state $|1\rangle$. So, if a qubit state is given by $|\psi\rangle$, then a superposition state of this qubit can be written as

$$|\psi\rangle = \alpha|0\rangle + \beta|1\rangle$$

1.5

Where α and β are complex numbers called the *amplitudes* of the state of the qubit.

The easiest way to envisage the principles of superposition is to think about a beam of light that is subjected to a polarizing filter which would polarize the light *vertically*, let's say, in the state $|0\rangle$. Now consider a second polarizer held with its axis *horizontal* to the vertically polarized beam emerging from the first polarizer (i.e., forcing a *horizontal polarization*); in this case, there is no longer any light to be seen emerging from the second polarizer. This is because the horizontal axis of polarization is at 90° to the axis of vertical polarization, which the beam of light already possessed due to passing through the first polarization filter, which enforces the horizontal state. The horizontal state, in our case, as per equation 1.5, is state $|1\rangle$. The state $|1\rangle$ is orthogonal to the vertical polarized light and the two polarizers cancel the remainder of emergent light to give absolutely nothing. Now, if the axis of the second polarizer would be tilted to another angle, say 30°, then we would have some light emerging from the whole set-up.

In another example, we consider an atomic qubit based on an electron exhibiting its UP or DOWN spin and affected by a magnetic field. The electron is *not spinning* in a traditional sense; the word *spin* describes its *angular momentum*, which can be quantized to mean UP or DOWN (1 or 0). Before we measure the qubit to see whether it is a 0 or a 1, it is in a superposition state, which means it can be in a fraction of a 0 or a 1 or, in other words, weighted combinations like 0 – 37%, 1 – 63% (the α and β parameters). Physicists have not yet developed a visualization of physical reality that underlies spin. But they can describe spin mathematically and predict its behavior in lab experiments.

At any given moment, the qubit may be in some proportion of spin biased toward one or zero. Only when we measure the qubit does it collapse into one of the definite states, $|0\rangle$ or $|1\rangle$, as depicted in Figure 1-4.

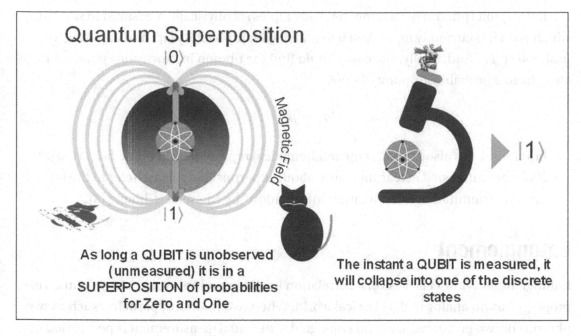

Figure 1-4. *Quantum superposition and effects of measurement—superposition collapses when measured to give a discrete state*

The states $|0\rangle$ and $|1\rangle$ can be represented in column vector forms as follows.

$$|0\rangle = \begin{pmatrix} 1 \\ 0 \end{pmatrix}$$

and

$$|1\rangle = \begin{pmatrix} 0 \\ 1 \end{pmatrix}$$

Hence, the superposition is described by taking a linear combination of the state vectors for the 0 and 1 paths, so the general path state is described by a vector that is analogous to equation 1.5.

$$|\psi\rangle = \alpha \begin{pmatrix} 1 \\ 0 \end{pmatrix} + \beta \begin{pmatrix} 0 \\ 1 \end{pmatrix} \qquad 1.6$$

If the qubit (photon or electron in the examples) is physically measured to see which path it is currently in, we find it in path 0 with probability $|\alpha|^2$, and in path 1 with probability $|\beta|^2$. And, finally, since we should find the photon in *exactly one path*, we must have, from a probabilistic point of view,

$$|\alpha|^2 + |\beta|^2 = 1$$

1.7

Equation 1.7 is also known as the mathematical representation of the *Born rule* [3].

If you are interested in learning more about superposition, please refer to the book "Quantum Computation and Quantum Information" by Nielsen & Chuang [3].

Entanglement

Entanglement is a special case of a correlation between multiple quantum systems. This property has no analog in the classical world. When two (or more) particles, such as two photons or two electrons, are in an entangled state and a measurement is performed on *one* of them, then that measurement can impact the behavior of the same measurement on the other particle(s) instantaneously, *independent* of the physical distance between them. The unintuitive nature of this phenomenon is so strange that Albert Einstein called it "spooky action at a distance" since there is no known explanation yet as to why it occurs. When we measure one qubit, the act of measurement collapses its state and simultaneously collapses the state of the other entangled qubit(s).

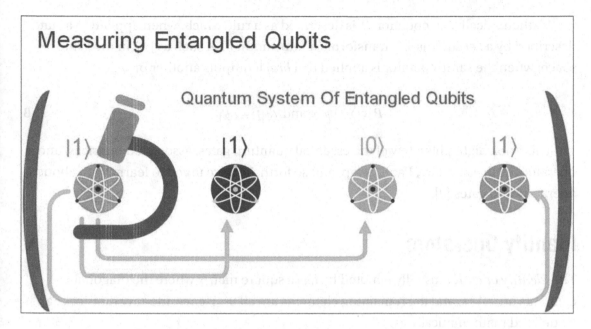

Figure 1-5. *Measuring a single qubit causes its superposition to collapse a definite state, causing all other entangled qubits to collapse*

As shown in Figure 1-5, this phenomenon enables us to deduce the state of the other entangled qubit(s) no matter how far the physical distance between them is. Entangled qubits become a system with a single quantum state.

Quantum Operators and Gates

Quantum gates are, like classical reversible gates, logically reversible, but they differ markedly on their universality properties. Whereas the smallest universal classical reversible gates must use three bits, the smallest universal quantum gates need only use two bits. Some universal classical logic gates, such as the Toffoli gate, provide reversibility and can be directly mapped onto quantum logic gates. Quantum logic gates are represented by unitary matrices. Any unitary operator acting on a two-dimensional quantum system or a qubit is called a *one-qubit quantum gate.* In the quantum circuit model, the logical qubits are carried along *wires*, and quantum gates act on those qubits. A quantum gate acting on n-qubits has the input qubits carried to it by n wires, and other n wires carry the output qubits away from the gate. Hence, quantum gates can be represented by $2^n \times 2^n$ matrices with orthonormal rows. Single qubit gates such as Hadamard gates are represented by 2×2 matrices.

Mathematically, an operator $\hat{P}$ is described as a rule which, when applied to a state described by a *ket* such as $|\psi\rangle$, transforms it into another ket such as $|\psi'\rangle$ in the same space; when the same operator is applied to a *bra*, it outputs another bra.

$$\hat{P}|\psi\rangle = |\psi'\rangle \text{ and } \langle\phi|\hat{P} = \langle\phi'| \qquad\qquad 1.8$$

This book highlights a few of the essential quantum gates, associated matrices, and operators such as unitary, Pauli, Swap, and so forth. You are urged to learn more about operators and gates [3].

Identity Operators

The *identity operator*, usually denoted by I, is a square matrix where the diagonal elements are all 1s, and the remaining elements are all 0s. Hence, identity matrices can be defined mathematically as

$$I_{nn} = 1 \text{ and}$$

$$I_{nm} = 0, \text{ when } n \neq m \qquad\qquad 1.9$$

An example of a 2×2 identity matrix is as follows.

$$\begin{pmatrix} 1 & 0 \\ 0 & 1 \end{pmatrix}$$

Unitary Operators

An operator is considered unitary if it produces an identity matrix as output when multiplied by its own adjoint. A unitary operator has its own inverse equal to its adjoint.

The inverse of an operator U is denoted by U^{-1}. This satisfies the following relationship.

$$UU^{-1} = I \qquad\qquad 1.10$$

where I is the identity operator.

For U to be unitary,

$$UU^\dagger = U^\dagger U = I \qquad\qquad 1.11$$

and

$$U^{-1} = U^\dagger \qquad\qquad 1.12$$

A *unitary transformation* transforms the matrix of an operator in one basis to a representation of the same operator in another basis. For example, for a two-dimensional matrix a change in basis from $|r_i\rangle$ to $|e_i\rangle$ is given by

$$U = \begin{pmatrix} \langle e_1|r_1\rangle & \langle e_1|r_2\rangle \\ \langle e_2|r_1\rangle & \langle e_2|r_2\rangle \end{pmatrix}$$

It is important to note that the unitarity of quantum evolution implies that quantum operations are *reversible,* except for those operators used for measurement. This reversibility is a consequence of restricting attention to closed systems. Any irreversible classical computation can be efficiently simulated by a reversible classical computation. The same holds for quantum computation.

The Pauli Group of Matrices and Gates

The Pauli group of matrices are a set of one qubit quantum operators that give rise to Pauli *gates* (applied to quantum circuits) and span the vector space formed by all one-qubit operators. The importance of Pauli gates in quantum computing is immense as any one-qubit unitary operator can be expressed as a linear combination of the Pauli gates.

It is considered that there are four different Pauli operators, including the Identity operator (however, some authors omit the identity operator). There are quite a few different notations employed in various literature to depict Pauli operators. In this book, and for context, I, X, Y and Z are consistently used. However, for reasons of completeness, all the notations are mentioned because Pauli operators are defined on a computational basis.

$$I \equiv \sigma_0 \equiv \begin{pmatrix} 1 & 0 \\ 0 & 1 \end{pmatrix} \qquad X \equiv \sigma_1 \equiv \sigma_x \equiv \begin{pmatrix} 0 & 1 \\ 1 & 0 \end{pmatrix}$$

$$Y \equiv \sigma_2 \equiv \sigma_y \equiv \begin{pmatrix} 0 & -i \\ i & 0 \end{pmatrix} \quad Z \equiv \sigma_3 \equiv \sigma_z \equiv \begin{pmatrix} 1 & 0 \\ 0 & -1 \end{pmatrix} \qquad 1.13$$

It is of interest to note that the NOT gate is often identified with the Pauli X gate as both have the same matrix representation and the same effect on a basis vector (i.e., they invert the basis vectors). For example, the column notation of the zero state is

$$|0\rangle = \begin{pmatrix} 1 \\ 0 \end{pmatrix}$$

If you apply the NOTE or the X gate to it, you have

$$X(\equiv NOT)|0\rangle = \begin{pmatrix} 0 & 1 \\ 1 & 0 \end{pmatrix}\begin{pmatrix} 1 \\ 0 \end{pmatrix} = \begin{pmatrix} 0*1+1*0 \\ 1*1+0*0 \end{pmatrix} = \begin{pmatrix} 0 \\ 1 \end{pmatrix} = |1\rangle \qquad 1.14$$

Similarly, the X operator, when applied to $|1\rangle$, would invert it to $|0\rangle$ giving the same effect as the NOT gate (for this reason, the X gate is also known as the *bit-flip* gate). You are encouraged to verify this.

The Z operator is also known as *the phase flip operator* as its effect is to rotate a state vector about the Z axis by π radians or 180 degrees. The Pauli gates X, Y, and Z correspond to rotations about the X, Y, and Z axes of the Bloch sphere, respectively.

Phase Gates

The phase shift operator or rotation gate causes the state $|0\rangle$ to remain unchanged but rotates the $|1\rangle$ state by a defined angle or phase θ.

$$R_\theta \equiv \begin{pmatrix} 1 & 0 \\ 0 & e^{i\theta} \end{pmatrix} \qquad 1.15$$

Using *Euler's identity* in equation 1.15, if we set $\theta = \pi$, we recover the Pauli Z gate because $e^{i\theta} = \cos(\pi) + i.\sin(\pi) = -1$.

If we substitute $\theta = \pi/2$ in equation 1.15, we get $e^{i\theta} = i$, which in turn gives another operator called S.

$$S \equiv \begin{pmatrix} 1 & 0 \\ 0 & i \end{pmatrix} \qquad 1.16$$

where the S operator rotates the original state by 90° or $\frac{\pi}{2}$ radians about the Z axis.

The T operator rotates the original state by 45° or $\pi/4$ radians about the Z axis. The T gate is also known as the $\pi/8$ gate because the $e^{i\pi/8}$ can be factored out, leaving the diagonal components with an absolute phase of $|\pi/8|$.

$$T \equiv \begin{pmatrix} 1 & 0 \\ 0 & e^{\frac{i\pi}{4}} \end{pmatrix} = e^{\frac{i\pi}{8}} \begin{pmatrix} e^{-\frac{i\pi}{8}} & 0 \\ 0 & e^{\frac{i\pi}{8}} \end{pmatrix} \qquad 1.17$$

More detailed discussions on types of rotation and parameterized gates can be found in the textbook by Nielsen and Chuang [3].

Cartesian Rotation Gates

Previously, you saw properties of few phase gates. The following are some additional rotation gate representations that are commonly used in QC and QML.

$$RX_\theta \equiv \begin{pmatrix} \cos\dfrac{\theta}{2} & -i\sin\dfrac{\theta}{2} \\ -i\sin\dfrac{\theta}{2} & \cos\dfrac{\theta}{2} \end{pmatrix} \qquad 1.18$$

$$RY_\theta \equiv \begin{pmatrix} \cos\dfrac{\theta}{2} & -\sin\dfrac{\theta}{2} \\ \sin\dfrac{\theta}{2} & \cos\dfrac{\theta}{2} \end{pmatrix} \qquad 1.19$$

$$RZ_\theta \equiv \begin{pmatrix} e^{-i\theta/2} & 0 \\ 0 & e^{i\theta/2} \end{pmatrix} \qquad 1.20$$

Hadamard Gate

Hadamard gates create superposition of states. The Hadamard gate has the following matrix representation.

$$H \equiv \frac{1}{\sqrt{2}} \begin{pmatrix} 1 & 1 \\ 1 & -1 \end{pmatrix} \qquad 1.21$$

It maps the computational basis states as follows.

$$H|0\rangle = \frac{1}{\sqrt{2}}\begin{pmatrix} 1 & 1 \\ 1 & -1 \end{pmatrix}\begin{pmatrix} 1 \\ 0 \end{pmatrix} = \frac{1}{\sqrt{2}}\left[\begin{pmatrix} 1 \\ 0 \end{pmatrix} + \begin{pmatrix} 0 \\ 1 \end{pmatrix}\right] = \frac{1}{\sqrt{2}}(|0\rangle + |1\rangle) \qquad 1.22$$

Similarly, $H|0\rangle = \frac{1}{\sqrt{2}}(|0\rangle - |1\rangle)$. The operation rotates the qubit state by π radians or 180 degrees about an axis diagonal in the $x - z$ plane.

Another property of the Hadamard gate is that it is self-inverse (i.e., $H = H^{-1}$); hence,

$$H\left[\frac{1}{\sqrt{2}}(|0\rangle + |1\rangle)\right] = |0\rangle$$

and

$$H\left[\frac{1}{\sqrt{2}}(|0\rangle - |1\rangle)\right] = |1\rangle$$

$$1.23$$

This operation is equivalent to applying an x-gate followed by a $\frac{\pi}{2}$ rotation about the Y axis.

CNOT Gate

The CNOT gate, also known as a *controlled not* gate, is a two-qubit gate. A *controlled* gate in quantum computing involves using a *control*, or C, qubit as the first input qubit and the second qubit as the target qubit. The first input to the CNOT gate acts as the control qubit: If the control qubit is in a state $|0\rangle$, then the gate does not do anything to the target qubit; but if the state of the control qubit is $|1\rangle$, then the gate applies the NOT or X operator (equation 1.14) to the target qubit. In other words, the CNOT gate forces *entanglement* on the two qubits in the realm of quantum computing.

The possible input states to a CNOT gate are as follows: $|00\rangle$, $|01\rangle$, $|10\rangle$, $|11\rangle$. The action of the CNOT gate on these states are as follows.

$$|00\rangle \Rightarrow |00\rangle$$

$$|01\rangle \Rightarrow |01\rangle$$

$$|10\rangle \Rightarrow |11\rangle$$

$$|11\rangle \Rightarrow |10\rangle$$

A circuit representation of the CNOT gate is given by

The matrix representation of the controlled CNOT gate is given by

$$CNOT \equiv \begin{pmatrix} 1 & 0 & 0 & 0 \\ 0 & 1 & 0 & 0 \\ 0 & 0 & 0 & 1 \\ 0 & 0 & 1 & 0 \end{pmatrix} \qquad 1.24$$

The outer product representation of the CNOT is

$$CNOT = |00\rangle\langle 00| + |01\rangle\langle 01| + |10\rangle\langle 11| + |11\rangle\langle 10| \qquad 1.25$$

The CNOT gate is useful for preparing entangled states. It is self-adjoint; applying it for a second time reverses its effect.

SWAP Gate

SWAP gates are also two-qubit gates. A *SWAP* operator reverses the states of bits in an input qubit state. For example, it can take in the state $|10\rangle$ and reverse it to $|01\rangle$. *SWAP* operators are represented by the following matrix.

$$SWAP \equiv \begin{pmatrix} 1 & 0 & 0 & 0 \\ 0 & 0 & 1 & 0 \\ 0 & 1 & 0 & 0 \\ 0 & 0 & 0 & 1 \end{pmatrix} \qquad 1.26$$

Density Operator

The last important operator in this book, among other operators encountered, is the *density operator*. In a closed, pure quantum system, it is simple to calculate the probability of finding a certain state $|x\rangle$. It is the square of its amplitude $|\alpha|^2$ (reference equation 1.7). The density operator of a *pure* single state $|\psi\rangle$ is given by its outer product.

$$\rho = |\psi\rangle\langle\psi| \qquad\qquad 1.27$$

The trace of the square of a density operator of a *pure* system *only* is 1: $Tr(\rho^2) = 1$

If the density of a closed pure system is time-dependent, then the time evolution of the density operator is given, in terms of the unitary operator, by

$$\rho(t) = U\rho(t_0)U^\dagger \qquad\qquad 1.28$$

However, in an environment where there is a *statistical distribution of mixed* quantum systems (for example, n_1 of system-1 and n_2 of system-2), a general description of the density operator can explain the nature of such a mixture. This is done by first formulating density operators for each state, applying appropriate weights to these states, and then summing up all the possibilities.

Hence, let $|\psi_i\rangle$ represent n possible states in an ensemble in a mixed quantum system, where $i \in \{1...n\}$, and p_i is the probability that a member is in the corresponding state $|\psi_i\rangle$. Therefore, in this case, the density operator for the whole mixed quantum system is given by

$$\rho = \sum_{i=1}^{n} \rho_i p_i = \sum_{i}^{n} p_i |\psi_i\rangle\langle\psi_i| \qquad\qquad 1.29$$

The density operator is Hermitian (i.e., $\rho = \rho^\dagger$; trace of a density operator is 1), and the density operator is always positive for any state vector.

Hamiltonian

The Hamiltonian of a system is one of the most important concepts in quantum mechanics, quantum computing, and quantum machine learning. Some of the most important algorithms in quantum optimization, machine learning, variational algorithms depend on Hamiltonians. A *Hamiltonian* is the total energy of a system.

It provides the physics connection between classical and quantum mechanics. Each system has its own Hamiltonian function. It is the *form* of this function rather than the value which determines how a system evolves with time. The Hamiltonian operator is one of the most widely used operators in quantum mechanics. Hence understanding the process of constructing the Hamiltonian function can be of utmost importance in QML, as you see in the later chapters.

Time Evolution of a System

The time evolution of a physically isolated or "closed" quantum system is given by the Schrödinger equation as follows.

$$i\hbar \frac{\partial}{\partial t} |\psi\rangle = H |\psi\rangle \qquad 1.30$$

where H is the *Hamiltonian* of the system. Since the Hamiltonian corresponds to the total energy of the system, the possible energies that the quantum systems may have been given by the eigenvalues of the operator H. The state of the system after a time t is given by

$$|\psi(t)\rangle = e^{-\frac{iHt}{\hbar}} |\psi(0)\rangle \qquad 1.31$$

And the time evolution of the quantum state of the system is described by the *unitary operator*.

$$U = e^{-\frac{iHt}{\hbar}} \qquad 1.32$$

No-Cloning Theorem

In 1982, Wootters and Zurek [4] derived what is now known as the *no-cloning theorem*, which is one of the most fundamental properties of quantum systems. This states that it is impossible to create an *exact perfect* copy of an arbitrary quantum system that belongs in a specific Hilbert space. This adds to the challenge of programming a quantum computer as opposed to a classical one. Hence, as per the no-cloning theorem, there is no valid quantum operation that can map an arbitrary state $|\psi\rangle$ to $|\psi\rangle |\psi\rangle$.

Fidelity

Fidelity is a measure of "how close" one state in a quantum system is to another. Fidelity is described as the amount of statistical mixture or overlap between two quantum distributions.

Unlike in classical systems, telling two quantum sequences of bits apart is not possible in general unless they are orthogonal to each other. Therefore, performing projective measurements on two states are not productive if those two states are not orthogonal. This property restricts measurement to be conducted as a *positive operator-valued measure* or POVM. Fidelity is a measure of distinguishability of two different states giving the best measurement to distinguish the two states.

Fidelity measures the overlap between two states. Hence, if $\rho_1 = |\psi\rangle\langle\psi|$ and $\rho_2 = |\phi\rangle\langle\phi|$ are the density operators of the two different distributions, then the fidelity of the system is given by

$$\mathcal{F}(\rho_1, \rho_2) = Tr\left[\sqrt{\sqrt{\rho_1}\rho_2\sqrt{\rho_1}}\right] = |\langle\phi|\psi\rangle| \qquad 1.33$$

Equation 1.33 implies that $|\langle\phi|\psi\rangle|^2$ is the probability of the system being in state $|\phi\rangle$, given that it is already known to be in state $|\psi\rangle$.

The following are the basic properties of fidelity.

1. Fidelity of two pure states is symmetric (i.e., $\mathcal{F}(\rho_1, \rho_2) = \mathcal{F}(\rho_2, \rho_1)$) in an argument in the range {0, 1}.

2. Fidelity is invariant when a unitary operator acts on it.

$$\mathcal{F}\left(U\rho_1 U^\dagger, U\rho_2 U^\dagger\right) = \mathcal{F}\left(\rho_1, \rho_2\right) \qquad 1.34$$

3. Fidelity equals 1 if and only if $\rho_1 = \rho_2$.

4. Tensor product of the density operators for fidelity is given by

$$\mathcal{F}\left(\rho_1 \otimes \rho_2, \upsilon_1 \otimes \upsilon_2\right) = \mathcal{F}\left(\rho_1, \upsilon_1\right)\mathcal{F}\left(\rho_2, \upsilon_2\right) \qquad 1.35$$

Complexity

The *complexity* of a problem refers to the study of classes of problems, such as NP-hard or PSPACE. Computational exercises involve the engagement of resources from the perspective of algorithms and time of computation to solve a specific problem. In order to estimate the overall computational cost, and so forth, you need to analyze the resources required to run an algorithm (on a computing system), which are needed to solve the problem.

A problem can be any computational problem; for example, categorizing cats and dogs using a range of images. To explain the difference between the two animals, we may define the sizes and shapes of cats and dogs as constraints or labels. Then we pick an algorithm, which is generally independent of the hardware, to do the work for us. The algorithm, as we run it, requires two main computational resources: run time and memory.

The time a computer takes to run an algorithm is the runtime of that algorithm on that specific computational platform. As per the fundamentals of classical computation, a solution for the target of a database search can be obtained by verifying items in the database until the one desired item is found. This exercise requires a worst-case scenario of time complexity $O(n)$.

A quantum complexity theory [5] tries to answer the question about speedups and efficiency of quantum computers vs. their classical counterparts in relation to runtime. In this book, the word *complexity* refers to the asymptotic runtime complexity of a quantum system. Chapters 2 and 3 expand more on complexity.

Grover's Algorithm

Originally, Grover's search algorithm was created to do an unsorted search in a quantum database for a particular element over an unstructured environment of N entries with no repeated elements. The only way to find a specific element in a database where the elements are stored in a random order is to do an exhaustive search. Grover's algorithm can be generalized to search databases with repeated elements. Grover's algorithm is also optimal for a multiplicative constant (i.e., it is *not possible* to improve its computational complexity). If the database consists of database entries of N elements

$\{x_1, x_2..., x_n\}$, then Grover's algorithm utilizes quantum properties to query the database $O\sqrt{(n)}$ times using $O(\log N)$ storage space to find a marked element x_0 with high probability. Grover's algorithm can be implemented with $O\left(\sqrt{N}\log N\right)$ *universal gates.* Statistically, Grover's algorithm needs approximately an order of $\sqrt{2^n}$ tries to find a marked object in a quantum database. For example, if the bit string is 4 bits, then in a classical context, a search algorithm would require $2^4 - 1 = 15$ attempts to the correct marked entry; in comparison, Grover's algorithm would require $\sqrt{2^4} = 4$ attempts. As the bit strings get larger, the improvement offered by Grover's algorithm becomes more significant.

Grover's algorithm is based on *amplitude amplification,* a technique that is described in later chapters of this book.

To explain the mechanisms of Grover's algorithm, let us consider a function $f(x)$ with n inputs in a space where $N = 2^n$ and $x = \in \{0, 1\}^n$. Let the function $f(x)$ be such that $f(x) = 1$, if and only if $x = x_0$ for a particular x_0. Therefore,

$$f(x) = \begin{cases} 1, & x = x_0 \\ 0, & x \neq x_0 \end{cases}$$

1.36

The challenge is to find x_0 with as few numbers of evaluations as possible. The function $f(x)$ is known as the oracle and x_0 is called the *marked element.*

In practice, the oracle may be treated as a black box, and the details of its inner mechanisms may be ignored. The oracle is modeled as a gate that takes an n bit input and produces a 1-bit output. The oracle is *not a valid quantum gate* since it is not unitary and irreversible.

Firstly, the n input states of $|0\rangle_1 \otimes |0\rangle_2 \otimes |0\rangle_3... \otimes |0\rangle_n \equiv |0\rangle^{\otimes n}$ are all initialized to $|0\rangle$.

As a second step, Hadamard gates are applied to each of those $|0\rangle$ input states so that there is a total of $H^{\otimes n}$ gates which transform each of the $|0\rangle^{\otimes n}$ inputs into states of superposition. If $|\psi\rangle$ is the superposition state for all possible $|x\rangle$, then

$$|\psi\rangle = \frac{1}{\sqrt{2^n}} \sum_{x = \in \{0,1\}^n} |x\rangle$$

1.37

Application of the Grover algorithm utilizes the Grover operator, also knowns as the Grover Diffusion operator G.

The Grover operator is given by

$$G = (2|\psi\rangle\langle\psi| - I)O \qquad\qquad 1.38$$

The implementation of Grover's algorithm involves two Hadamard gates that require n operations each. A unitary gate controls the conditional phase shift, and whereas the overall complexity depends on the application, the time complexity is efficient as only one call is needed per iteration. It is found that Grover's algorithm is efficient to the point where it requires to evaluate $f(x)$ less than N times to find x_0. It evaluates in order of $\frac{\pi}{4}\sqrt{(n)}$, which is asymptotically optimal. There is a quadratic gain in time complexity in this quantum scenario as compared to the classical context.

Grover's algorithm is generally implemented as follows.

Algorithm: Grover's Algorithm

Input: $N = 2^n$ and f; initial state $|0\rangle^{\otimes n}$

Output: x_0 with probability $\geq \left(1 - \frac{1}{N}\right)$

Step 1. Apply two-level quantum computer with $n + 1$ qubits

Step 2. Initialize superposition by applying Hadamard gates: $H^{\otimes n} | 0\rangle^{\otimes n}$.

Step 3. for $O\sqrt{(n)} = \frac{\pi}{4}\sqrt{(n)}$ times do

       Apply the Grover operator G

Step 4. Measure the system in the computational basis.

An example of the implementation of Grover's algorithm with a five-qubit circuit is shown in Figure 1-6. The On boxes shown in green indicate the qubit's probability is in the On state if measured. In this case, the probability that they are on is 100% except for the middle qubit, which shows Off. The oracle, as discussed before, is an evaluation in order of $\frac{\pi}{4}\sqrt{(n)}$ which is asymptotically optimal.

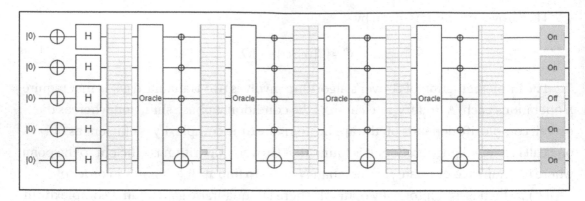

Figure 1-6. *Circuit representing Grover's algorithm for five qubits.The first register is initially in state |0⟩. The H in the boxes represent the Hadamard gates implementing superpsoition. In this case the expected iterations are in the order of $\frac{\pi}{4}\sqrt{5}$, represented in the simulation image as "Oracle". [6]*

Shor's Algorithm

Shor's algorithm was fundamental in demonstrating the power and importance of quantum computation. It can be used to factor prime numbers, which means that it can break encryption code on a practical quantum computer (a large enough quantum computer was yet to be built when this book was written).

Shor's efficient factoring algorithm consists of a quantum and a classical part. The former one is a quantum-based solution to the so-called *order-finding problem*. Because this algorithm hides the seminal idea, which allows factoring a large number N in $\mathcal{O}(\log_2 N)^3$ steps or gates instead of the best-known classical method requiring asymptotically $\mathcal{O}[\hbar \cdot (\log_2 N)^{1/3}(\log_2 \log_2 N)^{2/3}]$, where $\hbar$ is a constant often taken as 0.9.

This means that the best-known algorithm for factoring prime numbers on a classical (non-quantum) computer requires an amount of time that is fundamentally exponential with the size of N. This is why Shor's algorithm is considered important; a quantum computer running it would only require polynomial (as opposed to exponential) time. Since the security of the popular RSA cryptosystem is based on the difficulty of prime factorization, Shor's algorithm has attracted a lot of interest from the privacy and data security communities. This algorithm got the attention of a lot of people!

The order-finding part is the *only* quantum part of Shor's algorithm; the rest involves classical calculations. A simple way to state Shor's algorithm is that it uses order finding to find the factors of some odd integer N.

Shor's algorithm factors prime numbers by reducing the problem into an *order* or *period* finding exercise. If x and N are positive integers which are co-prime to each other, that is, they do not have any common factors and $<N$, then the *order* of x is the smallest positive integer r in the relationship given by

$$x^r = 1 \bmod N \tag{1.39}$$

The purpose of Shor's algorithm is to find r which is the period of x^r when N is the number to be factored.

The fundamentals of Shor's algorithm originated as a result of the number theory, as follows.

$$f(r) = x^r \bmod N \tag{1.40}$$

where $f(r)$ is a periodic function, and x is co-prime to N (i.e., they do not have a common factor, and N is the number to be factored).

Heisenberg's Uncertainty Principle

The key to understanding the depth of challenges of quantum computing and quantum machine learning is embedded in realizing the Heisenberg's uncertainty principle.

The main difference between the classical world and the quantum world was pointed out by Werner Heisenberg (1901–1976). According to classical laws of nature, if the initial position, velocity, and all other forces acting on a body are known, then the future path (position and velocity) of this physical system can be predicted exactly with absolute accuracy with the help of Newton's laws. Since this prediction is accurate, classical physics is *deterministic*.

Heisenberg's uncertainty principle, in its original form [7] states: If the x-component of the momentum of a particle is measured with an uncertainty Δp_x, then its x-position, cannot, at the same time, be measured more accurately than $\Delta x = \hbar / (2\Delta p_x)$; where $\hbar$ is the Planck's constant. Hence, there is a natural limit of precision in the measurement outcome of quantum physical systems. The uncertainty relations in three dimensions can be mathematically represented as follows.

$$\Delta x \Delta p_x \geq \hbar / 2 \,;\, \Delta y \Delta p_y \geq \hbar / 2;\, \Delta z \Delta p_z \geq \hbar / 2 \tag{1.41}$$

As per Heisenberg's principle, whereas it is possible to *simultaneously* measure *either* the position or momentum of a particle accurately, it is *impossible* to measure both these observables to any degree of *arbitrary accuracy* when the measurement of these two are done *at the same time*. A measurement of a quantum system for its location cannot be done without inducing a large momentum to the system; the position measurement of a quantum particle cannot be done passively. The act of measurement disturbs the inherent nature of the particle position and gives it momentum. The uncertainty principle is an inherent nature of matter alone as it is embedded in the wave nature of its property.

It is very important to realize that the uncertainty principle applies to a physical observation and associated measurement and *any interaction between classical and quantum systems* regardless of any specific observer. This principle profoundly impacts quantum optics, photonics, hardware design efforts for various forms of quantum computers, and innumerable other devices and processes that we benefit from.

Learning from Data: AI, ML, and Deep Learning

Learning from data is used in situations where analytical and deterministic results are not available. This basic fact has led to the marriage of data science to the world of quantum mechanics, where determinism is largely absent.

Situations in which determinism fails surround us in our daily lives. For example, if we show a photo of a zoo to a five-year-old child and ask if there is an elephant in it, we will likely get a correct answer. However, if we ask a twenty-year-old what the definition of an elephant is, we may get an inconclusive answer. The description of an elephant was explained to us via mathematical definitions; we learned what an elephant is by looking at them or photos of them. In other words, we learned from data.

The term *artificial intelligence* was coined in 1956. Since then, artificial intelligence (AI) has occupied people's imagination, particularly science fiction fans. A. C. Clarke's HAL the computer in *2001: A Space Odyssey* and C-3PO in *Star Wars* are two examples. Machine learning (ML) originated from the early thinkers of AI as they tried to simplify the algorithmic approaches to solve real-life problems over the years which included decision tree learning, inductive logic programming, clustering, reinforcement learning, and Bayesian networks among several others.

The terms *AI*, *ML*, and *deep learning* are often used interchangeably. However, when most people talk about AI, they are referring to ML. Fundamentally, ML uses algorithms to learn from data and then make a deterministic forecast or prediction about the system or environment represented by the dataset. Hence, rather than repeated manual coding of software routines with a specific set of instructions to accomplish a particular task, the machine is "trained" by using large amounts of data and algorithms to empower it to learn how to perform the specific task. Figure 1-7 shows the differences between deep learning (DL), AI, and ML.

Artificial intelligence is a science similar to physics or chemistry. The subject leverages mathematical and statistical theories and looks at the processes to create intelligent programs or machines that can proactively solve problems on their own, which has always been considered a human prerogative. Artificial Intelligence is often categorized as weak AI, strong AI, and super-intelligent (strong) AI.

Machine learning is a subset of artificial intelligence (AI) that uses mathematical and statistical insights to endow a computational system with the ability to automatically learn and improve its efficiency as it learns from experience without any need for further manual programmatic tuning. Tasks in ML, as in AI, can be achieved using various algorithms which can be used to solve various corresponding problems.

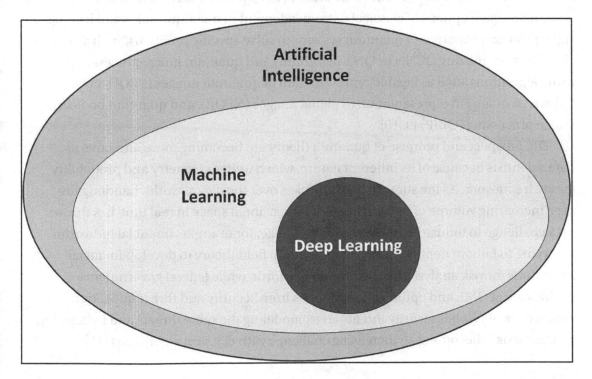

Figure 1-7. Relationship between AI, ML, and DL

Deep learning is a subset of machine learning, which largely uses neural networks to analyze different problems within a structure that is similar to the human neural system.

Chapters 2 and 3 expand more on algorithms relevant to these theories.

Quantum Machine Learning

Computational intelligence uses machine learning algorithms to solve optimization problems. One of the largest challenges of ML is handling data in multiple dimensions. Principles of superposition, entanglement, and contextual sensitivity have proven to be strong proponents of offering solutions for hard problems. Quantum mechanics have been useful in solving problems where the systems under investigation show contextual behavior. Natural quantum mechanical properties have proven to be an ideal match in such cases [8].

In recent years, *quantum neural networks* (QNN) and *quantum convolutional neural networks* (QCNN) have become areas of deep interest to researchers. These areas investigate options of classical data encodement in quantum states and leverage superposition properties of quantum systems to solve specific problems, such as image classification utilizing QCNN or QNN algorithms and quantum image processing (QIP) using algorithms such as flexible representation of quantum images (FRQI), novel enhanced quantum representation of digital images (NEQR), and quantum boolean image processing (QBIP) [9, 10].

The prospect and promise of quantum theory are becoming more attractive to data scientists because of its inherent nature, which unites geometry and probability in one framework. As the surge of big data takes over the digital world, handling the ever-increasing volume of data in different dimensional space in real time has thrown up a challenge to industries and governments alike; for example, financial industries are trying to fathom depths of QML and quantum field theory to develop financial algorithms for risk analysis, forecasting, and so forth, while federal governments are looking at QML and optimization theories from security and threat mitigation perspective. As global security threats arise, modeling the same threats and forecasting on that basis is becoming an increasing challenge with classical algorithms [11].

This book mostly focuses on QC treatment of classical datasets obtained as observables from classical systems, such as texts, images, and financial data, leveraging a classical quantum interface. As you progress through the book, you see various options for QC and QML tasks; that is, not all forms of quantum computers available do every job the same way at the same efficiency. To this end, we leverage Google's Cirq, OpenFermion, and TensorFlow Quantum, D-Wave's tools, Xanadu's PennyLane and Strawberry Fields, and others.

You look at several leading machine learning algorithms boosted by physics as you progress through this book, finishing with cutting-edge works such as quantum walks and QBoost.

Setting up the Software Environment

Before you become quantized and start to hack based on the laws of physics, it is good practice to set up your basic software environment. As you go through the book, we install various other libraries on top of our basic environment.

My operating system is Linux Ubuntu 18.04 LTS. The code should be supported in Python 3.6.x and up. For up-to-date support of TensorFlow Quantum, Cirq, and so forth, Python 3.7+ is recommended. There are instructions for installing Python on Windows and macOS on websites for your choice of Python installation. Most of the code in this book is based on Jupyter notebooks. You should download the relevant libraries for your OS; hence, the code is independent of the OS.

Python can be downloaded from various sites in various flavors. The following are two examples.

- The default Python download site is at www.python.org/downloads/. Here, there are choices and instructions for downloading and installing Python supported on Windows, Linux, and macOS.

- My preference is Anaconda Python (Anaconda3) at www.anaconda. com/products/individual, which offers free Python supported on various operating systems, as shown in Figure 1-8.

Anaconda Installers

Windows 🪟	MacOS	Linux 🐧
Python 3.8	Python 3.8	Python 3.8
64-Bit Graphical Installer (466 MB)	64-Bit Graphical Installer (462 MB)	64-Bit (x86) Installer (550 MB)
32-Bit Graphical Installer (397 MB)	64-Bit Command Line Installer (454 MB)	64-Bit (Power8 and Power9) Installer (290 MB)

Figure 1-8. *Versions supported for Anaconda Python*

The installation of Python on Ubuntu Linux is done using its native packaging system.

```
apt-get install
```

macOS offers MacPorts or HomeBrew as packaging systems. Anaconda has a packaging system called Conda. Python uses `pip` or `pip3` for Python 3.x installation.

You need to install a Jupyter notebook to use the code in this book. If you want access to the Python IDE, the free PyCharm for non-commercial use can be downloaded from `www.jetbrains.com/pycharm/download/#section=linux`.

Anaconda Python's individual free distribution bundle gets downloaded with a Jupyter notebook and its free IDE called Spyder, as shown in Figure 1-9.

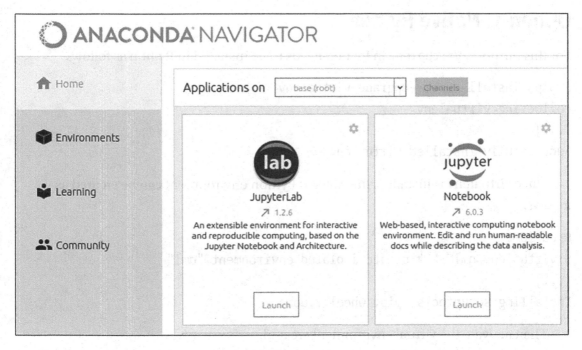

Figure 1-9. *Anaconda Python dashboard offering bundles of Jupyter, Spyder, and others*

If Anaconda is downloaded on Windows, then package upgrade can be done by opening the Anaconda PowerShell from **Start ➤ Anaconda3 ➤ Anaconda PowerShell Prompt** and running the conda commands. Anaconda installation instructions are at `https://docs.anaconda.com/anaconda/user-guide/tasks/install-packages/`.

Once Python is installed, it is recommended that a permanent workspace directory is created for QML code, for example (feel free to change the name or path to something more meaningful).

```
$ export QML = "$HOME/qml"   # path and directory
$ mkdir - p $QML
```

It is strongly recommended that a separate isolated environment is created for all the coding work in this book. This prevents accidental overlap between other projects and coding versions in the same environment.

Option 1: Native Python

For this purpose, `virtualenv` in Python needs to be installed in Python as follows.

```
$ pip3 install -user -upgrade virtualenv
collecting virtualenv
[...]
Successfully installed virtualenv
```

Once virtualenv is installed, the isolated Python environment can be created as follows.

```
$ cd $QML
$ virtualenv qml    # create isolated environment "qml"
[...]
Installing setuptools, pip, wheel...done
```

Then activate the virtual environment created.

```
$ source qml/bin/activate #activates the isolated env
(qml) sgangoly@ubuntu:~$
```

To exit the isolated system to revert to your global Python environment, enter the following `deactivate` command.

```
(qml) sgangoly@ubuntu:~$ deactivate
```

Option 2: Anaconda Python

Using conda package manager to create the isolated virtual environment is easy.

```
$ conda create --name qml       # create isolated environment "qml"
[...]
$ conda activate qml            # activate isolated environment
(qml) sgangoly@ubuntu:~$
```

To exit the isolated system to revert to your global Python environment, type in the following `conda deactivate` command.

```
(qml) sgangoly@ubuntu:~$ conda deactivate
```

As long as the environment is active, any package installed in the isolated virtual environment will not interact with any other packages or libraries installed in any other part of your system.

Installing the Required Packages and Libraries

You need to install a few necessary packages and libraries to set up our base system. As you progress through the book, you install additional libraries and packages when required. Use pip3 if you are using a global Python environment and pip if inside an isolated virtual Python 3.x+ environment, such as "qml" in the preceding example. If using an Anaconda install, use the conda command. Ubuntu environments may need the sudo command.

For native Python (Option 1)

```
$ pip3 install -upgrade jupyter numpy pandas scipy scikit-learn matplotlib
Collecting jupyter
[...]
```

For Anaconda (Option 2)

```
$ conda install -c anaconda scikit-learn
$ conda install -c anaconda matplotlib
```

If the installation is successful without any errors or warnings, try out opening a Jupyter notebook with the following command (see Figure 1-10).

```
$ jupyter notebook
[...]
[I 21:26:32.372 NotebookApp] The Jupyter Notebook is running at:
[I 21:26:32.372 NotebookApp] http://localhost:8888/?token=2c99c1f3108242781
947a02d6ff692187f19af7565c355cb
```

```
[I 21:26:32.371 NotebookApp] Serving notebooks from local directory: /home/sgangoly
[I 21:26:32.372 NotebookApp] The Jupyter Notebook is running at:
[I 21:26:32.372 NotebookApp] http://localhost:8888/?token=2c99c1f3108242781947a02d6ff6921{
7f19af7565c355cb
[I 21:26:32.372 NotebookApp]  or http://127.0.0.1:8888/?token=2c99c1f3108242781947a02d6ff{
92187f19af7565c355cb
[I 21:26:32.372 NotebookApp] Use Control-C to stop this server and shut down all kernels {
twice to skip confirmation).
[C 21:26:32.417 NotebookApp]

    To access the notebook, open this file in a browser:
        file:///home/sgangoly/.local/share/jupyter/runtime/nbserver-2172-open.html
    Or copy and paste one of these URLs:
        http://localhost:8888/?token=2c99c1f3108242781947a02d6ff692187f19af7565c355cb
     or http://127.0.0.1:8888/?token=2c99c1f3108242781947a02d6ff692187f19af7565c355cb
```

Figure 1-10. *Jupyter notebook: Terminal output and browser interface*

Once the Jupyter notebook browser opens, click the New tab, and select Python 3, as shown in Figure 1-10. This opens a second Jupyter notebook browser with your Python 3.x installation. You are ready to code!

If you do not want to install anything, you are welcome to try out Google's Colab at `https://colab.research.google.com`. It is built on a Jupyter notebook and natively supports Python. In addition to being coding-friendly, it also offers support for GPU and TPU (Google's TensorFlow Processing Unit).

Quantum Computing Cloud Access

The following are guidelines for setting up free (but limited) public cloud access to various *real quantum hardware* systems if you are interested in getting your hands dirty with *real* qubits. Most of these QC environments are based on superconducting qubits, except for Xanadu, which is industry-first in creating QC with semiconductor photonic qubits and has very recently opened cloud access to their QC platform.

Cloud access to the QC environments is quite easy. It is only a matter of registering with them and getting access. Besides offering cloud-based access, almost all QC vendors provide options for simulators that you can try algorithms on, such as Rigetti's QVM. The following are websites for some providers that offer access to a limited number of qubits via the cloud.

- Rigetti: `www.rigetti.com`

- IBM Q Experience: `https://quantum-computing.ibm.com`

- Xanadu: `https://xanadu.ai/request-access`

- D-Wave Leap: `https://cloud.dwavesys.com/leap/login/ ?next=/leap/`

- Amazon (AWS) has recently started a cloud offering called Braket. Customers can use Amazon Braket to run their quantum algorithms on their choice of quantum processors based on different technologies, including systems from D-Wave, IonQ, and Rigetti. Both simulated and quantum hardware jobs are managed through a unified development experience, and customers pay only for the compute resources used. To get started with Amazon Braket, visit `https://aws.amazon.com/braket`

Summary

The adaptation of quantum computing and machine learning algorithms in industry has seen a historic rise in recent years. Chapter 1 covered some rudimentary fundamentals required as basics for a venture into the world of QML. Appendix A covers some vital mathematical background as a reference, such as tensor products and Fourier transforms. The next chapter covers the essentials of classical machine learning concepts as a launching pad for translation into the quantum domain.

CHAPTER 2

Machine Learning

Machine learning (ML) emerged as a subfield of research in artificial intelligence and cognitive science. The three main types of machine learning are *supervised learning*, *unsupervised learning*, and *reinforcement learning*. There are several others, including *semisupervised learning, batch learning*, and *online learning*. Deep learning (DL) is a special subset that includes these various types of ML. It extends them to solve other problems in artificial intelligence, such as reasoning, planning, and knowledge representation.

What is machine learning? In 1959, Arthur Samuel defined machine learning as "the field of study that gives computers the ability to learn without being explicitly programmed." ML is a subject that focuses on teaching computers how to learn without any need for task-specific programming in today's context. Machine learning explores the creation of algorithms that learn from and make predictions on data.

Note   This chapter looks at some rudimentary fundamentals of classical machine learning from an applied perspective, such as types of learning, variance, and bias. It indulges in hands-on exercises using publicly available data types. The code is downloadable from the book's website. ML is a massive field all by itself, and it is not possible to include all aspects of it in one chapter. We revisit classical concepts throughout the book, including in-depth error correction and feature engineering to fit quantum concepts in data-driven predictions.

Traditionally, handcrafting and optimizing algorithms to complete a specific task has been the way to build legacy software and address workloads on data manipulation and analysis. At a high-level, traditional programming focuses on a deterministic outcome for well-understood input data. In contrast, ML has the power to solve a whole class or range of problems for datasets whose input-output correspondence is not clearly understood.

41

© Santanu Ganguly 2021
S. Ganguly, *Quantum Machine Learning: An Applied Approach*, https://doi.org/10.1007/978-1-4842-7098-1_2

ML constitutes learning from data. One example of ML is a *spam filter*. A spam filter basically "learns" from examples of spam emails flagged by users, examples of regular *non-spam* emails, examples of known cases of widespread frauds created by spam (such as phishing emails, well-known frauds that ask for personal information, etc.), and other types of use cases.

A spam filter governed by ML learns from examples and applies what it has learned to suspected spam emails. However, this learning process and the subsequent algorithmic application of the results of the "learning" are probabilistic (i.e., it is not always perfect or deterministic). Hence, occasionally, you find legitimate emails in your spam folder and spam in your inbox. More examples the spam filter gets to "see," the more it trains itself to detect legitimate spam, and the better it performs at its job of protecting our mailbox.

One of the main reasons behind the success of ML over traditional programming in this type of task is the ever-increasing volume of data. Spam filters typically act on rules. In the past, we had to manually define rules to stop spams in our mailboxes where we had to define "from," "to," "subject," and so forth, as shown in Figure 2-1.

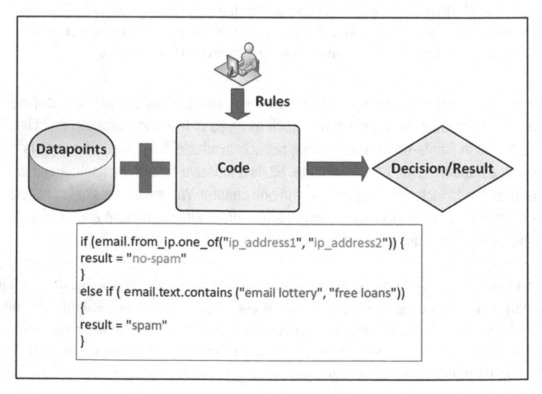

Figure 2-1. *The legacy method of manually entering rules for desirable results*

With the explosion of traffic and data in this age of big data in the modern world, the ability of humans to manually define rules by hand and keep up with the increasing volume of data has fallen well below the required efficiency level. Hence, the machines have taken over that task to define comprehensive rules to process large chunks of data and apply those same rules to provide solutions to problems. The data that the ML system uses to learn from is called a *training set*. In the example of the spam filter, flagging and filtering the spam or the *training data* to define the *performance* of the algorithm. The performance measure, in this case, can be the ratio of correctly or incorrectly spammed emails. The performance measure is called *accuracy* and this value, as demonstrated in some of the examples in this chapter, usually improves with training and time. The workflow of the application of ML for a spam filter is shown in Figure 2-2.

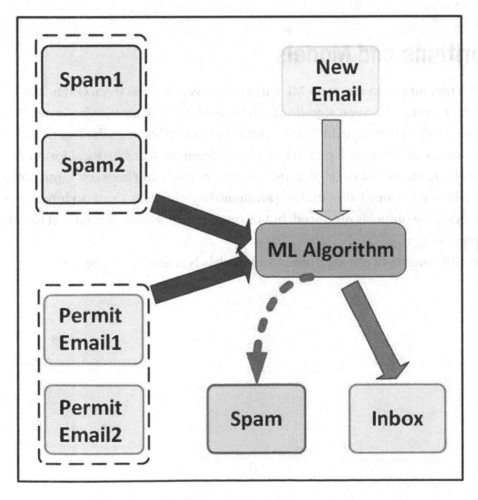

Figure 2-2. *Spam filtering with ML*

This chapter goes through the fundamentals of machine learning algorithms, principles of data-driven prediction, performance, and complexity. This chapter explores data-driven models and predictions and various learning algorithms from a classical perspective, followed by some hands-on examples. Subsequent chapters build on the concepts of classical machine learning powered by quantum physics.

Before we dive into some rudimentary and essential machine learning concepts, it may be beneficial to do a quick survey of some terminology involved in the subject.

Machine learning and big data: Until recently, most ML workloads were performed on single machines (computers). However, developments in GPUs (graphics processing units), Google's TPU (Tensor Processing Unit), and distributed computing now make it possible to run machine learning algorithms for big data at a massive scale, distributed among clusters.

Algorithms and Models

There is a plethora of literature on ML and its aspects of it these days. Often, it is hard to differentiate between algorithms and models as both terminologies are used interchangeably in literature. In this text, we differentiate between the two as follows.

Algorithms are methodologies defined by mathematical or statistical formulas.

Models are mathematical objects defining the relationship between input data and output, which is obtained after training an algorithm with data. Once an input goes into a model, the output is produced; but the model needs to be trained first to obtain a meaningful output.

The difference between algorithms and models is shown in Figure 2-3.

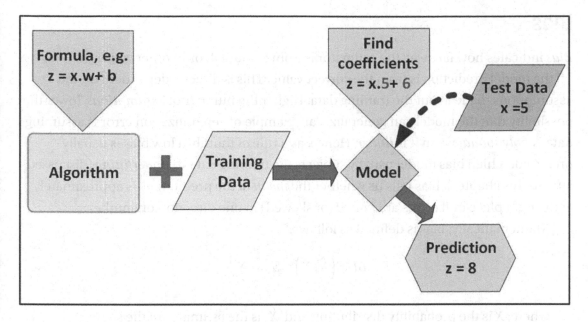

Figure 2-3. *Algorithms and models*

Different training data causes different models with different coefficients to be produced. An example of working algorithms is a housing price prediction where the price is based on the number of rooms, size of the plot, living space, and so forth. However, before the algorithm is employed and the results produced by it can be is trusted, there needs to be some proof that it works.

The "expected" values from an algorithm refer to building an *estimator*. All ML algorithms are estimators. An estimator tells us what to *expect* in the future but *does not actually predict* the future. Ideally, we would prefer to have an estimator which predicts the future perfectly! But that would be akin to having foresight or psychic vision. From a mathematical and scientific point of view, we create models and optimize them to extract improved accuracy to the best of our abilities.

The following are some basic properties of data and associated models which are looked at when measuring the success of an algorithm.

Bias

Bias indicates how far away the estimator is from the target, or in other words, how far off the model prediction is from the correct value. This is directly dependent on the assumptions made about the training data. Higher the number of *assumptions*, lower the possibility that the model can generalize; an example of generalization error is assuming data is *polynomial* when it is *linear*. Hence, as a rule of thumb, a low bias is usually preferred. A high bias model usually subjects the training data to *underfitting* (discussed later in this chapter). Bias tells us whether the *mean* of the predictions is approximately at the right place or if errors are *biased* (or skewed) in one direction or another.

Mathematically, bias is defined as follows.

$$BIAS\left(\hat{X}, X\right) \equiv \mathbb{E}\left[\hat{X}, X\right]$$

2.1

where, X is the probability distribution and $\hat{X}$ is the estimator of the same probability distribution.

From a glance at equation 2.1, bias shows how much an estimate may miss a target on average.

In summary, the following are the main properties of bias.

- One of the most common reasons for *bias errors* is simplification of assumptions by models to achieve a target function that is easier to learn.

- As a family of algorithms, parametric algorithms usually hit a high bias.

  o Learning is quick, hence less computational cost, so attractive

  o They are simple but not so flexible

  o Their performance is low on predicting complex data as those data characteristics are not a good fit for simplified assumptions.

Less number of assumptions made about the target function is one way to get to *low bias*. Some of those techniques are *decision trees*, *k-nearest neighbors* (KNN), and *support-vector machines* (SVM).

Variance

Variance is an indicator of how sensitive a prediction is to the training set. Variance is directly dependent on the spread of the data. Variance tells us how much wider or narrower the estimates are compared to the future predictions' actual values. A model consisting of many degrees of freedom, such as a polynomial model, usually exhibits a high degree of variance, which leads to *overfitting* (elaborated on later in this chapter) of the training data. Mathematically,

$$VAR\left(\hat{X}\right) \equiv \mathbb{E}\left[\left(\hat{X} - \mathbb{E}\left(\hat{X}\right)\right)^2\right]$$

2.2

In other words, variance is the *average of the squared differences from the mean*. Variance measures how spread out the associated data is. Variance has two broader varieties: *population variance* and *sample variance*.

Population variance is usually designed by sigma squared (σ^2). The following are the steps to compute the *population variance* of a dataset.

Population variance is mathematically given by

$$\sigma^2 = \frac{\Sigma\left(x - \mu\right)^2}{N}$$

2.3

where x denotes the individual data points, μ is the mean, and N is the *number* of data points in the entire *population*.

The following are the steps to calculate σ^2.

1. Calculate the mean of the dataset. For example, say, I need to calculate the number of cars seen on my street in one hour. To achieve that, we record observations of three cars, one car, two cars, six cars, and three cars. The mean is the average of the data, which in this case is $\frac{3+1+2+6+3}{5} = 3.0$.

2. The differences from the mean for each data point is given by

   $(3 - 3 = 0); (1 - 3 = -2); (2 - 3 = -1); (6 - 3 = 3); (3 - 3 = 0)$

3. The square of the differences gives the following: 0, 4, 1, 9, 0.

4. The average of the squared differences gives the final variance value.

$$\sigma^2 = \frac{0+4+1+9+0}{5} = 2.8$$

Sample variance is calculated slightly differently. Whereas population variance refers to the whole dataset, sample variance refers to a subset of that dataset. Sample variance is usually taken in chunks of well-defined dataset spaces, such as counting the number of cars seen each hour for five hours in a row. In this case, we may wish to count the cars each hour with the chunks of data grouped per hour. Then we would divide the *squared differences from the mean* by $N - 1$. In other words, the sample variance s^2 is given by

$$s^2 = \frac{\Sigma(y-\bar{m})^2}{n-1} \qquad\qquad 2.4$$

where y denotes the individual data points, $\bar{m}$ is the mean, and n is the *number* of data points in the *sample* space.

Variance is the amount that *the estimate of the target function will change* if *different training data* were used [15]. As related to variance, the target function is calculated from training data and hence influenced by training data. In an ideal case, the change in the target function should be minimal if the training dataset is changed. A good algorithm should address the hidden underlying mapping between the input and the output variables.

A *low variance* is indicated when the change in estimates of the target function is *small* with respect to the change in the training dataset, for example, parametric algorithms such as linear regression and logistic regression.

High variance is indicated when the change in estimates of the target function is high with respect to the change in the training dataset. Usually, nonparametric algorithms such as decision trees tend to have a high variance. Higher flexibility in an algorithm opens it up to higher variance errors.

Standard deviation is calculated as the square root of variance $\sigma^2 \Rightarrow \sigma$. Here, σ is the standard deviation.

Irreducible errors arise due to the noise in the data and unknown variables. Hence, cleaning of data is of utmost importance in any data science-related project.

This book has code available for download for the hands-on exercises. From this chapter onward, we follow the convention that code snippets annotated with the following icon are available to download.

The following is a quick exercise to visualize concepts of variance using the matplotlib Python library and distributed random data.

If you open the variance.ipynb code file in a Jupyter notebook using Python 3.7 or 3.8 (as described in Chapter 1), you see an example of the calculation of the variance and standard deviation of a hypothetical data spread for house prices in a residential area. The code is shown in Listing 2-1, and the corresponding histogram is shown in Figure 2-4.

Listing 2-1. variance.ipynb

```
%matplotlib inline
import numpy as np
import matplotlib.pyplot as plt
houseprice = np.random.normal(150.0, 50.0, 20000)
plt.hist(houseprice, 75)
plt.show()
```

The data is generated at random for a normal distribution. The distribution is centered at approximately 150 with a standard deviation of 50.0 for 20,000 data points.

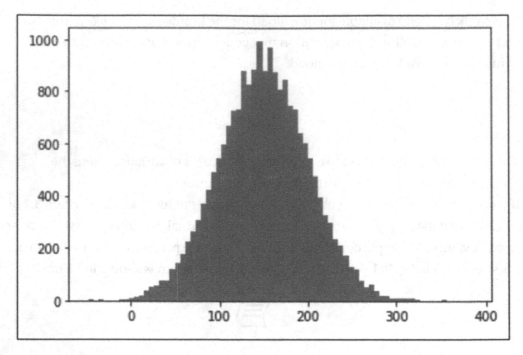

Figure 2-4. *Histogram for variance.ipynb*

The histogram shows a standard deviation of ~50.0 is around ~100 and ~200 with data centered around~150.

The standard deviation and variance of the data can be calculated using the `.std` and `.var` function calls, respectively, as shown in Figure 2-5.

Figure 2-5. *Standard deviation and variance from variance.ipynb*

For standard deviation, the value obtained is approximately 50.38, which is very close to what is specified in Listing 2-1. The variance is 2538.34, which is about 50.38^2 as we would expect since standard deviation is the square root of variance.

Bias vs. Variance Trade-off

Variance is directly related to a model's complexity. Variance typically increases if a model's complexity increases. However, increasing the complexity also reduces bias. As expected, reducing the complexity increases its bias and reduces its variance. Hence, when choosing a model, a trade-off between variance and bias is often required, and precise balancing of the two is needed.

The choice at the trade-off often comes down to overfitting vs. underfitting of data. A high degree of variance leads to overfitting, whereas a high-bias model subjects the training data to underfitting. Figure 2-6a shows an example of low bias. The zigzag graph has a low bias but high variance. In Figure 2-6b, the linear graph has low variance relative to the data, but high bias, related to error at each data point.

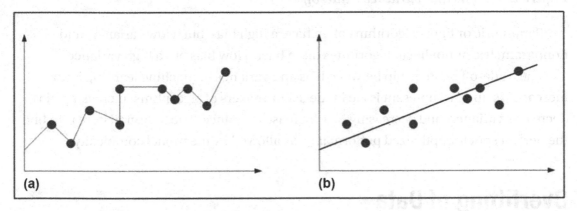

Figure 2-6. *Bias and variance trade-off*

The trade-off between variance and bias is important to reduce errors, which is the goal in any model. An error can be expressed in terms of variance and bias, as follows.

$$\epsilon = \beta^2 + \sigma^2 \qquad\qquad 2.5$$

where ϵ is error, β is bias, and σ^2 is variance.

Fewer assumptions about target function value by default enable more flexibility and hence low bias. On the other hand, fewer variations in target functions related to changes in training data give low variance and a less flexible but more stable algorithm. Figure 2-7 depicts the relationship between the two.

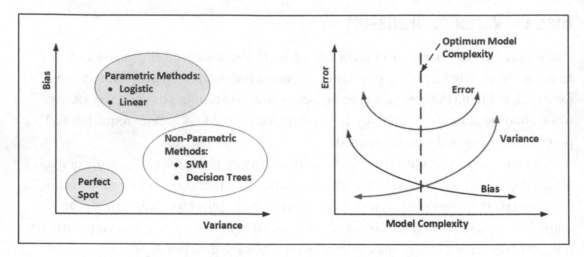

Figure 2-7. *Bias and variance trade-off*

Parametric or linear algorithms often have a high bias, but a low variance, and nonparametric or nonlinear algorithms often have a low bias but a high variance.

The trade-off relationship between bias and variance in machine learning is an inescapable and omnipresent issue for decision-makers of algorithms: increasing bias decreases variance, and decreasing bias increases variance. The holy grail here is to find the perfect spot for optimized performance as allowed by the model complexity.

Overfitting of Data

Suppose that in a soccer game, a striker scores a goal at the first scoring chance she gets. Can we then surmise that all strikers can score goals off every chance they get? We cannot make that assumption since more scoring chances end up not being goals than they do. Overestimating a predictive outcome based on an insufficient amount of data is called *overfitting* in ML. It is shown in Figure 2-8.

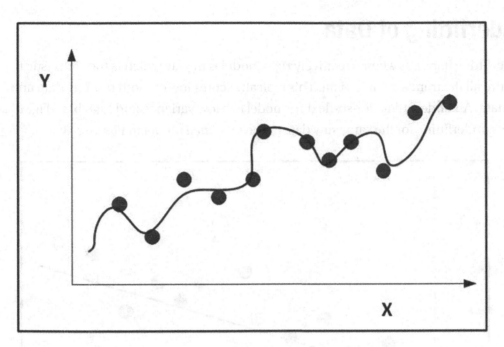

Figure 2-8. *Overfitting*

When a model is fitted too accurately to the training dataset, it appears "correct" at first glance, but because it fits the training data too "closely," it can exclude valid data that it had not "seen" before *while testing*. An example is a curvy function touching almost every data point in Figure 2-8 with very low or no error at all. In this case, the best-fit curve agrees with the training data with excellent fit but may perform poorly when evaluated using test data which the model has not seen before. Hence, the accuracy of the model's prediction suffers because of overfitting since it tends to *memorize* training data rather than *learning* from it. A model that is too flexible tends to display high variance and low bias.

Well-defined characteristics can detect overfitting. Usually, an overfitted model gives great results for training data but low performance on unseen, new test data. An example is 96% training accuracy and 68% testing accuracy. One resolution is to use an algorithm or model that better fits the data, such as k-fold cross-validation (explained later in this chapter).

Underfitting of Data

Underfitting happens when a relatively *rigid* model is in play which is too simplistic to capture all the trends in input data. This typically scores low on both training data and test data. An underfitting not-so-flexible model has low variance and high bias. Figure 2-9 shows underfitting for the same raw data as the *overfitted* model in Figure 2-8.

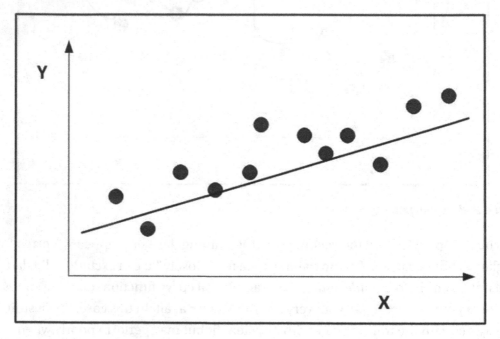

Figure 2-9. *Underfitting*

Ideal Fit of Data

Ideally, you want a model with low variance-based error and low bias-based error, as shown in Figure 2-10. Since, as shown in equation 2.5, the error is proportional to the square of bias and linearly proportional to variance. Hence, bias vs. variance trade-off is of high importance.

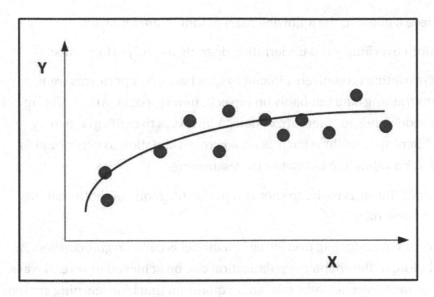

Figure 2-10. *Ideal fit*

In ML, one important desirable aspect is to achieve a "perfect spot" or "sweet spot" as a balance between the overfitting and underfitting models, as shown in Figure 2-11.

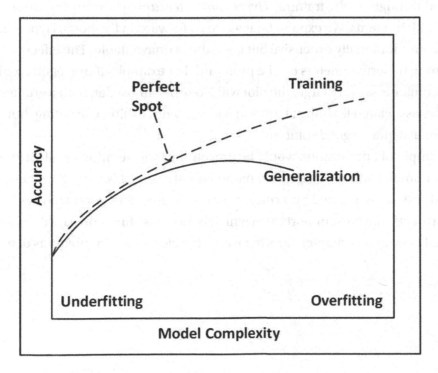

Figure 2-11. *The perfect or sweet spot*

To achieve a good fit, the following rules of thumb are followed.

- Both overfitting and underfitting degrade model performance.

- Overfitting is relatively difficult to spot because it performs well on training data but badly on unseen, new test data. An overfitting model tends to "memorize" data. A few ways to confirm are to try different algorithms such as k-fold cross-validation to verify or *hold back* a validation dataset or try resampling.

- Underfitting is easier to spot as it performs poorly on both training and test data.

The process of mitigating overfitting in a model is called *regularization*. As discussed before in the chapter, regularization can be achieved in several ways. One of the ways that more recent works relevant to quantum machine learning are looking at regularization is *optimization*, which is addressed in forthcoming chapters.

The amount of regularization applied during a learning process can be controlled with *hyperparameters*. A hyperparameter is a parameter related to a learning algorithm but not to a model. The hyperparameter is fixed before training the data and remains unchanged throughout the training. *Hyperparameter tuning* is a very important aspect of building ML systems. We expand on it as we go forward in the book. Hyperparameter tuning is computationally expensive but sometimes unavoidable. The effect of setting wrong hyperparameters can be profound. For example, if the regularization hyperparameter is set very large, the plot will likely be almost flat, with a gradient close to zero. The associated learning algorithm will likely not exhibit overfitting, but it will not produce an acceptable good solution.

An example of optimization would be preventing long iterations in algorithms by pruning parameter subsets to zero. If a parameter of a model is given by τ then, it is associated with a *cost* defined by *a cost function C*. The cost function C consists of several different aspects, the most important of which is *loss*. Loss functions are discussed in more detail later in this chapter. Loss measures the *closeness* of predictions of a model to its target.

Model Accuracy and Quality

Machine learning uses a scoring strategy analogous to school exam scores. *Accuracy* is defined by

$$accuracy = \frac{number\ of\ correctly\ classified\ cases}{total\ number\ of\ cases}$$

On the other hand, error is defined as the opposite of accuracy.

$$error = 1 - accuracy$$

It is intuitive from the definition that error is the ratio between *the number of incorrectly classified cases* and the *total number of cases*. This scheme of estimating accuracy and error is akin to rules of thumb leveraging approximation rather than employing precision-oriented algorithms.

In practice, there are well-defined ways to estimate accuracy and error and the relationship between the two. For example, some methods include *false positives* and *false negatives*. Regularization can be defined by the cost function as follows.

$$C(\tau) = regularizer + loss(\tau) \qquad\qquad 2.6(a)$$

where, *regularizer* is a penalty expression. Regularizers add constraints to training parameters. This involves a method of adding smaller coefficient vectors multiplied by a non-negative number:

$$C(\tau) = \alpha(\lambda) + loss(\tau) \qquad\qquad 2.6(b)$$

where λ is a small coefficient vector. α is a non-negative number. $\alpha(\lambda)$ is the penalty term (or the *regularizer*) and can be varied depending on how much weight we wish to attribute to it. If $\alpha = 0$, then the regularizer is zero, hence regularization is non-existent. As α gets larger, the parameters with larger norms are heavily penalized. The choice of a norm is usually taken as L1 or L2, which is the focus of future chapters in quantum computing and quantum machine learning.

Bayesian Learning

Bayesian learning (BL) [17] is an important approach to machine learning in that it has fundamentally different logic and uses probabilistic models. BL tries to translate learning into a computational problem. This perspective of learning utilizes integration rather than optimization.

Bayes' theorem states that given the probability of *B*, the probability of *A* is equal to the probability of *A* times the probability of *B* given *A* divided by the probability of *B*. Mathematically, this implies the following.

$$P(a|b) = \frac{P(a)P(b|a)}{P(b)}$$

2.7

where, *a* and *b* are subsets or values of random variables *A* and *B*, and

$$P(a|b) = \frac{P(a,b)}{P(b)}$$

2.8

where, $P(a, b)$ is the probability that both *a* and *b* take place. The term $P(a|b)$ is called the *posterior*, $P(b|a)$ is called the *likelihood,* and $P(a)$ is called the *prior*. If we are given a training dataset, *D*, with random variables *a* and *b*, the goal is to investigate the probability, $P(a, a|D)$. $P(a, a|D)$ is the probability of finding the happenings *a* , given the likelihood of *a* given *D*. Mathematically, this model is expressed as

$$P(a,a|D) = \int P(a,b|\tau)P(\tau|D)d\tau$$

2.9

where, τ is the model's parameter. $P(a, b|\tau)$ is the parametrized distribution of the model written as an unconditional probability. $P(\tau|D)$ is the probability that given the data, a certain desirable set of parameters are the correctly chosen ones. Conceptually speaking, the fundamental reason for training a dataset is to find an optimal parameter—in this case, represented by τ—for the given data. The unknown term, $P(\tau|D)$, can be calculated using Bayes' formula, which gives

$$P(\tau|D) = \frac{P(\tau)P(D|\tau)}{\int P(\tau)P(D|\tau)d\tau}$$

2.10

Here, the $P(\tau)$ addresses the previous assumption for the parameters, which leads to the best model *before* the model has seen the data. The prior $P(\tau)$ in Bayesian learning allows an *educated guess* at the parameter τ without seeing the data. As shown in equation 2.6, it can be related to the *regularizer* in certain cases. There are trade-offs between an integration-based approach and an optimization-based approach. Both have challenges when there is a lack of known structure and quickly become hard problems to solve.

Applied ML Workflow

We often get a huge set of data, sit back and wonder, "How on earth do I work with this, and how do I attack the problem?". There are some basic rules of thumb for working on ML problems. The following are the broad steps.

1.   Get the data.

2.   Explore the data. What is the data about?

3.   Build a model.

4.   Evaluate the model. Does the model work?

5.   Optimize the model.

6.   Deploy the model and monitor performance. If necessary, tune the model and try again.

Data exploration is a vital step to start the process of building an efficient ML model. This involves three main steps.

1.   Data collection

2.   Data cleaning

3.   Data processing

Whereas data collection is standard procedure, data cleaning can be one of the most time-consuming processes in the entire ML workflow. Cleaning dirty data, data filled with unintelligible special characters, spaces not well understood by algorithms, and various other sorts of noise has been the scourge of data engineers and scientists forever. These days, there are some out-of-the-box solutions for cleaning data; for example, the Google Cloud Platform (GCP)–hosted Dataprep tool.

The task of data processing can be tricky if the desired goal is ambiguous. An ML enthusiast needs to define a goal that she/he is trying to achieve by feeding a dataset to a model. These days there are specialized libraries available in Python and other languages which greatly facilitate data processing. Python specifically has NumPy, Pandas, and scikit-learn, which are leveraged throughout this book. The overall workflow for ML is shown in Figure 2-12.

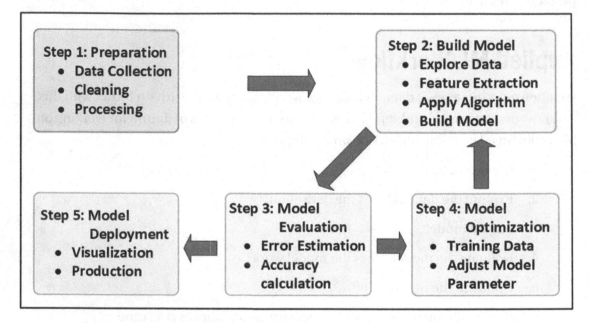

Figure 2-12. *Machine learning workflow*

The next step in data exploration is *model building*. This involves data inspection, extraction of feature sets from the data, choosing an appropriate algorithm that would serve the dataset best, and finally build the desired model.

The third step is *model evaluation*, which involves error estimation and accuracy calculation.

The fourth step is *model optimization,* where the model is fed more data, and parameters are tuned to obtain better performance. Once optimization is done, the output is fed back into steps 2 and 3 until the error and accuracy level are considered acceptable.

The fifth and last step is *model deployment* which involves production-level usage and visualization.

Validation of Models

Once we decide on a model and have built and tested it, we need to validate it. It is a mistake to validate the model with data that it has already seen. For example, if we reuse training data to test the model, then its prediction will be quite good since it was tested on data that it has already seen. Because the model's prediction on this dataset is a "correct" one, it delivers false confidence in the efficiency of the model's predictions. The same model, when tested with new data, will likely do much worse in predictions. Hence, a common practice is to separate the datasets for training and testing, some of which are discussed briefly next.

The Hold-Out Method

This method simply means that part of the data is subjected to "hold out" as a training set to evaluate models and select the best one; then test the model (i.e., validate it with the remainder of the data as a test set). This means the "test set" is chosen at random and reserved for testing.

Splitting the data in the hold-out method is a common practice in ML. As a rule of thumb, data is split randomly into 80% or 75%, or 70% training data and 20% or 25% or 30% testing data, as shown in Figure 2-13.

There is no set rule for this split. The split is done keeping in mind the nature of the data and the computational cost of training. Higher the volume of the training data, the more the computational cost. As a matter of discussion, if the split in the dataset would be, for example, 50% training and 50% testing, then that is *not* considered as "good" since the volume of training data is too low, raising a risk that the model may not have seen enough data to have trained efficiently. Similarly, a split of circa 95% training and 5% testing data is not desirable as then the testing data percentage is too low. This is not acceptable because if the final model is trained on a set too large, it is not ideal to test it against models trained on a much smaller dataset. This may be akin to selecting a Formula 1 car to participate in the Dakar Rally in the desert.

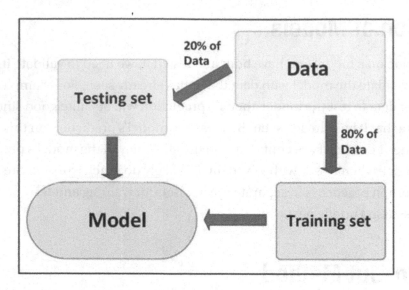

Figure 2-13. *Hold-out validation*

Although the hold-out method works for general-purpose data, it is not very efficient when some of the data may be biased. For example, consider an ML exam where a student gets 20 questions out of a question bank of 50, which is a mix of easy and difficult questions. Because in the hold-out method, the data is chosen randomly, we choose the questions randomly in this case. Let's say the student scores 80% in this particular examination. In this case, can we consider that his marks reflect his true knowledge of the subject? We cannot. This may be because most of the questions were easier due to the random nature of question selection. To get a true essence of his knowledge, we need to conduct/run more tests with the same student and similarly randomly chosen datasets and then take an average of the score.

As the training/testing split in the hold-out method is done at random, a model can learn well and deliver great predictions if a well-rounded dataset representing the entire data characteristic is chosen. Additionally, if the test set is also a relatively easy one, the model can deliver high accuracy. But the randomness of the choice of data makes the accuracy of the training and subsequent prediction somewhat unreliable. For example, we can pick a "weak" dataset for training that lacks data diversity—where the model does not learn much, and end up testing it against the remainder of the data, which includes most of the diversity unknown to the model giving a poor model performance. Hence, in the hold-out method, model accuracy can vary significantly based on the random split of data.

The Cross-Validation Method

How do we ensure a better, more efficient way to validate data-driven models? One solution is *cross validation*. The basics of cross validation are as follows: the model is trained on a training set and then evaluated once per validation set. Then the evaluations are averaged out. This method provides higher accuracy. However, because the training time is multiplied by the number of validation sets, this is more computationally expensive than the previous hold-out method. One of the cross-validation methods that are also common is *k-fold cross-validation.*

In a general case of choosing models, there is a chance that the training data is not representative of the whole dataset, in which case the dataset may be skewed, and overfitting may occur. k-fold cross-validation helps avoid this scenario as in this case, the model is trained using data in the training dataset. Then the model performance is evaluated using the data reserved for testing.

In k-fold cross-validation, the data is divided into 10 to 20 chunks of *k* equal sections, known as *k-folds*. Figure 2-14 shows a diagram of the k-fold cross-validation workflow.

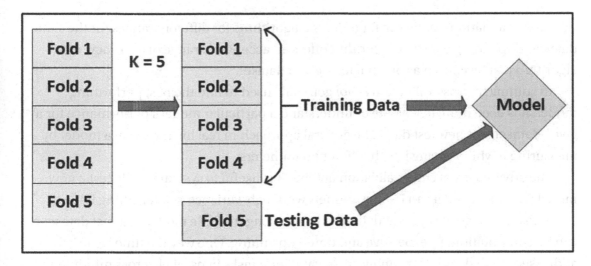

Figure 2-14. *k-fold cross-validation*

Usually, one of the folds, say fold- *i* (=5 in Figure 2-14), is reserved for *testing*. The other folds in Figure 2-14 are used for training. Once training is completed, the model is tested against fold- *i*. In this manner, after all the folds are cycled through with training and testing, the accuracies of predictions are compared.

An example of k-fold cross-validation is depicted in Figure 2-15, where k = 6.

Runs	Splits						Accuracy
1	Test	Train	Train	Train	Train	Train	80
2	Train	Test	Train	Train	Train	Train	82
3	Train	Train	Test	Train	Train	Train	88
4	Train	Train	Train	Test	Train	Train	90
5	Train	Train	Train	Train	Test	Train	85
6	Train	Train	Train	Train	Train	Test	92

Figure 2-15. *Six-fold cross validation*

This six-fold cross-validation has its data split into six chunks, one reserved for testing and the remaining five data folds reserved for training. The accuracy column shows a variation of accuracy between 80% to 92%. The average accuracy is given by

$$\frac{80+82+88+90+85+92}{6}=86.17$$

Cross-validation can be run for different algorithms for different ranges for the same dataset. Depending on which algorithm offers an acceptable range of accuracy, that algorithm can be chosen as optimal for a given dataset.

In summary, cross-validation is *not* generally used to find the best performing model. It is used more to validate or understand a particular model's performance for a given dataset and new test data. The general approach in real life is to tune a model by measuring it while varying it with different parameters.

The advantages of cross-validation are that it is useful to systematically test a new model with data, and it can identify models with high-variance or over-fitting issues.

A disadvantage of cross-validation is that running multiple models against datasets can be computationally expensive and time-consuming. One way the time issue is addressed these days is by running cross-validation tasks in parallel across multiple CPU/GPU cores or on a cluster.

As an exercise to have a feel of k-fold CV (cross-validation), let's look at the kFoldCV. ipynb code. The data used here is a publicly available dataset called *Iris*. This specific dataset consists of measurements of length and width of petal and sepal of 150 iris flowers. There are three classes of iris flowers with 50 instances each in the dataset: Iris virginica, Iris versicolor, and Iris setosa [13], as shown in Figure 2-16.

Figure 2-16. Iris setosa (by Radomil, CC BY-SA 3.0), Iris versicolor (by Dlanglois, CC BY-SA 3.0), and Iris virginica (by Frank Mayfield, CC BY-SA 2.0)

The code in Listings 2-2a, 2-2b, and 2-2c looks at a model which tries to predict the correct Iris species from its given attributes (width and length of petal and sepal).

Listing 2-2a. k-Fold Cross-Validation of Iris Dataset from kFoldCV.ipynb

```
#Import Libraries
import numpy as np
from sklearn.model_selection import train_test_split, cross_val_score
from sklearn import svm  #import (SVM) for classification
from sklearn import datasets
iris = datasets.load_iris() #load the data
# Data split into train/test sets with 25% for testing
X_train, X_test, y_train, y_test = train_test_split(iris.data, iris.target,
test_size=0.25, random_state=0)
# Train data for Linear Support Vector Classifier (SVC) model for
prediction
clf = svm.SVC(kernel='linear', C=1).fit(X_train, y_train)
# Performance measurement with the test data
clf.score(X_test, y_test)
```

The cross_val_score library of scikit-learn [14] allows us to evaluate the score by cross-validation. The data is input to the train_test_split() function, which facilitates splitting of the data into training and testing chunks. In this model, we start out by training 75% of the data and testing 25%, which is specified by test_size=0.25. As per conventions of k-fold CV, the test data is extracted at random (random_state=0) and reserved for testing.

This scheme gives us four datasets: X_train contains 75% of Iris measurement data, X_test contains 25% of Iris measurement data reserved for testing, y_train and y_test contains species corresponding to the measurements.

The model used is *support-vector classification* (SVC), which is part of support-vector machine (SVM). It is discussed later in this chapter. The model is named svc here. The score() function measures the performance of the SVC model. The SVC model is scored against the reserved test dataset. The score is given as

$$0.9736842105263158 \hspace{4cm} 2.11$$

However, since the dataset size is only 150, it is a fairly small dataset, with 75% of it used for training and 25% reserved for testing. Hence, there is a good chance that we are overfitting. K-fold cross-validation is a good way to verify that and see if we have a better fit.

In the next part of the kFoldCV.ipynb code, shown in Listing 2-2b, the SVC model is used to pass on the entire dataset (i.e., measurements of iris.data) and the target data iris.target to the function cross_val_score(). We are using cross-validation cv=5, which implies that it will use five different datasets and keep one dataset in reserve for testing. The datasets are chosen randomly.

Listing 2-2b. Continued: 5-Fold Cross Validation kFoldCV.ipynb

```
#cross_val_score a model, the data set, and the number of folds:
cvs = cross_val_score(svc, iris.data, iris.target, cv=5)
# Print accuracy of each fold:
print(cvs)
# Mean accuracy of all 5 folds:
print(cvs.mean())
```

This snippet of code returns a list of error values as output from each iteration (5 in total) or "fold" as follows.

$$[0.96666667 \quad 1. \quad 1. \quad 0.96666667 \quad 1. \quad]$$

These five values are then averaged to give an error metric for the k-fold CV for the Iris dataset.

$$0.9866666666666667 \qquad\qquad 2.12$$

If we compare the results of 2.12 and 2.11, we can immediately surmise that the results of the 5-Fold CV (2.12) are better than before; hence this is a definite improvement.

The usual approach here is to see if there can be an improvement over this result. To obtain the result of 2.6, a linear kernel (kernel='linear') was used. An approach would be to test the model and corresponding data behavior with a polynomial-based kernel that may better fit. Whether a linear fit is preferable over a polynomial one depends on the data characteristics. Hence, looking at the next part of the kFoldCV.ipynb code, shown in Listing 2-2c, we try out a kernel, which is *poly* for polynomial.

Listing 2-2c. Continued: K-Fold Cross-Validation with Polynomial Kernel from kFoldCV.ipynb

```
svc = svm.SVC(kernel='poly', C=1).fit(X_train, y_train)
cvs = cross_val_score(svc, iris.data, iris.target, cv=5)
print(cvs)
print(cvs.mean())
```

Once the polynomial kernel is run, it gives the following as five-fold error values and corresponding average.

$$[0.96666667 \quad 1. \quad 0.96666667 \quad 0.96666667 \quad 1. \quad]$$

$$0.9800000000000001 \qquad\qquad 2.13$$

A comparison between 2.13 and 2.11 shows that the polynomial kernel gives a value lower than the linear one. This indicates that the polynomial is probably overfitting. The lesson to learn here is that had we not used the k-fold CV for verification, and instead used a standard train/test split, we would not have realized that we were overfitting. This is a simple but good example of the importance of a k-fold CV.

A linear model is generally less flexible as it only has two degrees of freedom, whereas a high-degree polynomial is quite flexible due to its several degrees of freedom. The results of a k-fold cross-validation run are often summarized with the mean of the model skill scores. It is a good practice to include measuring the variance of the skill scores, such as the standard deviation or standard error. CV is popular because of its simplicity. It generally results in a less biased or less optimistic estimate of the model skill than other methods, such as a simple train/test split.

Regression

Regression is the study of how to best fit a curve to summarize data. For example, agricultural output is "usually" directly relevant to the amount of rainfall. Regression constitutes some of the most powerful and most widely studied algorithms of supervised learning. In this approach, we try to understand the data points by the curve that may best fit them. There are several types of regression algorithms. As we move into the domain of quantum machine learning, there will be more to say about regression and associated algorithms.

Linear Regression

As the name suggests, *linear regression* is a straight line fit to a set of observed data. An example is the height and weight of players on a soccer team. Figure 2-17 shows the data and the curve from this dataset.

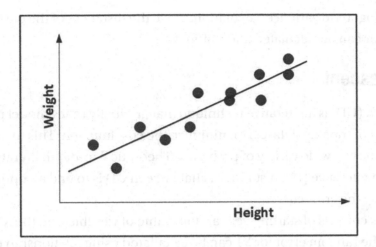

Figure 2-17. *Linear regression*

In Figure 2-17, the X axis shows the height, and the Y axis depicts the weight. Once the data points are plotted, the coach may look at the plot and think that while most players' height and weight match, some don't, and they may need to start a fitness regime to gain or lose some weight. If the coach drew a curve to best fit these data points, he would likely end up with a straight curve or a linear plot.

In this example, the curve's equation is given by a straight line, which is $y = mx + c$, where m is the gradient of the line and c is a constant. Linear regression can be modeled as a deterministic function [5].

$$f(x;\tau) = \tau_0 + \tau^\epsilon x \qquad\qquad 2.14$$

where, the parameters are contained in vector $\tau \in \mathbb{R}^N$ and $x \in \mathbb{R}^N$, where $\mathbb{R}$ is the space of real number and N is the dimension of the parameters. The bias τ_0 can be included in $\tau^\epsilon x$ as an extra dimension.

Learning in linear regression considers finding parameters τ which may fit function f in equation 2.14 to the training data for an efficient prediction.

Least Squares Technique

The *least squares technique* consists of using the square loss (explained later in the chapter). The errors generated in a straight-line fit are given by the distance of the individual data points to the line. We square up all these errors and calculate their sum. This way, if there are negative errors, then they do not cancel out the positive ones and give us a false positive from an error-generation perspective.

This is analogous to variance calculation. Once the distances of the data points are known, the error can be calculated and minimized.

Gradient Descent

Gradient descent (GD) is an iterative technique that gradually tunes model parameters for a specific set of training datasets to minimize the cost function. This technique finds optimal solutions to a wide variety of problems. Theoretical in-depth literature of GD can be found in reference [15]. A standard challenge in GD is to find an optimal value *without* knowing the error curve!

The process consists of starting at a random value of variable x on the X axis in Figure 2-18a. The random error for X1 can be calculated using relationship error = f(x1). The gradient of the error can also be calculated using the slope. Mathematically, we need to take the *derivative* of the error function to calculate the gradient.

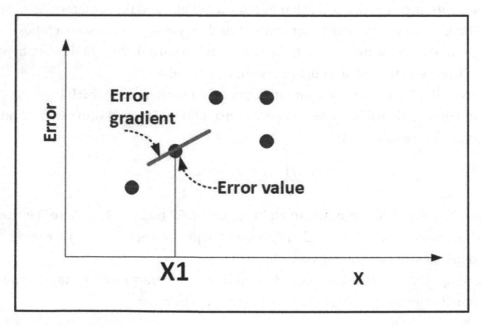

Figure 2-18a. *Gradient descent step 1*

The error derivative gives us an indication of which way the slope or gradient is inclined. At this point, a second data point x2 is calculated at random (if unknown), as shown in Figure 2-18b.

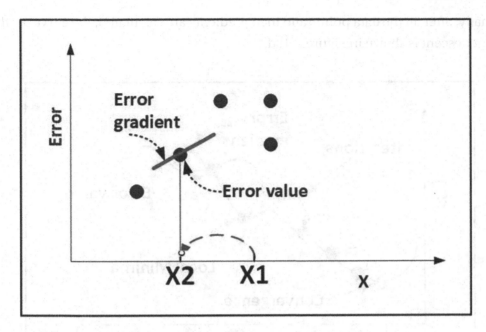

Figure 2-18b. *Gradient descent step 2*

Next, as shown in Figure 2-18c, the error for point x2 is calculated as error = f(x2), the error gradient is measured, and then we "guess" a value for point x3 and repeat the process.

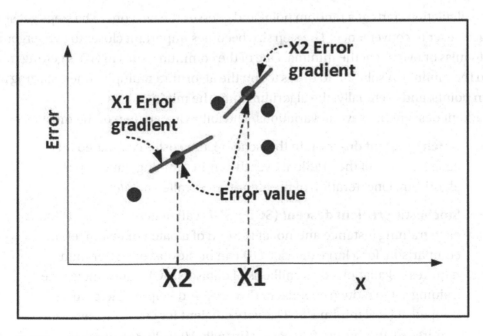

Figure 2-18c. *Gradient descent step 3*

Finally, after all the data points and their gradients are calculated, the curve for the gradient descent is shown in Figure 2-18d.

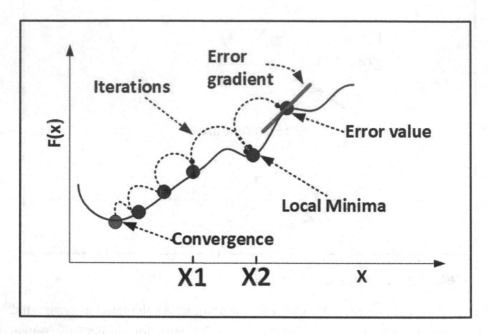

Figure 2-18d. *Gradient descent step 4*

GD calculation starts at a random point and guesses the next one. The steps get smaller as we get closer to convergence. The step size becomes important closer to convergence so as not to miss or overshoot the minima. One of the common issues in GD is getting trapped close to the minima. A solution to this is to run the algorithm multiple times, starting at random points, and eventually, the algorithm finds the minima.

Gradient descent has several variations. The following are two of the most important.

- **Batch gradient descent**: In this scheme, the cost is calculated for each iteration of the gradient over the entire training dataset per algorithm. One iteration of the algorithm is called a *batch*.

- **Stochastic gradient descent (SGD)**: SGD calculates coefficients for each training instance and not at the end of a batch of instances. In comparison, for a large dataset, GD can be slow as each iteration requires calculations over millions of datasets. SGD randomizes the training set to reduce coefficient diversity and helps avoid getting trapped at local minima. SGD is very efficient for large datasets as it requires relatively very few passes (usually 10 to 20) to converge.

Nonlinear and Polynomial Regression

Polynomial regression is used in more general cases of regression where the data distribution may not be well suited for a linear (straight line) fit. A polynomial fit allows more degrees of freedom. However, more degrees of freedom are not always great; depending on the dataset, it may cause overfitting.

At present, the most successful nonlinear regression models in ML are *neural networks*, which you look at in the next chapter.

Classification

Classification in machine learning algorithms is a basic task that deals with assigning a class label to examples from the problem domain.

There are various types of classification for predictive modeling in machine learning. This section focuses on the following types.

- Binary classifiers

- Multi-class classifiers

- Multi-label classifiers

- Imbalanced classifiers

Categories of uniquely labeled data offer the obvious benefit of characterizing data for specific tasks. The previous section addressed *regression*, which is largely about curve fitting to a dataset. You saw that the best fit curve is a function which inputs a data item and assigns a number. In contrast to regression, *classification* refers to an ML model which assigns discrete labels to its inputs. Classification is typically a supervised learning algorithm dealing with a *class* or discrete value. A typical input is a feature vector, and a typical output is a class.

A great example of *predictive modeling* is our first use cases of ML earlier in this chapter: the spam email filter. This basically involves classifying emails as "spam" or "not spam."

For a *binary classifier*, the output is *binary* (i.e., Yes/No, True/False, On/Off, etc.). From a *neural network* (discussed in the next chapter), only one neuron in the output layer can achieve this, and common activation functions are `sigmoid` or `logistic`. The following are examples of where binary classification works.

73

- Spam filter: Is it spam? Yes/No

- Transaction fraud prediction: Is it fraud? Yes/No

Binary classification tasks typically involve a class 0, which is considered the normal state, and a class 1, which is called the *abnormal state*. For example, "not spam" would be considered the normal state and "spam," the abnormal state. Certain algorithms, such as logistic regression and support-vector machines (SVM), support binary classification and do not *natively* support more than two classes. In its basic form, SVM does not support multiclassification [272]. It only does so via a process where the multiclassification problem is broken down into multiple binary classification problems.

Intuitively, a *multi-class classifier* is not binary. An example is classifying a number or digit into one of the first ten numbers {0 to 9}. From a *neural network* perspective, this requires *one neuron per class*. Hence, typically *more t*han one neuron is required for this type of classifiers. Another example of a multi-class classifier is optical object classification (face, plant, etc.). In a neuron networks treatment, the *softmax function* deals with multi-class classifiers.

Multi-label classification refers to tasks with two or more class labels, where one or more class labels may be predicted for each data point. An example is a photo with multiple objects in it and a corresponding model which may predict the presence of multiple known objects, such as car, elephant, human, and so forth. In binary and multi-class cases, a single class label is dealt with, whereas in multi-class, there are two or more class labels.

Imbalanced classification involves tasks where the number of examples in each class is *not* equally distributed. Typically, imbalanced classification tasks are binary classification tasks. For example, much of the data points in the training dataset belong to class 0, and a minority of examples belong to class 1. Specialized techniques such as undersampling the majority class or oversampling of the minority class are sometimes used to manipulate the training dataset for ease of computational tasks. Similarly, specialized modeling algorithms are also used.

Data-Driven Prediction

Most ML algorithms have the same overall characteristic: they are fed input data and output responses to a query. In this age of big data, the inspection and evaluation of real-time raw data are of primary importance, especially in the finance industry and cybersecurity. When a financial body wishes to predict stock prices, they may look at time series values for a price *forecast.*

If the treatment of the data involves image analysis, then the content and corresponding sorting of the images are treated as *classification.* Usually, the data is considered the input, and the model's prediction is considered the *output.*

Examples of predictive models and their usage are varied and vast. One of them is *image recognition.* If we see an elephant, we recognize it as an elephant from optical stimuli and because we learned what an elephant is by looking at them or photos of them. However, it is not intuitively obvious how to program a computer to instinctively recognize an elephant. In ML, we present many images to the computer, and it learns, from those images, what an elephant "looks" like. Similarly, the computer can also be taught what a tiger looks like. Hence, after several such *training* sessions with images as input data, the computer learns how to classify elephants and tigers. The more images the system sees of elephants and tigers in various shapes, sizes, and varieties of postures, the more efficient it becomes in recognizing and classifying them.

The *forecasting* of *time series* data is an important predictive application. A set of data points recorded at various points of time constitutes time series data. The data points are collected at adjacent periods of time; hence, it is logical to consider the possibility that there is a correlation between the observations. If we consider the recorded values of the dollar for eight consecutive years covering two United States elections, we can find records of important macroeconomic variables such as gross domestic product (GDP), price of gold, or price of crude oil during the same time period. Since the price of the dollar is affected by other indicators and political change and we learn its price behavior through time, we may be able to predict how the dollar price varies in a given year if a certain political party wins the presidential election.

Another important prediction is related to medical or fault-finding where it is called a *diagnosis.* For example, if we tried to predict the occurrence of cardiac arrests based on people's cholesterol level, blood pressure, age, weight, smoking habits, or exercise regime, we would not be able to predict a heart attack with any certainty. Instead, we may be able to predict how likely it is to happen given those factors. This type of prediction is based on *logistic regression.*

Note   *Logistic regression*, even though it sounds as if it is a *regression* algorithm, is a classification algorithm. Logistic regression is considered a regression model in statistics adopted by machine learning researchers who used it as a classifier.

The following are some common predictive models.

- Linear regression

- Multivariate regression

- Polynomial regression

Complexity

The complexity of a problem refers to the study of classes of problems that involve an engagement of resources from the perspective of algorithms, time of computation, and so on. To estimate the overall computational cost, an analysis of resources is required. This is done by running an algorithm that is needed to solve the problem.

The time a computer takes to run an algorithm is the runtime of that algorithm on that specific computational platform. As per the fundamentals of classical computation, a solution for the target of a database search can be obtained by verifying items in the database until the one desired item is found. This exercise requires a worst-case scenario of time complexity $O(n)$.

The computational complexity of algorithms is a quantitative expression of a measure of how the algorithm's time, size, and associated resource requirements evolve with the growth in a dataset or problem size. An algorithm is designed to be applied to an unknown function or a set of data to determine properties of the function (such as period or minima), factor a number, or classify the data.

The number of inputs (i.e., size of the data) to the function could be anything from a single digit to trillions or more. There are a lot of algorithms, and we have the luxury to pick and choose and try them out until we find the algorithm that does the job. Once we select an algorithm that we think serves our purpose or goal, we need to decide on the response to the following question. How does the algorithm's running time behave with an increase in the number of inputs? This is known as the *sensitivity* of the algorithm to *size*.

As an algorithm's problem size grows, so does its running time. This is known as *time complexity*. The following are some of the varieties of time complexities that are addressed in the next chapters.

- Constant time complexity

- Polynomial time complexity

- Non-polynomial time complexity

Chapter 1 touched on computational complexity, which is addressed throughout the book. It is discussed in depth when we get to the complexity of quantum machine learning.

Confusion Matrix

Now that we have explored the fundamentals of classification, let's look at *true positive* (TP), *false positive* (FP), *true negative* (TN), and *false negative* (FN). When a classifier says that a data point is *X* and it is *X*, which is known as the *true positive*. A false positive would be a case when the classifier identifies a data point as *X*, but it really is a *Y*. A true negative is when a data point is classified as *not X*, and it *really is not*.

The most important classification metric is *accuracy*, which was discussed earlier in this chapter when we looked at variance and bias. Accuracy was defined, but let's expand on that definition here.

$$accuracy = \frac{number\ of\ correctly\ classified\ cases}{total\ number\ of\ cases}$$

$$= \frac{TPs + TNs}{total\ number\ of\ cases}$$

It is also of interest to data scientists and engineers to know how well a classifier can avoid false alarms. This is defined by the *precision* of a classifier.

$$precision = \frac{TPs}{TPs + FPs}$$

In case we do not wish to miss out on any TPs and wish to catch as many Xs as possible, then the measure for that success is known as the *recall*.

$$recall = \frac{TPs}{TPs + FNs}$$

The standard way to represent the TPs, FPs, TNs, and FNs visually is called a *confusion matrix*. For example, for binary classification, the **confusion matrix** is a 2 × 2 table shown in the following format.

	Classified as **YES**	Classified as **NO**
True **YES**	No. of TPs	No. of FNs
True **NO**	No. of FPs	No. of TNs

where, TP is true positive; FP is false positive; TN is true negative; and FN is false negative.

For example, if patients in a clinic are classified as correctly diagnosed with cancer, or wrongly diagnosed with cancer, or having cancer but not diagnosed, or do not have cancer and were correctly diagnosed as not having the disease, then the *confusion matrix* for the cancer diagnostic process would look like the following table.

Actual Condition	Predicted Condition	
	Predicted "Has Cancer"	**Predicted "Does Not Have Cancer"**
Has Cancer	Patients with cancer correctly diagnosed	Patients with cancer not diagnosed
Does Not Have Cancer	Patient without cancer falsely diagnosed as having cancer	Patients without cancer correctly diagnosed

The f-*score* or *f-measure* is the harmonic mean of *precision* and *recall*. The f-score is used because while precision and recall are very important measures, dealing with only one of them does not present us with the full scenario. F-score is a way to summarize both these values and present a more realistic measure of the classifier's performance.

Supervised Learning

A *supervised learning* ML algorithm is presented with input data samples or *training sets* that are *labeled* and outputs responses to specific questions. In this case, the training set includes the desired solution or the *labels*, which allows it to learn from the data. The goal here is to learn from the training set so that meaningful predictions can be made for new data not previously seen by the algorithm. Supervised learning is usually divided into two types: regression and classification.

The spam filter use case is an example of a supervised learning task called *classification*. The filter is trained with known spam samples throughout time. The more training data it sees, the better it gets at its task of classifying which mail is spam and which one is not. The algorithm would classify an email under the category of spam classification and "fraud" or "not fraud." Table 2-1 lists some of the most common supervised learning algorithms and associated example use cases.

Table 2-1. *Examples of Supervised Learning Algorithms*

Supervised Learning Methods	Example	Algorithms
Regression	Stock price prediction, House price prediction	Linear regression, ridge, lasso, polynomial
Classification	Spam: Fraud/Not Fraud, Cancer/Not Cancer	Logistic regression, SVM, KNN
Decision Tress	Predict stock prices (regression), classification (credit card fraud)	Decision trees, random forest

Similarly, if we use the Wisconsin Breast Cancer Dataset[1] [12] to train an ML model such as logistic regression or KNN, a prediction can be made on whether a patient suffers from cancer based on the classification of the features.

One of the main goals of supervised learning is to achieve low bias and low variance errors.

From Data to Prediction

The path to prediction from data in supervised models involves four basic steps.

As a first step, the data is cleaned, processed and a *model family* is chosen as a first guess. Since this is supervised learning, the data needs to be labeled, and the data features and definitions are known. As an example, the data can be represented by a linear weighted function. The model is trained by fitting parameters. The steps are shown in Figures 2-19 (a–d).

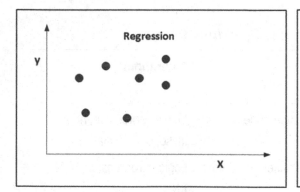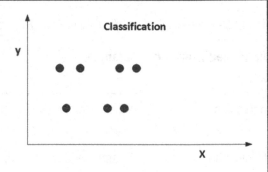

Figure 2-19a. *Step 1 for data-driven prediction*

---

[1]This refers to a publicly available dataset in the Wisconsin Breast Cancer database created by Dr. William H. Wolberg, a physician at the University of Wisconsin Hospital. This dataset contains 569 samples of malignant and benign tumor cells. Dr. Wolberg used fluid samples, taken from patients with solid breast masses and an easy-to-use graphical computer program called Xcyt, which can perform the analysis of cytological features based on a digital scan. The program uses a curve-fitting algorithm to compute ten features from each one of the cells in the sample and then calculates the mean value, extreme value, and standard error of each feature for the image. The dataset available at https://archive.ics.uci.edu/ml/datasets/Breast+Cancer+Wisconsin+(Diagnostic).

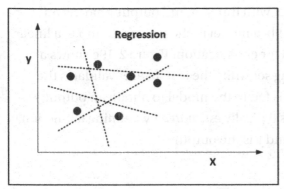
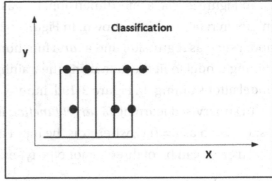

Figure 2-19b. *Step 2 for data-driven prediction*

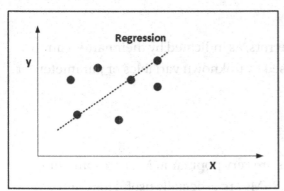
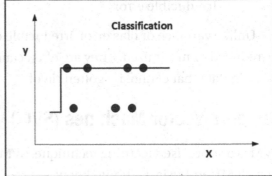

Figure 2-19c. *Step 3 for data-driven prediction*

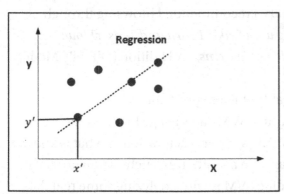
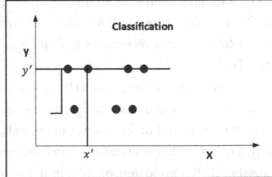

Figure 2-19d. *Step 4 for data-driven prediction*

In Figure 2-19a, a one-dimensional dataset with inputs x and output y having an unknown relationship is shown. In Figure 2-19b, a model is chosen; for example, a linear model such as regression and a *tanh* function for classification. Figure 2-19c shows a training model to fit the data using the training set while the test dataset validates the model after training. In Figure 2-19d, input x' is fed to the model to obtain an output y'.

In supervised learning, a *target function* is typically *estimated* by defining a function first up, such as $y = f(x)$, where x is the input and y is the output.

An error can be of three major type types.

- Variance error

- Bias error

- Irreducible error

Unlike variance or bias error, irreducible errors, as indicated by their name, cannot be reduced or minimized. They are often caused by unknown variables or parameters or noise in data that cannot be gotten rid of.

Support-Vector Machines (SVM)

SVM is a supervised learning technique. SVMs are very popular in ML and something that all ML enthusiasts should know as a tool. SVMs are especially useful for the classification of small to medium-sized datasets which exhibit complexity. This section focuses on the core functionalities of SVM. These ML algorithms are mathematically robust and have been worked on since the 1990s. I recommended browsing through Aurélien Géron's *Hands-On Machine Learning with Scikit-Learn and TensorFlow: Concepts, Tools, and Techniques to Build Intelligent Systems*, 2nd Edition (O'Reilly Media, 2019) [15].

SVM is an overall easy way to have reasonably efficient predictions from a complex higher-dimensional dataset with multiple features. SVM can salvage higher dimensional support vectors and divide them categorically. The support vectors define what is known as *hyperplanes*. Different kernels can be leveraged to achieve this, such as polynomial or Gaussian RBF. A good approach to understanding SVM is with analyzing some real data. In this exercise, we work with the Iris dataset again and adapt libraries from `scikit-learn` [16].

SVM Example with Iris Dataset

A quick revisit to the Iris dataset reminds us that the publicly available dataset consists of measurements of length and width of petal and sepal of 150 iris flowers. There are three different classes of iris flowers of 50 instances, each included in the dataset: Iris virginica, Iris versicolor, and Iris setosa [13]. You saw usage of *support-vector classification* (SVC), a typical classification for SVM. The SVC class allows the addition of several polynomial features *without requiring to add them.* This helps avoid the pressure of enlarging dimensions of numbers of features from a computational cost perspective because we do not add any features.

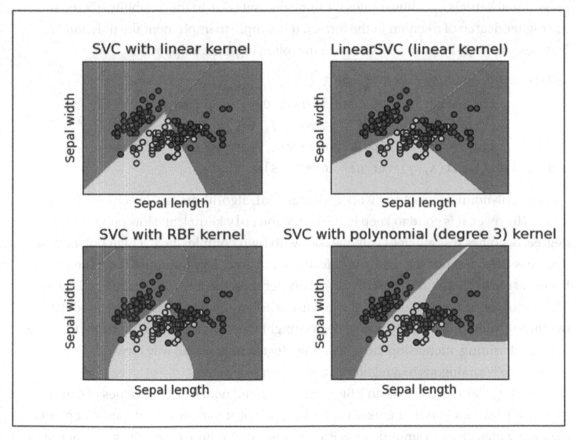

Figure 2-20. *SVM with Iris dataset (Courtesy of scikit-learn documentation [16])*

Figure 2-20 (the code to generate the graphics is available at `https://scikit-learn.org/stable/auto_examples/svm/plot_iris_svc.html`) shows sepal length and sepal width, with linear, polynomial, and Gaussian RBF kernels. The flexibility of the polynomial kernel against the linear ones is obvious if we inspect the graphics visually. The linear models have linear decision boundaries (intersecting hyperplanes). The nonlinear kernel models (polynomial or Gaussian RBF) have more flexible nonlinear decision boundaries with shapes that depend on the kind of kernel and its parameters.

An SVM model creation using `scikit-learn` inevitably involves specifying some hyperparameters such as C (see the following code snippet). A low or high C affects the margin definitions of the graphs. Previously, we discussed some advantages of polynomial kernels over linear ones for some datasets due to the flexibility offered by the higher degree of freedom in the former. It is simple to implement the polynomial features via the `poly` kernel, as shown in the following code snippet.

```
models = (svm.SVC(kernel='linear', C=C),
         svm.LinearSVC(C=C, max_iter=10000),
         svm.SVC(kernel='rbf', gamma=0.7, C=C),
         svm.SVC(kernel='poly', degree=3, gamma='auto', C=C))
models = (clf.fit(X, y) for clf in models)
```

The polynomial kernel can work with many ML algorithms and is not specific to SVMs. However, it is good to keep in mind that the `poly` kernel, at a low polynomial degree, becomes less efficient with datasets with high complexity. If a high polynomial degree is used, it tends to create many features at a great computational cost (i.e., it becomes slow). Note that LinearSVC is a lot faster than the SVC with `kernel = 'linear'`.

One of the issues to look out for when using polynomial kernels is if the model is overfitting, reducing the polynomial degree might help. On the other hand, if it turns out to be underfitting, increasing the polynomial degree may be the way forward. The gamma in the code is analogous to a regularization hyperparameter.

The SVC class with Gaussian RBF kernel is defined with different values of C and gamma. If gamma increases, it makes the bell-shaped part narrower, and a smaller gamma makes it wider, giving a smoother decision boundary. In these cases, the gamma acts as a regularizing hyperparameter. If the model is overfitting, gamma should be reduced; if it is underfitting, gamma should be increased.

Other types of kernels are less commonly used, such as the *string kernels* for classifying text or DNA sequences. As a rule of thumb, the linear kernel is tried first to deal with large training sets or large feature sets. For very large datasets, the Gaussian RBF kernel works well too.

As an exercise for SVM, it may be useful to look at visualization options along with classification. We utilize Python's `seaborn` library, which is built on top of `matplotlib`. The `seaborn` library provides a high-level data visualization interface to create a matrix representation called a *heatmap* to represent data in a two-dimensional form. The data values are represented as colors in the graph. The goal of the heatmap generated later in this section (see Figure 2-21) is to provide a colored visual summary of information. The SVM classification exercise code is downloadable from this book's website as `svmIris.ipynb` to be run on a Jupyter notebook and is shown in Listings 2-3 (a–d).

Listing 2-3a. SVM Classification of Iris Dataset: Importing Libraries and Observing Data

```
#import libraries
import pandas as pd
import matplotlib.pyplot as plt
import seaborn as sns
%matplotlib inline
#Get the data and define col names
colnames=["sepal_length_in_cm", "sepal_width_in_cm","petal_length_in_
cm","petal_width_in_cm", "class"]
#Read the dataset
dataset = pd.read_csv("https://archive.ics.uci.edu/ml/machine-learning-
databases/iris/iris.data", header = None, names= colnames )
#See the Data
dataset.head()
```

Listing 2-3a imports required libraries, including `seaborn`, downloads the Iris data, and opens the data header definition up in the following table format.

	sepal_length_in_cm	sepal_width_in_cm	petal_length_in_cm	petal_width_in_cm	class
0	5.1	3.5	1.4	0.2	Iris-setosa
1	4.9	3.0	1.4	0.2	Iris-setosa
2	4.7	3.2	1.3	0.2	Iris-setosa
3	4.6	3.1	1.5	0.2	Iris-setosa
4	5.0	3.6	1.4	0.2	Iris-setosa

ML algorithms work better with numbers than text. This means that if the data contains categorical data, we must encode it to numbers before they can be fitted to evaluate a model. Two popular techniques are *integer encoding* and *one-hot encoding*. A newer technique, called *learned embedding*, provides a useful middle ground between the other two methods.

- **Integer encoding**: Each unique label is mapped to an integer.

- **One-hot encoding**: Each label is mapped to a binary vector.

- **Learned embedding**: A distributed representation of the categories is learned.

Looking at the data format, it may be useful to change the class to numbers rather than the names of the flowers. The code snippet in Listing 2-3b changes the categorization format.

Listing 2-3b. SVM Classification of Iris Dataset: Change the Class to Numbers

```
#Use pandas to encode the categorized columns
dataset = dataset.replace({"class":  {"Iris-setosa":1,"Iris-versicolor":2,
"Iris-virginica":3}})
#Read the new dataset
dataset.head()
```

Iris setosa is changed to class 1, Iris versicolor is changed to class 2, and Iris virginica is changed to class 3 as a result of the code in Listing 2-3b, and as shown in the following sample output.

	sepal_length_in_cm	sepal_width_in_cm	petal_length_in_cm	petal_width_in_cm	class
0	5.1	3.5	1.4	0.2	1
1	4.9	3.0	1.4	0.2	1
2	4.7	3.2	1.3	0.2	1
3	4.6	3.1	1.5	0.2	1
4	5.0	3.6	1.4	0.2	1

Next, let's plot the heatmap using seaborn using the code in Listing 2-3c.

Listing 2-3c. SVM Classification of Iris Dataset: Generate Heatmap

```
plt.figure(1)
sns.heatmap(dataset.corr(), cmap="YlGnBu")
plt.title('Correlation between iris Classes')
```

Listing 2-3c generates the heatmap shown in Figure 2-21, which shows a correlation between classes of flowers and the length and width of their petals and sepals.

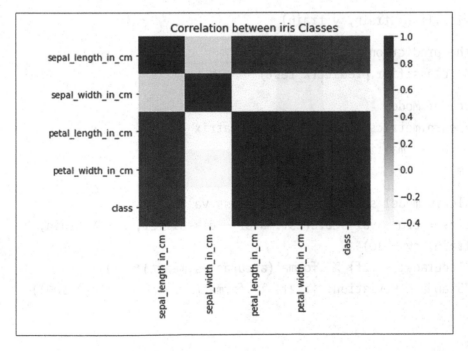

Figure 2-21. *SVM with Iris dataset: heatmap*

In the last step, in Listing 2-3d, the data is split, the SVM classifier is created, and the accuracy is calculated.

Listing 2-3d. SVM Classification of Iris Dataset: SVM Classifier and Accuracy code

```
# Split Data
X = dataset.iloc[:,:-1]
y = dataset.iloc[:, -1].values

from sklearn.model_selection import train_test_split
X_train, X_test, y_train, y_test = train_test_split(X, y, test_size = 0.25,
random_state = 0)

#Create the SVM classifier model
from sklearn.svm import SVC
classifier = SVC(kernel = 'linear', random_state = 0)
#Fit the model for the data

classifier.fit(X_train, y_train)

#Make the prediction
y_pred = classifier.predict(X_test)

#Accuracy of Model
from sklearn.metrics import confusion_matrix
cm = confusion_matrix(y_test, y_pred)
print(cm)

from sklearn.model_selection import cross_val_score
accuracies = cross_val_score(estimator = classifier, X = X_train,
y = y_train, cv = 10)
print("Accuracy: {:.2f} %".format(accuracies.mean()*100))
print("Standard Deviation: {:.2f} %".format(accuracies.std()*100))
```

The following output shows 98% accuracy and a 3.64% standard deviation. It also shows that there is only one misclassified data.

```
[[13  0  0]
 [ 0 15  1]
 [ 0  0  9]]
Accuracy: 98.18 %
Standard Deviation: 3.64 %
```

k-Nearest Neighbors

k-nearest neighbors (KNN) is one of the easiest to implement techniques in ML. KNN deals with structured, labeled data and groups them as per their *affinity* to each other. An example is the popularity of books. In Figure 2-22, the little circles represent the crime novels, whereas the squares are science-fiction books. The plotting appears to be based on distance somehow. If a new book whose genre is unknown (represented by a triangle) is published, the algorithm searches the training data for the nearest neighbor to the new example, *S*, and then classifies it. The KNN does not generate a line that separates the training data points.

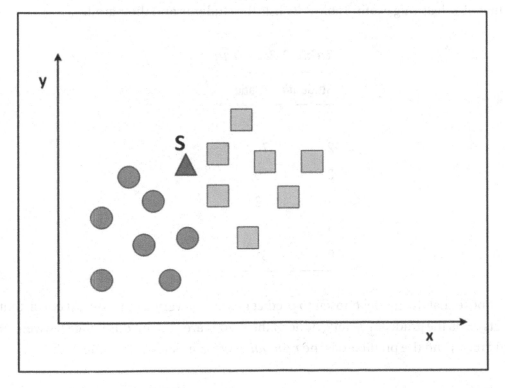

Figure 2-22. k-nearest neighbor

The pros of KNN are that it is easy to implement and works well with well-behaved and structured data. KNN is capable of correctly realizing arbitrarily complex dividing lines, also known as *hyperplanes*. A disadvantage of KNN is a single error in data points can give poor classification results. If a new, unknown point is immediately next to the point that is an outlier of another class, it is classified in the wrong class. This is another example of *overfitting*, where a fit has been made to random errors.

Error and Loss Functions

Earlier in this chapter, we looked at various errors generated by variance, bias, and their intermingled relationship in equations 2.5 and 2.6. *The loss function* is the computational function that computes errors. Prediction errors are directly related to the loss functions used to calculate them (i.e., different loss functions may calculate different errors for the same prediction).

Error and Loss Function for Regression

To understand how certain loss functions may work, let us consider six students' grades on a machine learning exam. Table 2-2a lists the students and their grades.

Table 2-2a. *Grades*

Student#	Grade
1	56
2	78
3	89
4	63
5	73
6	91

Suppose that the model chosen to predict grades is a very simple one. It consistently predicts 75 as the grade. In reality, none of the grades are 75. The difference between the actual grades and the predicted is the *residual* or *error,* as shown in Table 2-2b.

Table 2-2b. *Error/Residuals for Grades*

Student#	Grade	Prediction	Error or Residual
1	56	75	56 − 75 = −19
2	78	75	78 − 75 = 3
3	89	75	89 − 75 = 14
4	63	75	63 − 75 = −12
5	73	75	73 − 75 = −2
6	87	75	91 − 75 = 16

Table 2-2b shows that the sum of the errors is

$$(-19+3+14-12-2+16)=0$$

This means that the errors can cancel out, giving the impression that variance and bias are "well" while hiding in plain sight. However, suppose we square those errors to get rid of the negative signs and eliminate the cancellation of errors, thereby making it easier to inspect the pitfalls hidden in our "simple" model. Squaring the errors offers the additional advantage of amplifying the large deviations or "outliers," making them easier to detect. This leads to the *sum of squared errors* (SSE).

SSE is also known as the *residual sum of squares* (RSS) or *sum of squared residuals* (SSR). In this example, we are dealing with a toy dataset, but in reality, general data has labels and weights attached to them. Generally, SSE is given by E in the following expression.

$$E_{SSE} = \frac{1}{2}\sum_n \left(t^{(n)} - y^{(n)} \right)^2$$

2.15

where t datasets are the targets (or labels). The y are the outputs of the model. The (n) are the indices to mark the individual data which range across the samples. Hence $t^{(i)}$ is the index of the target for the ith training row vector.

The following is another way to mathematically define SSE.

$$E_{SSE} = \sum_n \left(x^{(n)} - \check{x}^{(n)} \right)^2$$

2.16

where x dataset are the actual values, $\check{x}$ are the predicted values. The (n) are indices to mark the individual data.

SSE is one of the simplest, all-purpose error metrics in ML.

There is also *mean squared error* (MSE), defined as follows.

$$E_{MSE} = \frac{1}{n} \sum_n \left(x^{(n)} - \check{x}^{(n)} \right)^2$$

2.17

MSE is fast and calculates gradients. But due to its being directly dependent on the dataset count, it can be somewhat sensitive to deviations in the dataset, and predictions that show large divergence from data can be penalized heavily.

The last important error for regression is *mean absolute error* (MAE), defined as

$$E_{MAE} = \frac{1}{n} \sum_n \left| x^{(n)} - \check{x}^{(n)} \right|$$

2.18

MAE is robust and generally not very affected by deviations. It is used when data is assumed to be corrupt. There are no set-in-stone rules to choose any regression error function. Of the three commonly used ones, trial and error are usually recommended for the best results.

Error and Loss Function for Classification

The following are some of the loss functions for prediction errors commonly used in classification tasks.

Binary class entropy measures the divergence of probability distributions between actual and predicted values and is commonly used in binary classification. This is mathematically given by

$$E_{BCE} = -\frac{1}{n} \sum_n \left[x^{(n)} \log \check{x}^{(n)} + \left(1 - x^{(n)} \right) \log \left(1 - \check{x}^{(n)} \right) \right]$$

2.19

In *categorical cross-entropy*, only one of many labels is predicted.

The following are some of the other loss functions.

- Negative log likelihood

- Margin classifier

- Soft margin classifier

- Sparse categorical cross-entropy

Unsupervised Learning

In unsupervised learning, the goal is to find the natural structure in the dataset which is unlabeled. The algorithm is fed with input data only and is expected to find meaningful structure by itself, with no external supervision or input. An example of this type of ML is finding clusters in a growing collection of texts in a library. This can be done based on changes in book topics across time, out-of-date discussions and themes, and authorship that may have become too sensitive for modern-day readers instead of target readers a couple of centuries back when they were originally written.

In unsupervised learning, there are no training and testing sets. All data points are training data points. The clusters are built from these data points.

Algorithms that address *visualizations* are also examples of unsupervised learning. An algorithm can be fed complex and unstructured data to obtain an output of a representation.

Clustering is the process of finding natural groupings in data. Human brains tend to naturally cluster data that we encounter. We classify them in our minds and organize them in certain categories as our brains simplify classifications by forming well-defined groups for them. Figure 2-23a illustrates a classification scheme for tea, coffee, milk, and juice. This helps you understand the data and with finding instances similar to what you have learned.

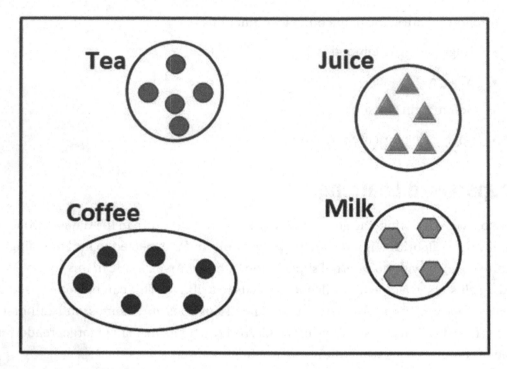

Figure 2-23a. *Clustering examples*

One significant application of unsupervised learning is in cybersecurity and *anomaly detection*, for example, credit card fraud, where the system, which has been shown normal transactions, tries to recognize any abnormal transaction pattern. Figure 2-3b illustrates how clustering can be used for fraud or anomaly detection.

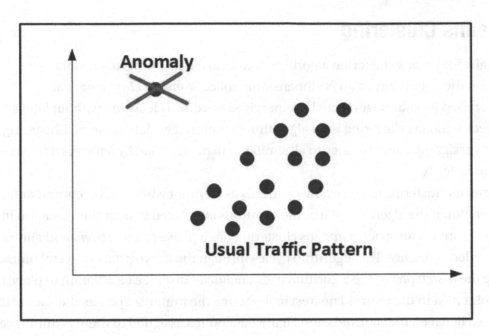

Figure 2-23b. *Anomaly detection with clustering*

Another great example of unsupervised learning in modern-day life is Google News, where an algorithm automatically groups news stories into specific sections. Table 2-3 lists some of the most common unsupervised learning algorithms and associated example use cases.

Table 2-3. *Examples of Unsupervised Learning Algorithms*

Unsupervised Learning Methods	Example	Algorithms
Clustering	Cluster virus data, visitor groups at a conference	Hierarchical clustering, k-means clustering
Dimensionality reduction	Reduction of dimensions in data	Principal component analysis (PCA)

k-Means Clustering

The main goal of any clustering algorithm is to assign data points to clusters that represent their similarity in an N-dimensional space. *k-means clustering* is an unsupervised learning algorithm. Unsupervised learning is learning without labels or targets. k-means clustering is an algorithm that produces data clusters. Clustering involves assigning cluster names to all similar data points. Usually, clusters are named 1, 2, 3, and so forth.

k-means clustering takes several centroids as an input where each centroid defines a cluster. When the algorithm starts, the centroids are placed in a random location in the data point vector space. k-means clustering has a phase called *assign,* and another phase called *minimize*. The algorithm cycles through these two phases several times. During the *assign* phase, the algorithm uses Euclidean distance calculation to place each data point next to the centroid nearest to it. During the minimize phase, the sum of the distances between the centroids and all the data points assigned to them is minimized by moving the centroids in a direction that brings them closer to the data points. This is called a *cycle*. The end of a cycle produces a *ready hyperplane*.

After one cycle ends, the next cycle starts by disassociating the data points from their centroids. Locationally, the centroids stay where they are. With that in place, a new association phase starts, which may decide to make assignments different than the one at the previous cycle. When a new data point is encountered, it is assigned to the closest centroid and inherits the name of that centroid as a label. A cluster in k-means is a region around a centroid.

In a theoretical setting, labels are not required in clustering. But the evaluation metrics become irrelevant if there are no labels to associate them. What will the metrics define if there is no element to define? Without labels, we cannot calculate true positives, false positives, true negatives, and false negatives. A detailed explanation of the usage of classification evaluation metrics in clustering, called the *external evaluation of clustering*, is given in [24].

However, there are times when we simply do not have access to labels and are forced to work without them. In this case, we can use an *internal evaluation of clustering*. Among several evaluation metrics available, the Dunn [25] evaluation metric is most popular. The focus of the Dunn coefficient is to measure the cluster density in an N-dimensional space. The quality of each cluster is evaluated by calculating the Dunn coefficient for each cluster.

Reinforcement Learning

In this case, the algorithms which make up the learning system is known as *agents*. The agents observe, select, and perform specific actions in return for rewards or penalties. An example of the application of reinforcement learning is how robots are taught to walk or perform tasks. They get "rewarded" for doing the correct work and "penalized" for getting them wrong. Another example is how sometimes people learn to play video games. They try out new things, such as clicking places, opening doors, getting ammunition or money, and eventually learning how to play. Similarly, the agents in reinforcement learning algorithm try out new things as it continues to learn from data-based observations and learns automatically. One of the areas of reinforcement learning that has become immensely popular is neural networks, which is explored in Chapter 3.

There are some very engaging media offerings where reinforcement learning is very well represented, including at `https://openai.com/blog/emergent-tool-use/`.

Summary

Machine learning (ML) has emerged as a subfield of research in artificial intelligence and cognitive science and has been adopted by businesses at a tremendous pace since 2014. Chapter 2 covered some fundamentals required as basics for a venture into the world of machine learning boosted by quantum mechanics. The next chapter covers the essentials of neural networks before moving onto quantum information processing concepts.

CHAPTER 3

Neural Networks

Chapter 2 looked at machine learning (ML), which emerged as a subfield of research in artificial intelligence. Deep learning (DL) is a subset of ML. It attempts to solve specific problems in areas of artificial intelligence, such as reasoning, planning, and knowledge representation.

Deep learning is centered on *artificial neural networks* (ANNs), which are powerful, scalable, and versatile. Google images use ANNs to classify billions of images at scale. From a mathematical perspective, a *neural network* (NN) can be looked at as a nonlinear regression model with a specific choice for the model function. A linear model is input to the nonlinear function, and then the expression gets nested several times. These models were derived from biological neural networks [18]. They represent graphics reminiscent of neurons connected by synapses.

Note   This chapter looks at the fundamentals of classical artificial neural networks from an applied perspective. It also looks at deep learning techniques such as learning, variance, bias, neural networks, and exposure to it via some hands-on exercises using publicly available data types. The code can be downloaded from the book's website. NN is a massive field all by itself, and it is not possible to include all aspects of the subject in one chapter. We revisit these concepts as we go through this book and look at how they fit in with quantum theory in data-driven predictions.

Artificial neural networks were inspired by the aspiration to map the human brain's logical functionalities to computing systems to create intelligent machines. Marvin Minsky was one of the founding fathers of artificial intelligence (AI). His PhD dissertation submitted to Princeton University in 1954 was titled "Neural Nets and the Brain Model Problem." In 1951, he built an Air Force Office of Scientific Research–funded machine,

© Santanu Ganguly 2021
S. Ganguly, *Quantum Machine Learning: An Applied Approach*, https://doi.org/10.1007/978-1-4842-7098-1_3

which saw the implementation of SNARC (Stochastic Neural Analog Reinforcement Calculator). When A. C. Clarke's legendary sci-fi book *2001: A Space Odyssey* was made into a film, he was an advisor to director Stanley Kubrick.

During classification, only a *forward pass* is made algorithmically. The forward pass is simply the sum of calculations that occur when an input travels through a neural network. The reverse of feedforward is known as *backpropagation*.

Building a neural network requires a structured approach and consists of the following overall components.

- A decision on the number of layers in a network

- Size of the input (usually the same as the number of neurons in the input layer)

- Number of neurons in hidden layers, if any

- Number of neurons in the output layer

- Initial values of weights

- Initial values for biases

Perceptron

Mathematically, the learning process in neurons consists of modifying weights and biases during training via backpropagation. The idea is to measure the network error when classifying and then minimize that error by modifying the weights. A *perceptron*, invented in 1962 by Frank Rosenblatt, is the building block of a neural network. A perceptron consists of a *binary threshold neuron*, which is defined as follows.

$$z = b + \sum_i w_i x_i \qquad 3.1$$

$$y = \begin{cases} 1, & z \geq 0 \\ 0, & otherwise \end{cases} \qquad 3.2$$

where, x_i are inputs, w_i are weights, b is the bias, and z is the *logit* or the *weighted sum*. The inputs and outputs are either real or binary numbers. Equation 3.2 defines the decision reached with nonlinearity. It is possible to absorb the bias in equation 3.1 as a weight making it possible to use only a rule which updates the weight. This is shown in Figure 3-1.

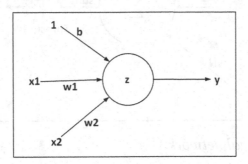

Figure 3-1. *Bias absorption*

To achieve this, input x_0 is added to equation 3.1 with a value of 1 and the bias as its weight.

$$z = b + \sum_i w_i x_i = w_0 x_0 + w_1 w_1 + w_2 x_2 + \dots$$

with $b = w_0 x_0$

In the preceding equation, b can either be x_0 while $w_0 = 1$ or b can be w_0 while $x_0 = 1$. The inputs never change and what we are trying to do is change the bias. Hence, the input is treated as a weight. This process is known as *bias absorption*.

The following describes the basics of the perceptron training rule.

1) A training scenario is chosen.

2) If the predicted output matches the output label, then no action needs to be taken.

3) If the perceptron predicts a 0 when it should have predicted a 1, the input vector is added to the weight vector.

4) If the perceptron predicts a 1 when it should have predicted a 0, the input vector is subtracted from the weight vector.

Usually, an NN consists of N inputs, say I_N and N weights as shown in Figure 3-2.

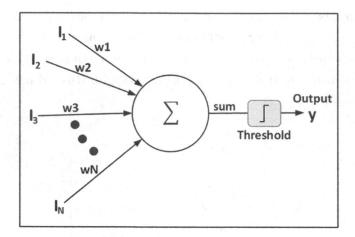

Figure 3-2. *Simple neural network*

In the simple NN shown in Figure 3-2, the inputs are *binary* (ON/OFF). In this model, the inputs and weights are summed, and the *threshold function*, which is usually a step function, gives the output. The following are the limitations of this model.

- The inputs and outputs are binary

- There is no automatic way to train the weights; hence, the setting of weights is done manually

In contrast to the simple NN shown in Figure 3-2, a perceptron model, shown in Figure 3-3, is more versatile because the inputs can be numbers and do *not have to be binary*. A perceptron model is the simplest form of feedforward neural network.

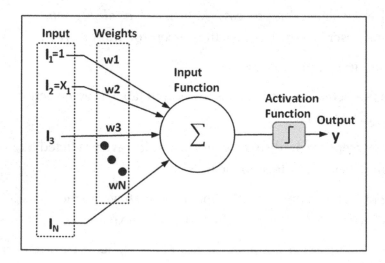

Figure 3-3. *Simple perceptron*

The generalized perceptron model shown in Figure 3-3 has inputs and weights, and then the bias absorption described in equations 3.1 and 3.2. If we revisit equation 3.1, the calculation of a perceptron involves a model function given by

$$z = f(x; w) = \phi(\mathbf{w}^\mathbf{T}.\mathbf{x}) \qquad 3.3$$

where $\mathbf{w}^\mathbf{T}$ is the matrix representation of the weights multiplied by the input vector $\mathbf{x}$. ϕ is a function dependent on the input and the weights. f is the same as the model function z. Here, the inputs and outputs are either real or binary numbers. The nonlinear function ϕ is referred to as the *activation function* (discussed later). The summation is done at the start of the process after the inputs are fed.

Toward the end of the process, the step function is applied. The following are common step functions.

$$z = \begin{cases} 0, & z < 0 \\ 1, & z \geq 0 \end{cases}, \quad sgn(z) = \begin{cases} -1, & z < 0 \\ 0, & z = 0 \\ 1, & z > 0 \end{cases} \qquad 3.4$$

where *sgn* is the *signum* function. When neurons in a layer are connected to every other neuron in the layer before it (i.e., the input neurons), the layer is *fully connected*.

A simple example of a perceptron is if one is going to watch a cricket match. The decision to go or not depends on several factors, or in this case, inputs.

- Weather: If it rains, then the cricket match will not take place

- The proximity of the venue to public transport as there is no car

- If a friend can also join to offer quality company

- If a favorite star player is participating in the game

The perceptron for the problem is shown in Figure 3-4.

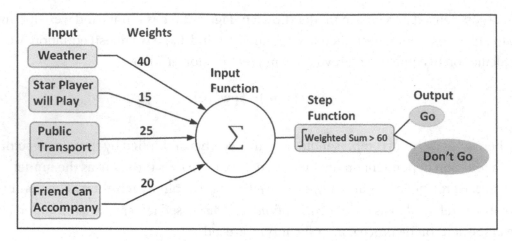

Figure 3-4. *Simple perceptron example of cricket match*

The output is calculated by assigning weights to each of the inputs. The output is binary (Yes/No) based on predetermined decision-making criteria that if the final score is higher than 60, then it's a Yes, otherwise it's a No. The weights affect the decision-making based on answers to questions like, "What if the friend cannot join?" or "What if it rains, but only for an hour?" or "What if the star player does not play?". Again, note that all manipulations of weights here are done manually.

Activation Functions

An activation function in a neural network defines how the weighted sum of the input is transformed into an output from a node or nodes in a layer of the network. To understand activation functions, let us suppose that we have a set of images of dogs and another set of images not containing dogs. Suppose that each neuron receives input from the value of a single pixel in the images. While the computer processes those images, we would like our neuron to adjust its weights and bias to have fewer and fewer images wrongly identified. This approach seems intuitive, but it requires a small change in the weights or bias to cause only a small change in the outputs. Because, if we have a big jump in output, we lose the "smoothness" in our learning curve and fail to learn progressively. The act of learning, to use the human analogy of kids' learning processes, is done in small steps (or, little by little). Unfortunately, the perceptron does not show this "learning by small steps" behavior. A perceptron is either a 0 or 1, and that's a big jump that does not help in learning, as shown in Figure 3-5.

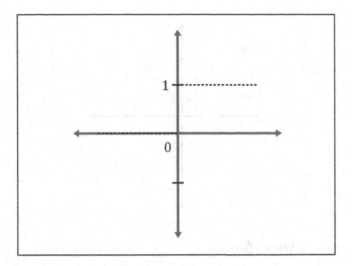

Figure 3-5. *Perceptron outputs*

To rectify this lack of progressive learning, what is required is "something" to smooth the abruptness of that step between 0 and 1 in Figure 3-5. Fundamentally, we need a function that progressively changes from 0 to 1 with no discontinuity. Mathematically, what is needed is a continuous function that allows us to compute the derivative. In mathematics, the derivative is how a function changes at a given point on a continuous function. For functions with input given by real numbers, the derivative is the slope of the tangent line at a point on a graph.

In our example of a cricket match in Figure 3-4, we defined a threshold as greater than 60 in an *activation function after* the neuron's output. This activation function was used to inject the "smoothness" in our cricket match example.

Without the activation function, the output of the perceptron would be linear and linear functions tend to be simple. They have challenges solving complex problems. Neural networks (NNs) are seen as *universal function approximators* because they can compute and learn almost any function. Hence, *nonlinearity* is required so that the NNs can compute complex functions. Hence, the need for the *activation functions*.

The following are some of the most common activation functions.

- **Linear:** In the linear output of the perceptron, the derivative of the function is zero since the function is linear. Linear activation functions are commonly used for *linear regression*. Hence, a neural network without an activation function is usually regarded as a linear regression model. Figure 3-6 depicts a linear activation function.

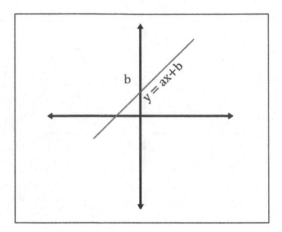

Figure 3-6. *Linear activation function*

- **Sigmoid**: The sigmoid function is one of the most widely used nonlinear activation functions that transform the values between the range 0 and 1. Mathematically, the Sigmoid function is continuous, and a typical sigmoid function, represented in Figure 3-7, is given by

$$\sigma(z) = \frac{1}{1+e^{-z}}$$

For input variations between $(\infty, -\infty)$, sigmoid produces small output changes in the range $(0, 1)$.

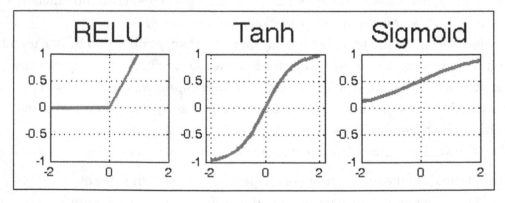

Figure 3-7. *ReLU, tanh, and sigmoid functions (Source [273])*

The sigmoid function has a well-defined non-zero derivative throughout; hence gradient descent can be used. Historically, the sigmoid function has been very popular. In recent times, ReLU functions are gaining popularity.

- **Tanh**: Tanh (see Figure 3-7) is another popular activation function mathematically defined as

$$\tanh z = 2\sigma(2z) - 1 = \frac{e^z - e^{-z}}{e^z + e^{-z}}$$

Tanh and sigmoid are closely related—tanh is considered a "stretched" sigmoid. Like sigmoid, tanh is S-shaped, continuous, and differentiable. Tanh is symmetric around zero and ranges from –1 to +1, whereas sigmoid ranges from 0 to +1.

- **ReLU**: A *rectified linear unit* is another nonlinear activation function that has gained popularity in deep learning. The main advantage of using the ReLU function over other activation functions is that it does not activate all the neurons simultaneously. ReLU (see Figure 3-7) is mathematically defined as follows.

$$\text{ReLU}(z) = \max(0, z)$$

ReLU is a simple function and has become popular because it helps address some optimization problems observed with the sigmoid. The function is zero for negative values and grows linearly for positive values, as shown in Figure 3-7. The ReLU is also very simple to implement, whereas the sigmoid's implementation is more complex.

For more information on other activation functions, read *Deep Learning* by Goodfellow et al. at `www.deeplearningbook.org` [274].

Hidden Layers

When we discussed SVM characteristics and discussed types of kernels, you saw that polynomial kernels offer more flexibility because of the curves in their decision boundaries compared to linear kernels. Similarly, in neural networks, a flexible decision boundary is often desired. A *hidden layer* is a way to achieve that.

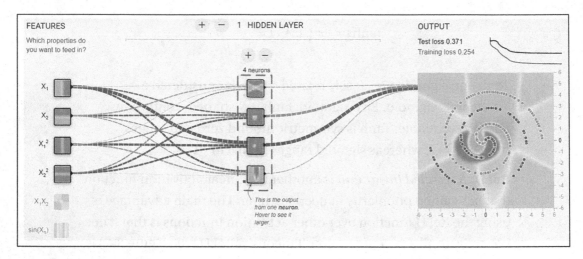

Figure 3-8. *NN with a single hidden layer (Created with TensorFlow Playground, ref [27])*

The marked red area in Figure 3-8 shows a *single hidden layer* in the feedforward network created with TensorFlow Playground [27]. Multiple hidden layers allow for more complex decision boundaries. Although depending on requirements, there can be several hidden layers in a NN; usually, one hidden layer is enough to solve most nonlinear problems. Recent research has shown that any function can be represented by a sufficiently large neural network with one hidden layer (Lin, Jegelka, 2018) [19]; however, training that network may be a challenge. As requirements and demands for feature sets grow across the industry, it has become common practice to consider more than one hidden layer.

Backpropagation

Backpropagation is the reverse of feedforward. A simple logistic regression network, strictly speaking, does not need any backpropagation. However, as more layers are added, the number of those calculations rises. The layers do not necessarily add to the complexity of calculation but add to the volume of those.

One of the rules of thumb in NN is *the backpropagation of errors is gradient descent*. Gradient descent is one of the most common training algorithms for feedforward networks, and mathematically, this is backpropagation, which is defined as

$$w_{ij}^{n+1} = w_{ij}^{n} - \eta \nabla E = w_{ij}^{n} - \eta \frac{\partial C(w_1, w_2)}{\partial w_{ij}} \qquad 3.5$$

where, the current weight w_{ij}^{n+1} is the updated weight, w_{ij}^{n} is the weight of the previous step down toward the direction of the steepest gradient, η is the learning rate, E is the cost function and C is the mathematical equivalent of E in a differential form with a derivative with respect to w_{ij} to obtain the gradient.

Hands-on Lab: NN with TensorFlow Playground

TensorFlow [20] was invented by Google to efficiently use graphical models in machine learning by leveraging tensors and quickly becoming a very popular ML library. TensorFlow was made a free, open source software library for fast machine learning tasks by Google. It can be used across a range of tasks but focuses on training and inference of deep neural networks. TensorFlow is a symbolic math library based on dataflow and differentiable programming. The documentation for TensorFlow is available at `www.tensorflow.org`. Due to its nature of utilizing graphical properties, many TensorFlow implementations run more efficiently on GPUs (graphical processing units) than CPUs. Google came out with TPUs (Tensor Processing Unit) to optimize TensorFlow's code performance. At the time this book was written, TensorFlow was at version 2.x.

Keras is another popular ML tool. Keras is a high-level API that allows us to build, evaluate, train, and execute all types of NNs. The original release of Keras is available at `https://github.com/keras-team/keras`. Multiple other implementations of Keras have been released. Three of the most popular open source deep learning libraries are TensorFlow, Microsoft Cognitive Toolkit (CNTK), and Theano. In this book, we use `tf.keras, which is` the Keras implementation bundled with TensorFlow.

To get a taste for neural networks and how they work, there is a resource to *experience* and visualize NN performance and architecture *before* we start detailed coding. It is called TensorFlow Playground [27].

The following steps simulate neural networks.

Step 1: Point your browser to `http://playground.tensorflow.org`. As shown in Figure 3-9, this should open the GUI for the playground that we use to *play* with some concepts. This simulator for neural networks helps users get used to the concepts of various NN algorithms.

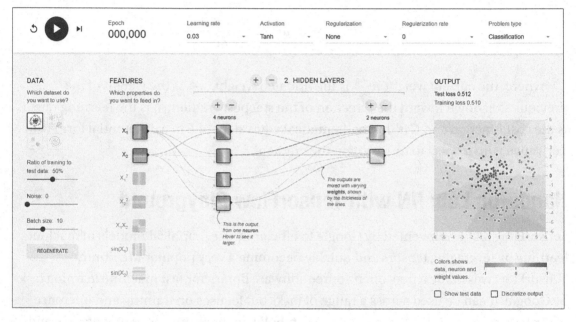

Figure 3-9. *TensorFlow Playground GUI*

For the next steps, refer to Figure 3-10. The corresponding steps are marked in dark blue numbers.

Step 2: Select the data.

Step 3: Select the features.

Step 4: Design the neural network.

Step 5: Adjust the parameters.

Step 6: Run.

Step 7: Inspect the results.

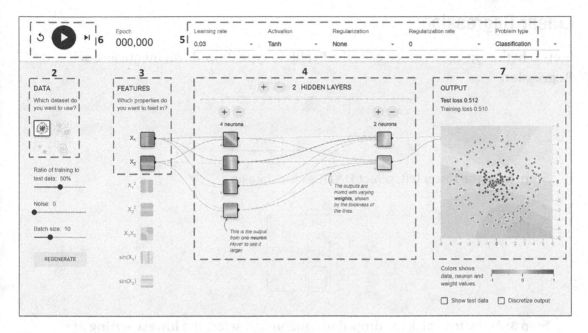

Figure 3-10. *TensorFlow Playground configuration*

Hidden Layers: Select **no Hidden Layers**. This should change the 2 beside the Hidden Layers caption on top to 0, and change the NN accordingly, as shown in Figure 3-11.

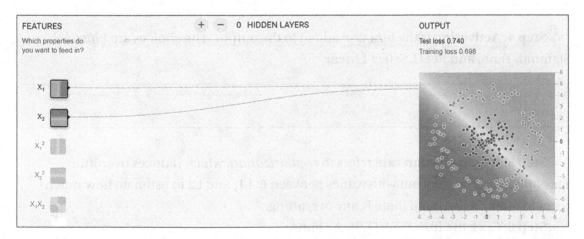

Figure 3-11. *TensorFlow Playground configuration with no hidden layers*

Observe the change in output by eliminating the hidden layer. The domain boundary in the output becomes linearly more well-defined with less complexity.

Linear Regression:

Step 1: In the top-right corner, click the **Problem type** drop-down menu, and select **Regression**.

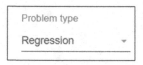

Step 2: On the lower left, select **DATA** as the dataset.

Step 3: At the top, click the drop-down menu and select the **lowest setting** of **Learning Rate**. Learning rate is the "step size" used for gradient descent.

Step 4: **Activation** is the function added to the output. The choices are Linear, Sigmoid, Tanh, and ReLU. Select **Linear**.

Learning rate	Activation	Regularization	Regularization rate	Problem type
0.00001	Linear	None	0	Regression

Step 5: **Regularization rate** refers to *regularization*, which reduces overfitting. This option can be tried out with values between 0, L1, and L2 to estimate how much regularization is needed if there is any overfitting.

Step 6: Click the RUN/START/PLAY button.

Step 7: Observe the OUTPUT curve. It appears to be converging slowly.

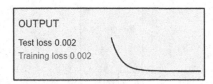

The reason for the delay is related to the value of the learning rate. The model appears to be learning too slow. The goal is to reach convergence as quickly as possible using as few epochs as possible. The **loss** referred to in the output is the **error**. In this case, it is the root mean squared error (RMSE).

Step 7: Click the Learning Rate drop-down menu, **increase** the learning rate, and hit RUN again. Is there a change in how fast the algorithm is running?

Step 8: Click the Learning Rate drop-down again, **increase** the learning rate to a **high value,** and hit **RUN** again. The output test loss shows up as **NaN** or *not a number*. A large value for learning rate causes overshoot.

To find the optimum learning rate, we need to determine the value for the learning rate, which takes us closest to zero using the fewest epochs.

Classification:

Step 1: To simulate classification, select the following parameters.

1.   Select **Classification** in the **Problem type** drop-down menu.

2.   Select Tanh in the **Activation** drop-down menu

3.   Select **0.01** in the **Learning rate** drop-down menu.

Learning rate	Activation	Regularization	Regularization rate	Problem type
0.01	Tanh	None	0	Classification

Step 2: Select the two-blob dataset.

Step 3: The output for the separated dataset might look like the following image.

Step 4: Zero loss may not be achievable, especially if some noise is introduced in the model

Step 5: As in the linear regression lab before, the challenge is to tune the *learning rate* to get to minimum loss in as few epochs as possible.

Readers are encouraged to play at the TensorFlow Playground and spend some time changing and adjusting parameters for hidden layers, regression, and classification while also adding new features on the left. As the number of hidden layers at the top is changed, the tool adds more hidden layers. Similarly, clicking the square boxes under FEATURES adds new features to the NN model.

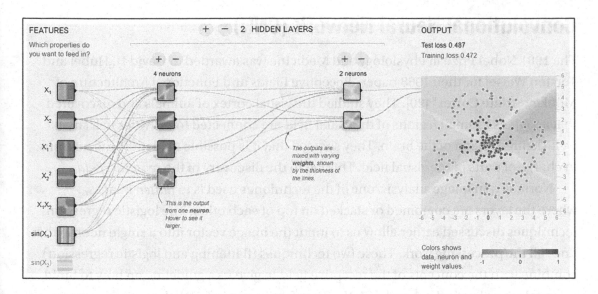

Change the dataset type under DATA, such as the tricky **Spiral** dataset to familiarize yourself with the behavior of various NN models with changes in parameters.

Next, let's venture into the world of neural network architecture.

Neural Network Architecture

We have already looked at the rationale, flexibility, and scalability of neural networks and how a perceptron is the building block for it. From an architectural point of view, there are several options to build a NN. The following are some of the more common ones that are used today.

- Convolutional neural network (CNN)

- Feedforward neural network (FFNN)

- Recurrent neural network (RNN)

Convolutional Neural Network (CNN)

The 1981 Nobel Prize in Physiology and Medicine was awarded to David H. Hubel and Torsten Weisel for their 1968 paper "Receptive Fields and Functional Architecture of Monkey Striate Cortex" [20]. They studied the visual cortex of animals and discovered that activities in small regions of the visual field are connected to activities in a small, well-defined region of the brain. They showed that it is possible to identify the neurons in charge of parts of the visual field. This led to the discovery of the *receptive field*.

Normally, in image analysis, one of the techniques used is to *flatten* images, where the layers are combined or stacked on top of each other. The logistic regression techniques discussed earlier allow us to input the image vector into a single neuron that does all the processing work. These two techniques (flattening and logistic regression) combine with the concepts of the receptive field to give us a guideline on how to build a convolutional neural network (CNN). For effective use in CNN, the logistic regression part uses a different activation function, but the structure remains the same.

Flattening allows the merging of images together into a single layer. This allows us to have all the visible information in one single layer. This is an important part of building NNs since the input of a fully connected layer consists of a one-dimensional array. The concept of flattening is illustrated in Figure 3-12.

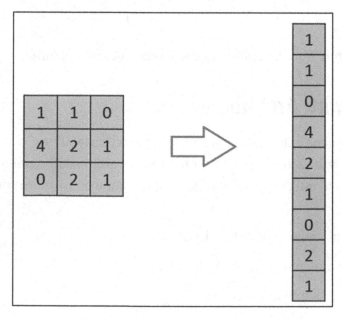

Figure 3-12. *Flattening (Courtesy of medium.com [21])*

A convolutional layer takes an image and passes a logistic regression over the whole image. Hence, the first input consists of components of the flattened vector, the second input consists of the next components, so and so forth. The process is illustrated in Figure 3-13.

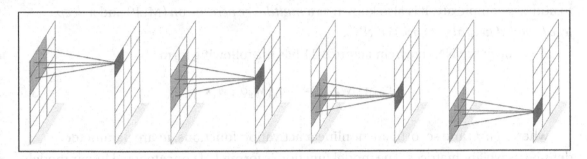

Figure 3-13. *Convolutional neural network*

A CNN has multiple layers. CNNs are designed specifically to handle images. Convolution layer neurons scan a particular area of the image, which is their *field of vision* (FOV), and pick up patterns in images, such as eyes, ears, nose, wheels, and so forth. For example, a CNN can be made to process a size 10 image with a 3 × 3 receptive field for all its three layers. Given that it has the property to randomly assign bias and weights, the first layer takes the 10X10 image and produces an image of size 8 × 8. This size is then given to the second CNN layer. This second layer can have its own receptive fields with randomly initialized bias and weights but for this example, let us assume it is also 3 × 3. This produces an output of 6 × 6, which is then passed on to the third layer with a third local receptive field. The third CNN layer produces a 4 × 4 image. This image is then flattened into a 16-dimensional vector and fed to a standard fully-connected layer with one output neuron and using a logistic function for nonlinearity. The logistic function gives an output between 0 and 1 and compares the output and the image label. It then calculates the error, backpropagates it, and repeats this process for every image in the dataset which has gone through the entire training of the network.

Training a CNN involves training the local receptive fields and the weights and biases of fully connected layers. The structural difference between logistic regression and a local receptive field is that a logistic function is used in the former. But, in the latter, any activation function can be used. One of the activation functions that is used widely is *rectified linear unit* or ReLU. A ReLU of an input x is given by the maximal value between x and 0 (i.e., it returns a 0 if the input is negative or otherwise the raw input). CNNs are used widely in computer vision applications.

Feedforward Neural Network

As discussed earlier, single-layer NNs can solve simple, linear problems but struggle to address complexity. To solve more complex problems, nonlinearity is introduced. A *feedforward neural network* (FFNN) is a way to implement the nonlinearity needed to address complexity. It is also known as a *multilayer perceptron* (MLP) and a *deep feedforward neural network* (DFNN).

A feedforward NN shown in Figure 3-11 has the following form

$$f\left(\mathbf{x},\mathbf{w_1},\mathbf{w_2},\ldots\right)=\ldots\phi_2\left(\mathbf{w_2}\phi_1\left(\mathbf{w_1}\mathbf{x}\right)\right) \qquad 3.6$$

where, ϕ_1, ϕ_2 and so forth are nonlinear activation functions. w_i are parameters defined as weight matrices. The model function is formed of concatenated linear models and functions. The concatenation can be repeated several folds. This is an example of the power of neural networks: they utilize a combination of the data processing power of linear algebra with the flexibility of nonlinear dynamics.

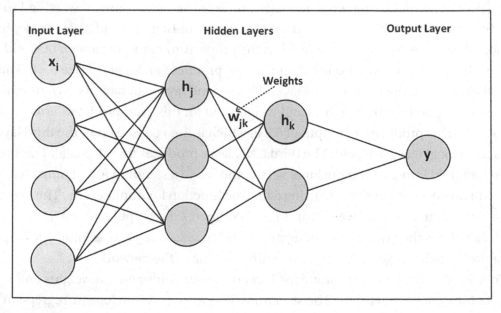

Figure 3-14. *Simple feedforward neural network with one hidden layer*

Figure 3-14 shows a feedforward NN with two hidden layers. A feedforward NN is made up by connecting multiple perceptrons in layers so that each unit of each layer is connected to units of the next layer. The first layer is made up of inputs x_i, the next layer is the hidden layer consisting of h_i units, and the last layer has the output y_i. There is an activation function to update each neuron, and each layer is updated after the previous one, thereby forming a *feedforward* neural network. The following is from equation 3.6, the model for the feedforward neural network with a single hidden layer of Figure 3-11.

$$f\left(\mathbf{x},\mathbf{w_1},\mathbf{w_2}\right)=\phi_2\left(\mathbf{w_2}\phi_1\left(\mathbf{w_1}\mathbf{x}\right)\right)$$

A feedforward NN can be combined with many possible *activation functions* (ϕ) such as *tanh, sigmoid,* and *ReLU*. Detailed discussion about the activation functions can be found in reference [15], and we expand on them as and when required for our purpose in upcoming chapters.

The following are typical characteristics of feedforward NNs.

- It consists of multiple layers.

- Each layer has several neurons.

- Each neuron is connected to neurons in the previous layer.

- Information flows through only one way; hence feedforward.

- It is composed of *input, output,* and *hidden* layers. There can be more than one hidden layer if needed.

The following are some rudimentary *sizing guidelines* when designing an FFNN.

- Input layer

  - Size is equal to the number of input dimensions.

  - It may need an additional extra neuron as a bias term.

  - For many sparse dimensions, wide and deep neural networks (with many layers) are often considered.

- Hidden layers

  - The size and number of hidden layers depend on the training sample, input features, outputs, and complexity of the problem.

  - Deep learning uses multiple hidden layers.

- Output layer

    - For regression, only a single neuron is needed.

    - For binary classification, only one neuron as output is binary (0/1, True/False...).

    - For multi-class classification, the softmax layer is used among other layers.

    - The sizing suggestion is one node per class label.

Hands-on Lab: Image Analysis Using MNIST Dataset

This lab exercise looks at image classification. We create and train a neural network model that, after the training, can predict digits from handwritten images with a high degree of accuracy. You should develop a basic understanding of how neural networks work and how TensorFlow works with Keras.

This lab is done using the MNIST dataset. MNIST is a set of 70,000 small images of digits handwritten by high school students and employees of the US Census Bureau. Each image is labeled with the digit it represents. This set has been studied so much that it is often called the "Hello, World" of machine learning. Whenever people develop a new classification algorithm, they are curious to see how it performs on MNIST and anyone who learns machine learning tackles this dataset sooner or later.

Figure 3-15 describes the task: create and train a model that takes an image of a handwritten digit as input and predicts the class of that digit; that is, it predicts the input image's digit or class.

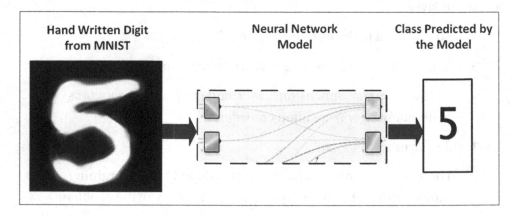

Figure 3-15. *Image classification task*

Open the Jupyter notebook file named TF_Keras_imageClassifiaction.ipynb available for download from the book's website.

Step 1: Import TensorFlow and verify the version.

```
import tensorflow as tf
tf.logging.set_verbosity(tf.logging.ERROR)
print('Using TensorFlow version', tf.__version__)
```

Step 2: Import the MNIST data.

```
mnist=tf.keras.datasets.mnist
(x_train, y_train),(x_test, y_test)= mnist.load_data()
```

Step 3: Shape the data.

```
print('x_train:', x_train.shape)
print('y_train:', y_train.shape)
print('x_test:', x_test.shape)
print('y_test:', y_test.shape)
```

The following shows the output.

```
x_train: (60000, 28, 28)
y_train: (60000,)
x_test: (10000, 28, 28)
y_test: (10000,)
```

Step 4: Look at an image.

```
from matplotlib import pyplot as plt
%matplotlib inline
plt.imshow(x_train[0], cmap='binary')
plt.show()
```

The following shows the output.

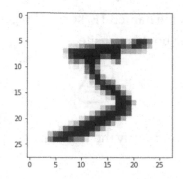

Step 5: Display the labels.

```
y_train[0]
```

Step 6: In this step we shall leverage hot encoding. Each label is converted to a list with 10 elements and the index element to the corresponding class is set to 1, rest are all set to 0 such as

original label	one-hot encoded label
5	[0, 0, 0, 0, 0, 1, 0, 0, 0, 0]
7	[0, 0, 0, 0, 0, 0, 0, 1, 0, 0]
1	[0, 1, 0, 0, 0, 0, 0, 0, 0, 0]

```
from tensorflow.keras.utils import to_categorical
y_train_encoded = to_categorical(y_train)
y_test_encoded = to_categorical(y_test)
```

Step 7: Validate the shape.

```
print('y_train_encoded:', y_train_encoded.shape)
print('y_test_encoded:', y_test_encoded.shape)
```

The following shows the output.

```
y_train_encoded: (60000, 10)
y_test_encoded: (10000, 10)
```

Step 8: Display the encoded labels.

```
y_train_encoded[1]
```

The following shows the output.

```
array([1., 0., 0., 0., 0., 0., 0., 0., 0., 0.], dtype=float32)
```

Step 9: Refer to Figure 3-16, which shows the neural networks architecture. As explained in the discussion on perceptrons and bias absorption, the diagram represents the equation.

$$z = b + w1 * x1 + w2 * x2 + w3 * x3$$

where the $w1$, $w2$, $w3$ are weights. b is an intercept term, otherwise called *bias*.

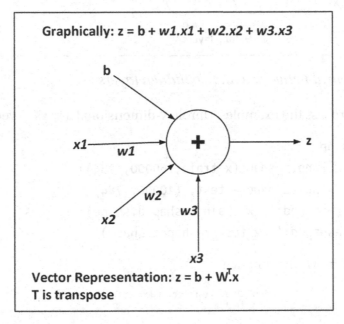

Figure 3-16. *Neural network architecture. The T means transpose. This gives the dot product for the result we wish to have*

Step 10: Build the neural network.

We have a single neuron with 784 features. This constitutes a neural network with two hidden layers, as shown in Figure 3-17.

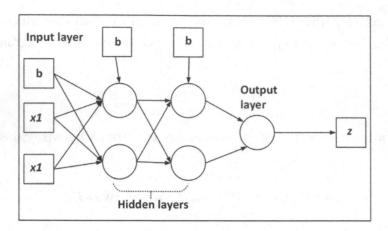

Figure 3-17. *Neural network with two hidden layers*

Step 11: Preprocess the examples. Unroll *N*-dimensional arrays to vectors.

```
import numpy as np
x_train_reshaped = np.reshape(x_train,(60000, 784))
x_test_reshaped = np.reshape(x_test, (10000, 784))
print('x_train_reshaped:', x_train_reshaped.shape)
print('x_test_reshaped:', x_test_reshaped.shape)
```

The following shows the output.

```
x_train_reshaped: (60000, 784)
x_test_reshaped: (10000, 784)
```

Step 12: Create a model.

```
from tensorflow.keras.models import Sequential
from tensorflow.keras.layers import Dense
model = Sequential([
    Dense(128, activation='relu', input_shape=(784,)),
    Dense(128, activation='relu'),
    Dense(10, activation='softmax')
])
```

Step 13: Activation Function: ReLU (Rectifier Linear Unit). To impose an activation function in the node, the first step is to define a linear sum of the inputs.

$$Y = W.X + b$$

The second step is to define the activation function output.

$$Z = f(Y)$$

Figure 3-18 shows the function.

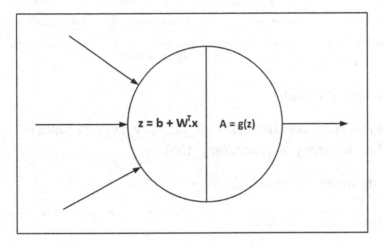

Figure 3-18. *Neural network with two hidden layers*

Compile the model.

```
model.compile(
    optimizer='sgd',
    loss='categorical_crossentropy', metrics=['accuracy']
)
model.summary()
```

The following shows the output.

Layer (type)	Output Shape	Param #
dense (Dense)	(None, 128)	100480
dense_1 (Dense)	(None, 128)	16512
dense_2 (Dense)	(None, 10)	1290

Total params: 118,282
Trainable params: 118,282
Non-trainable params: 0

Step 14: Train the model.

```
model.fit(x_train_norm, y_train_encoded, epochs=3)
```

The following shows the output.

```
Epoch 1/3
60000/60000 [==============================] - 6s 103us/sample - loss: 0.0727 - acc: 0.9787
Epoch 2/3
60000/60000 [==============================] - 6s 102us/sample - loss: 0.0638 - acc: 0.9818
Epoch 3/3
60000/60000 [==============================] - 6s 99us/sample - loss: 0.0569 - acc: 0.9835
<tensorflow.python.keras.callbacks.History at 0x7ff92032f940>
```

Step 15: Evaluate the model.

```
loss, accuracy = model.evaluate(x_test_norm, y_test_encoded)
print('Test Set Accuracy %', accuracy*100)
```

The following shows the output.

```
10000/10000 [==============================] - 1s 55us/sample - loss: 0.0877 - acc: 0.9697
Test Set Accuracy % 96.96999788284302
```

Step 16: Predict.

```
preds = model.predict(x_test_norm)
print ('Shape of predictions:', preds.shape)
```

The following shows the output.

```
                    Shape of predictions: (10000, 10)
```

Step 17: Plot the results.

```
plt.figure(figsize=(12, 12))
start_index = 0
for i in range(25):
    plt.subplot(5, 5, i+1)
    plt.grid(False)
    plt.xticks([])
    plt.yticks([])
    pred = np.argmax(preds[start_index+i])
    gt = y_test[start_index+i]
```

```
        col = 'b'
    if pred != gt:
        col = 'r'
            plt.xlabel('i={},pred={}, gt={}'.format(start_index+i,
            pred, gt), color = col)
    plt.imshow(x_test[start_index+i], cmap= 'binary')
plt.show()
```

The following shows the output.

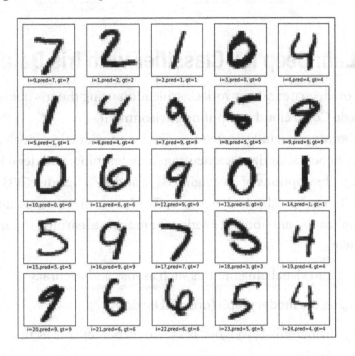

Step 18: Plot the results.

```
plt.plot(preds[8])
plt.show()
```

The following shows the output.

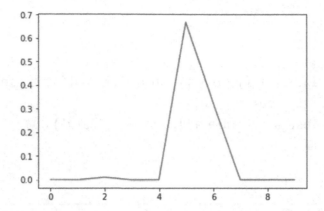

Hands-on Lab: Deep NN Classifier with Iris Dataset

This lab focuses on deep neural network classification using the Iris dataset. For this lab, we leverage Google Colab cloud computing environment.

Google Colaboratory is a fully hosted free service provided by Google, allowing us to load and run notebooks for data science, machine learning, and deep learning. The environment also offers options of leveraging GPU and TPU besides CPU.

This lab uses TensorBoard, which is TensorFlow's visualization toolkit [23]. It provides the visualization and tooling needed for machine learning experimentation, such as the following.

- Tracking and visualizing metrics such as loss and accuracy

- Visualizing the model graph (ops and layers)

- Viewing histograms of weights, biases, or other tensors as they change over time

- Projecting embeddings to a lower dimensional space

- Displaying images, text, and audio data

- Profiling TensorFlow programs

Step 1: Access Google Colaboratory at https://colab.research.google.com. The link should open the following page.

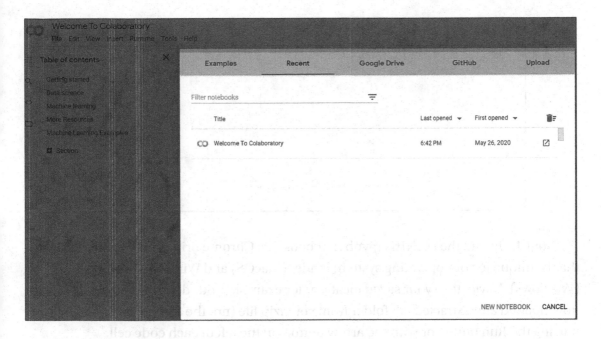

Step 2: Click the Welcome to Colaboratory link for access to the broader environment. Watch the introductory video to become familiar with the Google Colab environment.

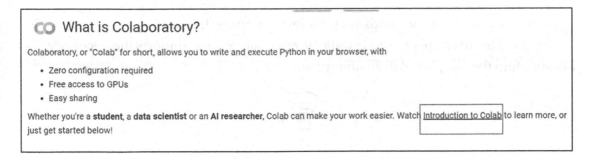

Step 3: Go to the top-left corner and click **File ➤ Upload Notebook**. You see the following screen.

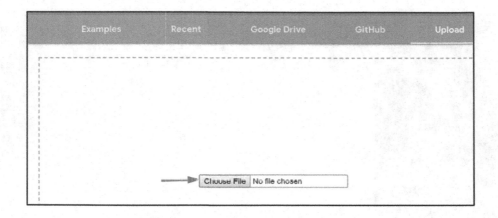

Step 4: Upload the DNNIris.ipynb notebook file: Chrome brings up a file requester that is unique to your operating system: Finder (macOS) and Windows Explorer (Windows). Navigate to your saved location (for example, your desktop). Select the downloaded and extracted lab folder from your .zip file (not the .zip file itself). Then run it using the Run button or with the arrow button on the left of each code cell.

Next, we proceed to the hnads-on exercise by running the code:

Step 1: The first snippet of code tests the environment and imports the necessary libraries into the Google Colab environment.

```
%matplotlib inline
import numpy as np
import pandas as pd
import matplotlib.pyplot as plt
try:
  # %tensorflow_version only exists in Colab.
  %tensorflow_version 2.x
except Exception:
  pass
```

```
import tensorflow as tf
from tensorflow import keras
```

Step 2: Read the data.

```
import os
data = "https://archive.ics.uci.edu/ml/machine-learning-databases/iris/
iris.data"
data_local = keras.utils.get_file(fname=os.path.basename(data),
                                                origin=data)
iris = pd.read_csv(data_local)
iris.columns = ['SepalLengthCm', 'SepalWidthCm', 'PetalLengthCm',
'PetalWidthCm','Species']
iris
```

The output lists a sample of the Iris dataset.

	SepalLengthCm	SepalWidthCm	PetalLengthCm	PetalWidthCm	Species
0	4.9	3.0	1.4	0.2	Iris-setosa
1	4.7	3.2	1.3	0.2	Iris-setosa
2	4.6	3.1	1.5	0.2	Iris-setosa
3	5.0	3.6	1.4	0.2	Iris-setosa
4	5.4	3.9	1.7	0.4	Iris-setosa
...	...	...	...	...	...

Step 3: Shape the data.

```
input_columns = ['SepalLengthCm', 'SepalWidthCm', 'PetalLengthCm',
'PetalWidthCm']
x = iris [input_columns]
y = iris[['Species']]

print (x.head())
print('-----')
print (y.head())
```

The output shows the selected columns.

```
     SepalLengthCm  SepalWidthCm  PetalLengthCm  PetalWidthCm
0            4.9           3.0            1.4           0.2
1            4.7           3.2            1.3           0.2
2            4.6           3.1            1.5           0.2
3            5.0           3.6            1.4           0.2
4            5.4           3.9            1.7           0.4

- - - - -
         Species
0  Iris-setosa
1  Iris-setosa
2  Iris-setosa
3  Iris-setosa
4  Iris-setosa
```

Step 4: Encode the labels. The output labels are strings like Iris-setosa or Iris-virginica. These are called *categorical variables*. They need to be changed to numbers, like we did for SVM. This is called *encoding*.

```
from sklearn.preprocessing import LabelEncoder
encoder =  LabelEncoder()
y1 = encoder.fit_transform(y.values) ## need y.values which is an array
print(y1)
```

Step 5: Split the data into train/test: 80% for train and 20% for test.

```
from sklearn.model_selection import train_test_split
# 'y1' (encoded labels)
x_train,x_test, y_train,y_test = train_test_split(x,y1,test_
size=0.2,random_state=0)
print ("x_train.shape : ", x_train.shape)
print ("y_train.shape : ", y_train.shape)
print ("x_test.shape : ", x_test.shape)
print ("y_test.shape : ", y_test.shape)
```

The following shows the output.

```
x_train.shape :  (119, 4)
y_train.shape :  (119,)
x_test.shape :  (30, 4)
y_test.shape :  (30,)
```

Step 6: Build the deep NN model. Since this is a classifier, the neural network is going to look like the following.

Neurons in Output layer = output classes (3 here) Output activation is 'softmax'

1. Neurons in Input layer = input dimensions (4 here)

2. Neurons in Output layer = output classes (3 here)

3. The output activation is 'softmax'.

4. Use the Adam optimizer, an optimization algorithm that can be used instead of the classical stochastic gradient descent procedure to update network weights iterative based in training data.

```
input_dim = len(input_columns)
output_clases = 3
print ("input_dim : ", input_dim, ", output classes : ", output_clases)
model = tf.keras.Sequential([
            tf.keras.layers.Dense(units=64, activation=tf.nn.relu,
            input_dim=input_dim),
            tf.keras.layers.Dense(units=32, activation=tf.nn.relu),
            tf.keras.layers.Dense(units=output_clases,  activation=tf.
            nn.softmax)
            ])
# loss = 'sparse_categorical_crossentropy'
model.compile(loss='sparse_categorical_crossentropy',
                optimizer=tf.keras.optimizers.Adam(),
                metrics=['accuracy'])
print (model.summary())
tf.keras.utils.plot_model(model, to_file='model.png', show_shapes=True)
```

The following NN model is the output of the code snippet.

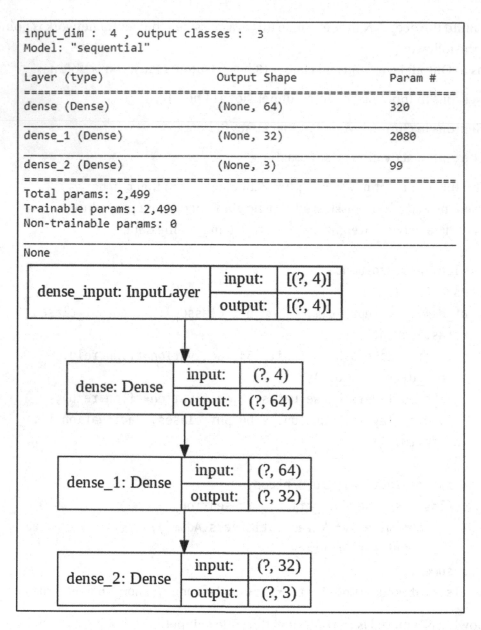

```
input_dim :   4 , output classes :   3
Model: "sequential"
```

Layer (type)	Output Shape	Param #
dense (Dense)	(None, 64)	320
dense_1 (Dense)	(None, 32)	2080
dense_2 (Dense)	(None, 3)	99

```
Total params: 2,499
Trainable params: 2,499
Non-trainable params: 0
```

None

dense_input: InputLayer	input:	[(?, 4)]
	output:	[(?, 4)]

dense: Dense	input:	(?, 4)
	output:	(?, 64)

dense_1: Dense	input:	(?, 64)
	output:	(?, 32)

dense_2: Dense	input:	(?, 32)
	output:	(?, 3)

Step 7: The TensorBoard GUI may come up at the first instance with an HTML error. **Refresh the page** using the refresh icon on the top-right corner of the TensorBoard GUI.

```
import datetime
import os
import shutil
app_name = 'classification-iris'
tb_top_level_dir= '/tmp/tensorboard-logs'
tb_app_dir = os.path.join (tb_top_level_dir, app_name)

tb_logs_dir = os.path.join (tb_app_dir, datetime.datetime.now().
strftime("%H-%M-%S"))
print ("Saving TB logs to : " , tb_logs_dir)
#clear out old logs
shutil.rmtree ( tb_app_dir, ignore_errors=True )
tensorboard_callback = tf.keras.callbacks.TensorBoard(log_dir=tb_logs_dir,
histogram_freq=1)
## This will embed Tensorboard right here in jupyter
%load_ext tensorboard
%tensorboard --logdir $tb_logs_dir
```

The following shows the output in TensorBoard view.

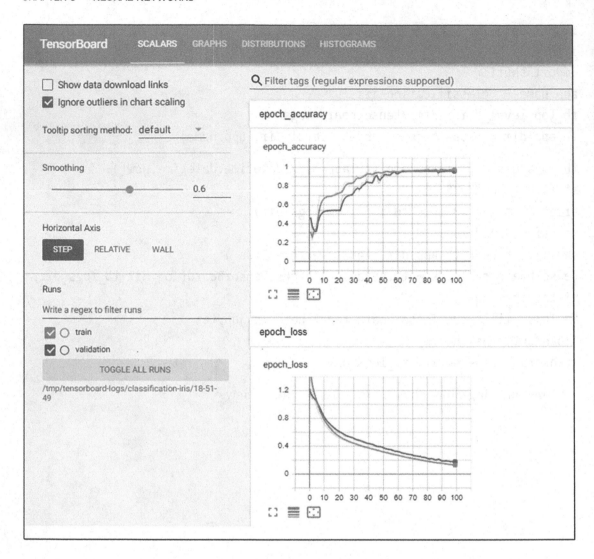

Step 8: Train the data (this may take a while depending on value of the epoch).

```
%%time
epochs = 100  ## experiment 100, 500, 1000
print ("training starting ...")
history = model.fit(
            x_train, y_train,
            epochs=epochs, validation_split = 0.2, verbose=1,
            callbacks=[tensorboard_callback])
print ("training done.")
```

The following shows the output.

```
3/3 [==============================] - 0s 16ms/step -
training done.
CPU times: user 3.73 s, sys: 250 ms, total: 3.98 s
Wall time: 4.54 s
```

Step 9: Plot the history.

```
%matplotlib inline
import matplotlib.pyplot as plt
plt.plot(history.history['accuracy'], label='train_accuracy')
plt.plot(history.history['val_accuracy'], label='val_accuracy')
plt.legend()
plt.show()
```

The following shows the output.

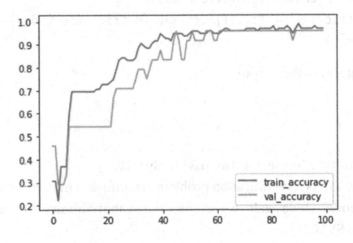

Compare this plot with the one from TensorBoard for accuracy of training and validation.

Step 10: Obtain the prediction.

```
np.set_printoptions(formatter={'float': '{: 0.3f}'.format})
predictions = model.predict(x_test)
predictions
```

In the output of step 10, for each test input, the softmax layer produces three numbers. These numbers are probabilities. If they are added, the total is 1.0. We want to choose the output that has the highest probability.

For example (0.03086184, 0.33362046, 0.6355177) means

- Class 1 has a probability of 0.03 or 3%.

- Class 2 has a probability of 0.33 or 33%.

- Class 3 has a probability of 0.63 or 63%.

Hence, we choose the class with highest probability as prediction: class 3

Step 11: Evaluate the model.

```
metric_names = model.metrics_names
print ("model metrics : " , metric_names)
metrics = model.evaluate(x_test, y_test, verbose=0)
for idx, metric in enumerate(metric_names):
    print ("Metric : {} = {:,.2f}".format (metric_names[idx],
    metrics[idx]))
```

The following shows the output.

```
model metrics :  ['loss', 'accuracy']
Metric : loss = 0.12
Metric : accuracy = 1.00
```

Accuracy shows very close to 1.0 as loss is only 0.12.

Step 12: Since this is a classification problem, a confusion matrix is an efficient way to evaluate the model. Once again, we leverage the seaborn library for visualization same as we did in SVM.

```
from sklearn.metrics import confusion_matrix
import seaborn as sns
cm = confusion_matrix(y_test, y_pred, labels = [0,1,2])
cm
import matplotlib.pyplot as plt
import seaborn as sns
plt.figure(figsize = (8,5))
sns.heatmap(cm, annot=True, cmap="YlGnBu", fmt='d').plot()
```

```
from sklearn.metrics import classification_report
from pprint import pprint

pprint(classification_report(y_test, y_pred, output_dict=True))
```

The following is the heatmap.

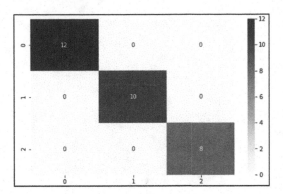

The following are the metrics from the confusion matrix.

```
{'0': {'f1-score': 1.0, 'precision': 1.0, 'recall': 1.0, 'support': 12},
 '1': {'f1-score': 1.0, 'precision': 1.0, 'recall': 1.0, 'support': 10},
 '2': {'f1-score': 1.0, 'precision': 1.0, 'recall': 1.0, 'support': 8},
 'accuracy': 1.0,
 'macro avg': {'f1-score': 1.0, 'precision': 1.0, 'recall': 1.0, 'support': 30},
 'weighted avg': {'f1-score': 1.0,
                  'precision': 1.0,
                  'recall': 1.0,
                  'support': 30}}
```

Going back to the discussion on confusion matrixes in Chapter 2, what can you conclude from this output?

Summary

Chapter 3 looked at artificial neural networks, which maps the human brain's logical functionalities to computing systems to create intelligent machines. Based on neural networks, deep learning solves specific problems in artificial intelligence, such as reasoning, planning, and knowledge representation. The next chapter covers the essentials of quantum information processing concepts.

CHAPTER 4

Quantum Information Science

Clouds come floating into my life, no longer to carry rain or usher storm,
but to add color to my sunset sky.

—Rabindranath Tagore

Quantum information is concerned with utilizing the special features of quantum mechanics for the processing and transmission of information. Quantum information refers to data that can be physically stored in a quantum system. Quantum information theory is the study of how such information can be encoded, measured, and manipulated. *Quantum computation* is a subfield of quantum information. It is often used synonymously with quantum information theory, which studies protocols and algorithms that use quantum systems to perform computations.

To quote Prof. John Preskill, quantum computing guru at Caltech, "Information is something that is encoded in the state of a physical system." Traditionally, computations have been carried out on a physically realizable device. Hence, it is an inescapable fact that the study of computing information processes should inherently be related to studying the underlying physical processes.

Note   This chapter focuses on addressing fundamentals of quantum information science such as quantum parallelism, Deutsch's algorithm, hands-on exercises with Qiskit and Google's Cirq, and associated theories. The chapter ends with introducing quantum computing systems as a precursor to the various platforms that we address. This should prepare you for a deeper dive into quantum algorithms and applications of machine learning in future sections of the book.

© Santanu Ganguly 2021
S. Ganguly, *Quantum Machine Learning: An Applied Approach*, https://doi.org/10.1007/978-1-4842-7098-1_4

There are notable examples of how physics has evolved related to the usage and manipulation of information. Landauer's principle, proposed in 1961, stated that the erasure of information is essentially a dissipative process. For example, we can consider a hypothetical "atom in a box" situation where an atom is placed in a gas-filled box with a horizontal partition in the middle. If the atom is moved to the bottom half of the partition, it is erased, irrespective of whether it started at the top or the bottom. The partition separating the top and bottom halves is removed suddenly, and a piston compresses the one-atom gas until the atom is confined in the bottom half. The act of compressing the gas causes a reduction in the entropy of the gas. The *change* in entropy *S* of a monoatomic ideal gas is given as follows [28].

$$\Delta S = Nk \left(\ln \frac{V_f}{V_i} \right)$$

where N is the number of atoms, k is Boltzmann's constant, V_f is the final volume after compression, and V_i is the initial volume. As the entropy varies, there is an associated change of heat from the "box" to the environment. If we consider the system to be isothermal (i.e., a system where the temperature remains constant at T, then a work $W = kT \ln \left(\frac{V_f}{V_i} \right)$ is done on the box. If used to erase the information, W must be performed at a cost, perhaps in the form of an electricity or gas bill.

As per Landauer's principle, if there is a finite supply of batteries, there should be a theoretical limit to computational time. However, as our understanding of the physics of information continued to evolve, came forth the theory by Charles Bennett. He stated in 1973 that computations could be performed using only *reversible* steps. In principle, it requires no dissipation and no power expenses; hence, *reversible computation*.

According to the theories, there is no need to pay to perform computation. However, in practice, the "irreversible" computers used today dissipate a huge amount of energy, orders of magnitude higher than the estimated $kT \ln \left(\frac{V_f}{V_i} \right)$. As computing components continue to get smaller and smaller, it becomes more and more important to *beat* Landauer's principle so that the components do not corrode or melt in the generated heat; hence, a reversible computation may be the only option left for engineering. As such, *quantum information* evolved as the interface of physics and computer science today.

Quantum Information

Classical information theory stems from the fundamentals of "information is physical," and instinctively, it is instructive for us to ponder upon the physical interpretation of information. After having largely conquered most details of classical information, we must consider that the laws of quantum mechanics fundamentally govern the universe.

Quantum physics is where true randomness resides as opposed to its classical counterpart, which is largely deterministic. In quantum theory non-commuting observables cannot simultaneously have precisely defined values described by Heisenberg's Uncertainty Principle (Chapter 1). In addition, if a measurement is performed on variable x, then that act of measurement influences the outcome of subsequent measurement of an observable y, if x and y do not commute. This means the act of obtaining information about a physical system disturbs the physical system; it is not a limitation to which classical physics is subjected to.

Information is represented by the configuration of matter and energy; hence the way that information can be stored, transmitted, and processed is determined by the laws governing physical systems. At the scales accessible to humans, the relevant physical laws are those of quantum mechanics. Hence, quantum information is the study of how quantum theory affects our ability to conceptualize and process information. Figure 4-1 is a diagrammatic representation of quantum information processing.

This led to the study of how complex quantum systems are understood and managed in terms of their information-carrying properties. The increasing understanding of information science in the quantum domain has opened the door to novel applications not possible with conventional "classical" information theory, such as *perfectly* secure communication, the efficient simulation of quantum systems, and ultrafast quantum computers. Recent years have seen an explosion of interest and progress in bringing quantum information into the real world. It has also offered insights into the fundamental nature of physics, computation, and information. Outside of the laboratory, the first commercial quantum devices have appeared on the market.

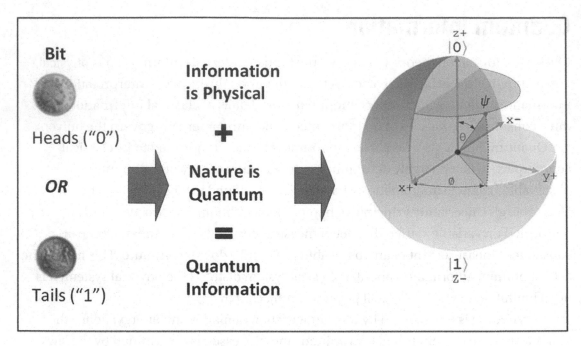

Figure 4-1. *Concept of quantum information processing*

The trade-off between creating a disturbance in the physical system to acquire information is due to the randomness of a quantum system. The property that obtaining information from a quantum system also causes a disturbance related to the *no-cloning theorem,* which states that quantum information cannot be copied with perfect fidelity. If it were possible to perfectly copy quantum information, that would enable us to measure an observable property of the copy without disturbing the original and hence allow us to circumnavigate the "disturbed when measured" property. But, the no-cloning theorem prohibits this act. As we know, nothing prevents the copying of classical information perfectly. This property has helped computing user and their larger ecosystem for decades when anyone has tried to back up a database or hard disk or even tried to copy-paste simple text.

Quantum Circuits and Bloch Sphere

A quantum computing circuit is a network of quantum logic gates where each gate performs some specific unitary transformation on one or more qubits. Analogous to classical computing, quantum circuits are constructed using options of different varieties of operations. This section expands on the basics of quantum gates mentioned in Chapter 1. It performs some experiments with quantum circuits as a warm-up to coding with quantum computing libraries in Python. The circuits are built using quantum gates that act as unitary operators on one or two qubits at a time. These circuits can be run using the Python notebook called qCircuitCirq.ipynb for this chapter. This code runs on the Google Cirq quantum simulator and includes a quantum teleportation experiment.

A quantum state space is constructed of complex vector space called *Hilbert space* (discussed in Chapter 1). In quantum information science, it is always assumed that the Hilbert space $\mathcal{H}$ are finite-dimensional. Since $\mathcal{H}$ is finite-dimensional, a basis can be chosen to represent vectors in this basis as finite column vectors and represent operators with finite matrices. The Hilbert spaces of interest for quantum computing typically have dimension $d = 2^n$, for some positive integer n. A qubit state can be represented u without any loss of generality, as follows.

$$|\psi\rangle = \cos\frac{\theta}{2}|0\rangle + \sin\frac{\theta}{2}e^{i\phi}|1\rangle \qquad 4.1$$

where, as usual, kets $|0\rangle$ and $|1\rangle$ are the orthonormal basis, and $0 \leq \theta \leq \pi$ and $0 \leq \phi \leq 2\pi$ and i indicate the imaginary part of the state vector. This is generally referred to as the *computational basis* without being specific about the physical realization of this basis. In quantum information science, computational visualizations are helpful in mapping this state to a Bloch sphere, as depicted in Figure 4-2.

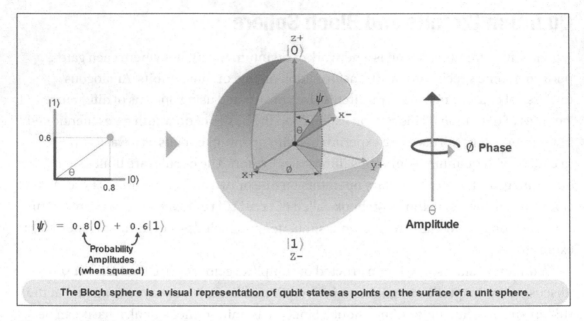

The Bloch sphere is a visual representation of qubit states as points on the surface of a unit sphere.

Figure 4-2. *Bloch sphere*

The Bloch sphere is an abstract 3-D representation with the state of a qubit represented by a point on the sphere. In terms of the Bloch sphere, the basis states $|0\rangle$ and $|1\rangle$ are the North and South poles, respectively. The antipodal points where the X axis intersects the surface of the sphere correspond to the following orthonormal states

$$|+\rangle = \frac{1}{\sqrt{2}}(|0\rangle + |1\rangle)$$

4.2

$$|-\rangle = \frac{1}{\sqrt{2}}(|0\rangle - |1\rangle)$$

4.3

Equations 4.2 and 4.3 are analogous to equations 1.19 and 1.20. In other words, referring to Chapter 1, $H|0\rangle = |+\rangle$ and $H|1\rangle = |-\rangle$. These states are known as *Hadamard basis* since they can be obtained from the computational basis by applying a 2×2 unitary matrix called the *Hadamard transformation, H.*

$$\begin{pmatrix} |+\rangle \\ |-\rangle \end{pmatrix} = H \begin{pmatrix} |0\rangle \\ |1\rangle \end{pmatrix} = \frac{1}{\sqrt{2}} \begin{pmatrix} 1 & 1 \\ 1 & -1 \end{pmatrix} \begin{pmatrix} |0\rangle \\ |1\rangle \end{pmatrix} = \frac{1}{\sqrt{2}} \left[\begin{pmatrix} 1 \\ 0 \end{pmatrix} \pm \begin{pmatrix} 0 \\ 1 \end{pmatrix} \right] = \frac{1}{\sqrt{2}}(|0\rangle \pm |1\rangle)$$

4.4

The Hadamard transformation is related to Pauli matrices. Referring to equation 1.13, which give the mathematical representation of the Pauli matrices, we see that the antipodal points where the Y axis intersects the surface of the sphere correspond to the following orthonormal states.

$$|i\rangle = \frac{1}{\sqrt{2}}\big(|0\rangle + i|1\rangle\big) \qquad 4.5$$

$$|-i\rangle = \frac{1}{\sqrt{2}}\big(|0\rangle - i|1\rangle\big) \qquad 4.6$$

Figure 4-3 shows some of the qubit states on the Bloch sphere.

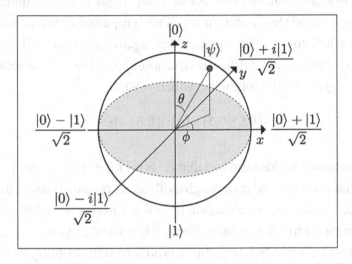

Figure 4-3. *The Bloch sphere with single-qubit states (Source [31])*

Superposition on Bloch Sphere

Superposition was defined in Chapter 1. A state space constructed from qubits carries substantially more information than a counterpart state space represented by *classical* binary bits, since the qubit state is specified in terms of two continuous real parameters θ and ϕ. In other words, the classical description only tells us whether we are in the northern hemisphere or the southern hemisphere, while the qubit state vector distinguishes every point on the surface of the Bloch sphere.

The qubit state is often represented as a superposition of the basis states $|0\rangle$ and $|1\rangle$. If measurements are performed on a large number N of identical single-qubit states, then the output is 0 for approximately $N\cos^2\dfrac{\theta}{2}$ times and 1 for approximately $N\sin^2\dfrac{\theta}{2}$ times. There is a marked difference between how probabilities affect the quantum world as opposed to its classical counterpart.

It is possible to start a physical process on a quantum computer—be that of the gate or annealing variety—with $|0\rangle$ as the initial state and then have the state *evolve*. This evolution of state can also be done by using a Hadamard operator H, similar to the GHZ circuit we experimented with previously. If this was tried on a *classical* system, the first step, *classical reasoning*, says that we have a 50% chance of being in the state $|0\rangle$ and 50% chance of flipping over to state $|1\rangle$. Now suppose that the second step of the process is subjected to evolution again by the Hadamard operator. Here, the classical reasoning says that if we were still in state $|0\rangle$ after the first step, there is a 50% chance that we will stay in $|0\rangle$ after the second step; furthermore, if we happened to flip to state $|1\rangle$ after the first step, there is a 50% chance that the state flips again on the second step, ending up back in $|0\rangle$. Hence, classical reasoning says that after applying H^2, the total probability of being in state $|0\rangle$ is given by the following, which is 50%.

$$(0.5)(0.5)+(0.5)(0.5)=0.5$$

However, if we now consider the fact that H^2 is an *identity operator* (i.e., equals 1), then we realize that this physical process gives $|0\rangle$ as an output 100% of the time. The difference here is that the computation of probabilities arising from quantum measurement can factor in interference effects. It is possible to simulate the interference classically with classical wave behavior, but it is an inherent property that quantum physics has built into it.

The limitations of classical reasoning exposed in the example show that we should be careful in how we think about a quantum superposition state. This is because the qubit state $|-\rangle = \dfrac{1}{\sqrt{2}}(|0\rangle - |1\rangle)$ is *not* in both the states $|0\rangle$ and $|1\rangle$ at the same time. It is not in either one of those states. It is instead in a different physical state, called the $|-\rangle$ state. In some physical realizations of qubits, the Hadamard basis is just as good a basis for measuring states as the computational basis. If the basis used to measure is switched, then the $|-\rangle$ state is not a superposition state. There is always a choice of measurement basis for any state in the Hilbert space where that state is one of the basis states, and hence, not a superposition. As such, superposition is not meaningful unless the quantum system indicates that some measurement bases are preferred over some others.

Quantum Circuits with Qiskit

In March 2017, IBM announced the release of IBM Quantum Experience, a platform connected to real quantum processors obtained with superconductive materials. It is possible to use this quantum computing platform via cloud access and execute programs on quantum hardware systems. Qiskit allows the management of real and simulated quantum circuits. The programming can be done via a graphical interface or CLI (command-line interface). The quantum computing environment also offers programmability via API (application programming interface): a set of classes, functions, and data structures for interfacing with devices and simulators and running experiments. The framework of Qiskit has the following main areas as per their API documentation [275].

- Qiskit (Terra)

- Qiskit simulator (Aer)

- Qiskit experiments (Ignis)

- Qiskit application modules (Aqua)

- Qiskit IBM Quantum (Provider)

Programs written in Qiskit follow a fixed workflow based on three high-level steps.

1. Build a circuit.

2. Execute the program.

3. Analyze the results.

Qiskit manages three main objects [275] to execute the workflow: the provider, the backend, and the job. The provider gives access to a group of different backends to choose from to suit our specific algorithm. The AER provider gives access to different simulators, while the IBM provider allows access to a real qubit environment via the IBMQ module.

The backends are responsible for running quantum circuits and returning results. They take a qobj as input and return a BaseJob object. Each execution is identified by a unique address, accessible by the Job object, which finds out the state of execution at a given point in time (for example, if the job is queued, running, or has failed). The Qiskit workflow is depicted in Figure 4-4.

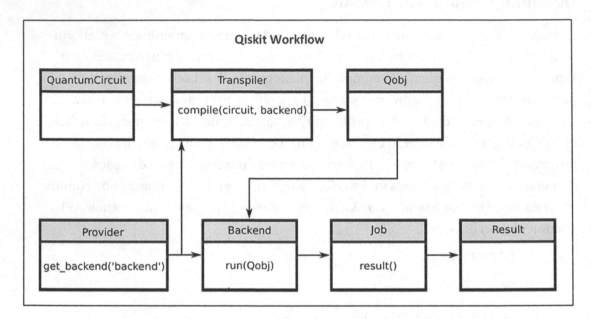

Figure 4-4. *Qiskit workflow*

Qiskit offers options to simulate quantum state spaces with the Bloch sphere. To achieve correct visualization and enable proper operation on Jupyter notebook, Qiskit Terra version 0.16.x is required along with Python 3.6 or later. This can be installed via the following command in a Linux environment.

```
$ pip install qiskit[visualization]
```

If you already have Qiskit installed and try the code without the visualization updates offered by Qiskit Terra 0.16.x, you may encounter the following error.

```
VisualizationError - input is not a valid N-qubit state.
```

If you already have Qiskit installed with an older version of Terra and encounter this error, update qiskit[visualization]with the -U option.

If you have an Anaconda environment on Windows, open an Anaconda prompt in the command-line window (`cmd.exe`) and type the following.

```
conda install qiskit[visualization] -U
```

This should start the update, download, and installation of all the relevant files shown in Figure 4-5.

```
(base) C:\Users\santagan>pip install qiskit[visualization] -U
Collecting qiskit[visualization]
  Downloading qiskit-0.23.2.tar.gz (4.1 kB)
Collecting qiskit-terra==0.16.1
  Downloading qiskit_terra-0.16.1-cp37-cp37m-win_amd64.whl (7.8 MB)
     |████████████████████████████████| 7.8 MB 1.3 MB/s
Collecting qiskit-aer==0.7.2
  Downloading qiskit_aer-0.7.2-cp37-cp37m-win_amd64.whl (24.0 MB)
     |████████████████████████████████| 24.0 MB 3.3 MB/s
Collecting qiskit-ibmq-provider==0.11.1
  Downloading qiskit_ibmq_provider-0.11.1-py3-none-any.whl (195 kB)
     |████████████████████████████████| 195 kB 6.8 MB/s
Collecting qiskit-ignis==0.5.1
  Downloading qiskit_ignis-0.5.1-py3-none-any.whl (204 kB)
```

Figure 4-5. *Qiskit Terra update and install in Anaconda on Windows*

Once the Qiskit packages are successfully installed, we can try some visualization exercises for quantum states using the Bloch sphere. This explains the interpretation of the vectors on the 3-D Bloch surface. We start with importing the `Qiskit` library and visualizing a sample Bloch sphere using the `plot_bloch_vector()` function shown in Listing 4-1a for the Jupyter notebook `qiskitBloch.ipynb` file. `qiskit.__qiskit_version__` shows the working versions of all Qiskit components and APIs.

Listing 4-1a. Bloch Sphere Simulation in `qiskitBloch.ipynb`

```
## import the libraries
import numpy as np
from qiskit import *
qiskit.__qiskit_version__

from qiskit.visualization import plot_bloch_vector
```

```
%matplotlib inline
plot_bloch_vector([0,1,0], title="Bloch Sphere")
```

This code generates version numbers and our sample Bloch sphere as output, as shown in Figure 4-6.

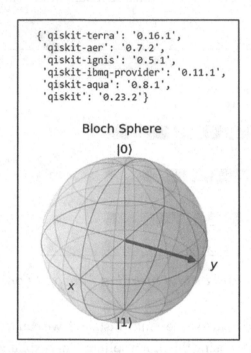

Figure 4-6. *Sample Bloch sphere and version numbers*

Note   Do not confuse qubit's state vector with its Bloch vector. The state vector is the vector which holds the amplitudes for the two states the qubit can be in. The Bloch vector is a visualization tool that maps the 2-D, complex state vector onto a real, 3-D space. A Bloch sphere is a geometric representation of a pure state. The Qiskit function that offers to plot blotch sphere is plot_bloch_vector(). It is important to note that each Bloch sphere represents a single qubit orientation (i.e., two qubits are represented by two Bloch spheres; three qubits need three Bloch spheres, and so on).

It is helpful to see how Qiskit handles gate definitions. We encountered GHZ states in Qiskit previously in this chapter. To do this, we first import all the necessary libraries as shown in Listing 4-1b and try out an X gate. X gate definitions were covered in chapter 1 when we looked at Pauli gates (equation 1.13). In this example, the plot_bloch_multivector() function is used, which takes a qubit's state vector and not the Bloch vector.

Listing 4-1b. Gate Programming in qiskitBloch.ipynb

```
from math import pi
from qiskit.visualization import import plot_bloch_multivector
#X-gate on a |0> qubit
qc = QuantumCircuit(1)
qc.x(0)
qc.draw()
```

Listing 4-1b gives the following circuit as output.

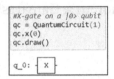

Here, we have performed an X or NOT gate operation on a qubit at state $|0\rangle = \begin{pmatrix} 1 \\ 0 \end{pmatrix}$ and flipped it to state $|1\rangle$. Let's verify the result of the X-gate operation on a Bloch sphere representation, as shown in Listing 4-1c.

Listing 4-1c. Bloch Representation of the X-operation qiskitBloch.ipynb

```
# Verify The result
backend = Aer.get_backend('statevector_simulator') # Simulate
out = execute(qc,backend).result().get_statevector() # Do the simulation,
returning the statevector
plot_bloch_multivector(out) # Display the output state vector
```

Listing 4-1c gives the following Bloch sphere representation as output.

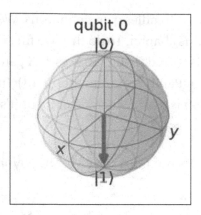

The output confirms that the state of the qubit is $|1\rangle$ as expected. This as a rotation by π radians around the X axis of the Bloch sphere.

The Qiskit library allows the addition of gates to circuits. For example, if we were to add Pauli Y and Z gates to our previously created X gate, we need the code snippet in Listing 4-1d.

Listing 4-1d. Y and Z gate Addition to X Gate `qiskitBloch.ipynb`

```
qc.y(0) # Do Y-gate on qubit 0
qc.z(0) # Do Z-gate on qubit 0
qc.draw()
```

This code snippet gives the following circuit as output.

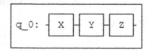

Qiskit allows measuring of only the Z basis. When a measurement is performed, it is not always a requirement to measure in any fixed computational basis (i.e., Z basis). the measurement of qubits can be done on any basis of our choice. For example, we can calculate the probability of measuring either $|+\rangle$ or $|-\rangle$ in the X basis.

$$p(|+\rangle) = |\langle+|q\rangle|^2 \, , p(|-\rangle) = |\langle-|q\rangle|^2$$

Once the measurement is performed, a qubit is guaranteed to be in one of these two states. Since Qiskit only allows measuring in the Z basis, it is required that we create our own using Hadamard gates (see Listing 4-1e).

Listing 4-1e. Hadamard `qiskitBloch.ipynb`

```
# Create Hadamard gate for the X-measurement function:
def x_measurement(qc,qubit,cbit):
    """Measure 'qubit' in the X-basis, and store the result in 'cbit'"""
    qc.h(qubit)
    qc.measure(qubit, cbit)
    qc.h(qubit)
    return qc

initial_state = [0,1]
# Initialise our qubit and measure it
qc = QuantumCircuit(1,1)
qc.initialize(initial_state, 0)
x_measurement(qc, 0, 0)  # measure qubit 0 to classical bit 0
qc.draw()
```

This code snippet gives the following circuit with H gate and M gate for measurement as output.

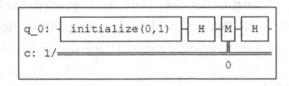

These exercises demonstrated that it is possible to create an X gate by sandwiching our Z gate between two H gates.

$$X = HZH$$

The calculation starts in the Z basis ⇒. Then the H gate switches the qubit to the X basis, after which ⇒the Z gate performs a NOT in the X basis, and finally ⇒ another H gate returns the qubit to the Z basis. Next, we perform verification on a Bloch sphere with the same syntax as Listing 4-1c and obtain the following output in Figure 4-7.

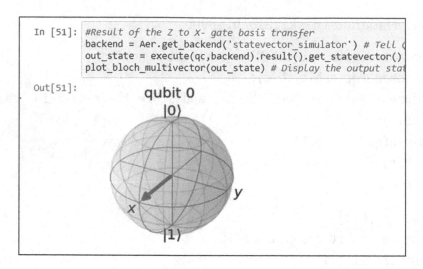

```
In [51]:  #Result of the Z to X- gate basis transfer
          backend = Aer.get_backend('statevector_simulator') # Tell (
          out_state = execute(qc,backend).result().get_statevector()
          plot_bloch_multivector(out_state) # Display the output sta
```

Out[51]:

Figure 4-7. *Verification of the X gate basis transfer*

If the Jupyter notebook cell is run again, you may see different results; however, the final state of the qubit is always $|+\rangle$ or $|-\rangle$. The qubit was initialized in the state $|1\rangle$. As a result of the measurement, it collapsed to states $|+\rangle$ or $|-\rangle$.

Qiskit allows visualization of data from a quantum circuit run on a real device or qasm_simulator. To facilitate this, there is a plot_histogram(data) function. As an example, a look at creating the Bell state with Qiskit. The computational basis of a system of two qubits can be converted to an orthonormal basis defined by the four Bell states.

Bell States with Qiskit

When a system is entangled, the individual component systems are linked together as a single entity. Any measurement that measures a part of the system—in our case, the first particle—is a measurement of the entire system. The wave function for the system then collapses, and both particles assume definite states. A popular example of a basis for a bipartite system is the *Bell basis* or the *Bell states*. The Bell states for two qubits are generally given by the following.

$$|\beta_{00}\rangle = \frac{1}{\sqrt{2}}(|00\rangle + |11\rangle)$$

4.7

$$|\beta_{01}\rangle = \frac{1}{\sqrt{2}}(|01\rangle + |10\rangle)$$

4.8

$$|\beta_{10}\rangle = \frac{1}{\sqrt{2}}(|00\rangle - |11\rangle) \qquad 4.9$$

$$|\beta_{11}\rangle = \frac{1}{\sqrt{2}}(|01\rangle - |10\rangle) \qquad 4.10$$

The $|\beta_{01}\rangle$ state is known as the *triplet state*. There are three triplet states. The other two are $|00\rangle$ and $|11\rangle$. $|\beta_{11}\rangle$ is known as the *singlet state*. A quantum circuit to build the Bell state is shown in Listing 4-1f, which prints the Bell measurement counts and then produces a histogram.

Listing 4-1f. Bell State `qiskitBloch.ipynb`

```
# quantum circuit to create a Bell state
bell = QuantumCircuit(2, 2)
bell.h(0)
bell.cx(0, 1)

meas = QuantumCircuit(2, 2)
meas.measure([0,1], [0,1])

# execute the quantum circuit
backend = BasicAer.get_backend('qasm_simulator') # the device to run on
circ = bell + meas
result = execute(circ, backend, shots=1000).result()
counts  = result.get_counts(circ)
print(counts)

from qiskit.visualization import import plot_histogram
plot_histogram(counts)
```

This produces the following output as measurement counts for Bell states.

```
{'00': 481, '11': 519}
```

The histogram in Figure 4-8 is the output of the `plot_histogram()` function called from `qiskit.visualization`.

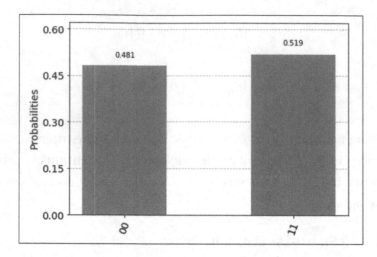

Figure 4-8. *Histogram for Bell state probabilities*

This section is a warm-up exercise to give insight into the operations of Qiskit for the upcoming chapters, where we indulge in defining more complex algorithms.

Quantum Circuits with Cirq

Quantum computers execute programs as quantum circuits based on unitary gate operations on qubits in gate-based quantum computers. A quantum computing circuit is a network of quantum logic gates where each gate performs some unitary transformation on one or more qubits. You have seen examples of some such gates in the previous section using Qiskit. This section explores similar insights using Cirq [30].

Cirq is an open source framework created by Google for programming quantum computers. It is "a Python software library for writing, manipulating, and optimizing quantum circuits, and then running them on quantum computers and quantum simulators." Cirq provides useful abstractions for dealing with today's noisy intermediate-scale quantum computers, where details of the hardware are vital to achieving state-of-the-art results. Cirq documentation is at `https://quantumai.google/cirq`. The live GitHub is at `https://github.com/quantumlib/Cirq`.

Cirq is installed via the following command in Python 3.6.x (preferred Python version in this book).

```
$ pip3 install cirq
```

```
(qml) sgangoly@ubuntu:~$ pip3 install cirq
Collecting cirq
  Downloading cirq-0.9.1-py3-none-any.whl (1.6 MB)
     |████████████████████████████████| 1.6 MB 892 kB/s
Requirement already satisfied: matplotlib~=3.0 in ./anaconda3/envs/qml/lib/pytho
n3.7/site-packages (from cirq) (3.3.3)
Collecting sortedcontainers~=2.0
```

Figure 4-9. *Cirq installation on Ubuntu 18.04 LTS. (qml) is the virtual environment name*

An example of the installation process is shown in Figure 4-9, where the package is installed in a virtual environment called qml. Chapter 1 describes ways to create virtual environments.

Once Cirq is installed, a Jupyter notebook session is opened from the same virtual environment, and a quick verification test is run in Listing 4-2a by printing Google's cirq.google.Bristlecone with the code in Cirq.ipynb. Bristlecone is a Google quantum computer, with which they achieved quantum supremacy. Every Bristlecone chip features 72 qubits.

Listing 4-2a. Bloch Sphere Simulation in Cirq.ipynb

```
import cirq
import numpy as np
print(cirq.google.Bristlecone)
```

This code produces the output in Figure 4-10, confirming that the Cirq installation was successful.

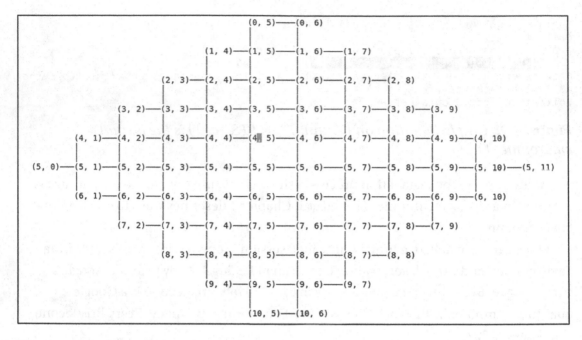

Figure 4-10. *Google Bristlecone in Cirq*

The circuit consists of a single qubit prepared initially in the |0⟩ state treated with the unitary X gate operation and then measuring the qubit again on the computational basis.

Listing 4-2b shows the steps to define an X gate.

Listing 4-2b. Generate a Qubit and a Circuit `Cirq.ipynb`

```
# Get a qubit and a circuit
qbit = cirq.LineQubit(0)
circuit = cirq.Circuit()
# Add an X gate: acts like the Pauli Matrix sigma_x
circuit.append(cirq.X(qbit))
```

In Listing 4-2c, the next step, we run a simulation with the `cirq.Simulator()` function.

Listing 4-2c. Simulation of Wave Function `Cirq.ipynb`

```
# Simulation that extracts the wavefunction

sim = cirq.Simulator()
result = sim.simulate(circuit)
```

```
print("\nBloch Sphere of the qubit in the final state:")
state = cirq.bloch_vector_from_state_vector(result.final_state,0)
print("x: ", np.around(state[0], 4), " y: ", np.around(state[1], 4),
" z: ", np.around(state[2], 4))
```

This code generates the following Bloch definition of the qubit at its final state as output.

```
Bloch Sphere of the qubit in the final state:
x:  0.0  y:  0.0  z:  -1.0
```

The next step, Listing 4-2d, adds a measurement gate to the circuit.

Listing 4-2d. Add Measurement on X Gate `Cirq.ipynb`

```
# Add a measurement gate at the end of the circuit:
circuit.append(cirq.measure(qbit, key="Final state"))
# Display the circuit:
print("\nCircuit:")
print(circuit)
```

This code produces the following circuit as output showing the measurement gate as *M*.

```
Circuit:
0: ──X──M('Final state')──
```

The next step of this experiment (see Listing 4-2e) uses the simulator and runs ten trial simulations to determine the final state of the qubit.

Listing 4-2e. Repeated Simulations `Cirq.ipynb`

```
# Invoke the Cirq quantum simulator to execute the circuit
simulator = cirq.Simulator()
# Simulate the circuit several times:
result = simulator.run(circuit, repetitions=10)
# Print the results:
print("\nResults of 10 trials:")
print(result)
```

The output of this code gives the following as a final state.

```
Results of 10 trials:
Final state=1111111111
```

This example uses a user-friendly feature of the Cirq simulator that extracts the quantum final state wave function. This is usually not possible on a real quantum computer, on which you must run the same circuit many times to get to the final state. In this example, the final state is $|1\rangle$.

Bell State Measurement with Cirq

You encountered Bell states and their definition in Qiskit earlier in this exercise. You should become familiar with the generation of Bell states and measurements in Cirq. It is possible to take two-qubit states in the computational basis and convert them into Bell states by using a Hadamard gate on the first qubit followed by a CNOT gate on the pair of qubits, as follows.

$$|00\rangle \Rightarrow |0\rangle = \frac{1}{\sqrt{2}}(|100\rangle + |11\rangle)$$ 4.11

$$|01\rangle \Rightarrow |\beta_{01}\rangle = \frac{1}{\sqrt{2}}(|01\rangle + |10\rangle)$$ 4.12

$$|10\rangle \Rightarrow |\beta_{10}\rangle = \frac{1}{\sqrt{2}}(|00\rangle - |11\rangle)$$ 4.13

$$|11\rangle \Rightarrow |\beta_{11}\rangle = \frac{1}{\sqrt{2}}(|01\rangle - |10\rangle)$$ 4.14

The unitary transformation shown in equations 4.11–4.14 in the Hilbert space of two qubits is a change of basis from the two-qubit computational basis to the Bell state basis. If the unitary operations are performed in the reverse order, it is possible to rotate from the Bell basis back to the computational basis. This inverse operation, followed by measuring the qubits, is known as *Bell state measurement*. The net effect is to project the two-qubit state onto one of the four Bell states using EPR pairs. The circuit is shown in Figure 4-11. Listing 4-2f is the code to initiate the exercise for the Bell state measurement.

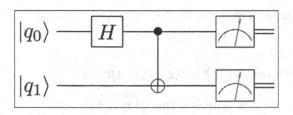

Figure 4-11. *Gate operations on Entangled EPR pair*

Listing 4-2f. Measurement on X Gate `Cirq.ipynb`

```
# Get two qubits and a circuit
qubit = [cirq.LineQubit(x) for x in range(2)]
circuit = cirq.Circuit()
# Add a Hadamard gate to qubit 0, then a CNOT gate from qubit 0 to qubit 1:
circuit.append([cirq.H(qubit[0]),
cirq.CNOT(qubit[0], qubit[1])])
# Run a simple simulation that extracts the actual final states
sim = cirq.Simulator()

result = sim.simulate(circuit)
print("\nBloch Sphere of the qubit 0 in the final state:")
state = cirq.bloch_vector_from_state_vector(result.final_state,0)
print("x: ", np.around(state[0], 4), " y: ", np.around(state[1], 4),
" z: ", np.around(state[2], 4))
print("\nBloch Sphere of the qubit 1 in the final state:")
state = cirq.bloch_vector_from_state_vector(result.final_state,1)
print("x: ", np.around(state[0], 4), " y: ", np.around(state[1], 4),
" z: ", np.around(state[2], 4))
```

This snippet gives the following output.

```
Bloch Sphere of the qubit 0 in the final state:
x:  0.0  y:  0.0  z:  0.0

Bloch Sphere of the qubit 1 in the final state:
x:  0.0  y:  0.0  z:  0.0
```

Then we add a measurement gate to the circuit, the same as in the X gate. This is shown in Listing 4-2g.

Listing 4-2g. Measurement on X Gate `Cirq.ipynb`

```
# Add a measurement at the end of the circuit:
circuit.append(cirq.measure(*qubit, key="Final state"))
# Display the circuit:
print("\nCircuit:")
print(circuit)
```

The following circuit is printed out as an output showing the measurement gates as *M* and the Hadamard gate as *H*.

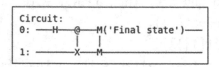

For the last step in the Bell states exercise, we run the simulation ten times, the same as in the previous exercise.

Listing 4-2h. Measurement on X Gate `Cirq.ipynb`

```
# Run the Cirq quantum simulator to execute the circuit:
simulator = cirq.Simulator()
# Simulate the circuit several times:
result = simulator.run(circuit, repetitions=10)
# Print the results:
print("\nResults:")
print(result)
```

The output is as follows.

```
Results:
Final state=1000101001, 1000101001
```

If Listing 4-4h is run several times, the results for each qubit vary randomly, but they are 100% correlated between the two qubits. The results from the quantum simulator are from 10 measurements in the computational basis of the final state, $|\beta_{00}\rangle$, which is the same EPR pair state as Alice and Bob. Notice that Cirq reports that the final state of both qubits is in the center of the Bloch sphere, not on its surface.

In summary, there are a few things to keep in mind about EPR pairs.

- Quantum randomness occurs between an eavesdropper (say Eve) and Alice and between Eve and Bob, but not between Alice and Bob.

- While Alice and Bob receive the same binary bitstring, it is causally impossible at that time for Eve to be in possession of their bitstring.

- While randomly generated, the bitstring that they share can eventually have large causal consequences.

This quantum weirdness is what Einstein called *spukhafte Fernwirkung*, meaning "spooky action at a distance."

Entropy: Classical vs. Quantum

Entropy is loosely defined as a way to quantify information in a signal. Briefly, entropy can be interpreted either as the *uncertainty* we have about the state of a given system *or* as the amount of information we gain after measuring the system.

Shannon Entropy

The key notion of classical information is *Shannon entropy*. Claude Shannon's entropy seeks to provide a statistical representation of information in a signal associated with given states $|\psi_i\rangle$, $i \in \{1...n\}$. Shannon's theory was based on finding the probability of seeing a given piece of information. This allowed scientists to characterize the amount of information gained from a signal.

Shannon quantified the amount of information gained by taking the base 2 logarithm of the probability of a given message. If the information contained in the message is denoted by I, and the probability of its occurrence by p, then

$$I = -\log_2 p$$

where, the negative sign ensures that the information content of the message remains positive. For example, if the probability of the stock market not crashing below 5% today is 0.9, then the information content of the news is I = −log$_2$ 0.9 = 0.152.

Formally, if $x \in \{x_1, x_2...x_n\}$ is a random variable with probability distribution $p \in \{p_1, p_2...p_n\}$, then $\sum_{i}^{n} p_i = 1$ and the Shannon entropy of x is given by

$$S_{SHE}(x) = -\sum_{i=1}^{n} p_i \log_2 p_i$$

4.15

Hence, if the probability of a given x_i is 0, the Shannon entropy is 0 $*$ log 0 = 0. The Shannon entropy estimates its minimum value as 0 if the state is pure and takes its maximum value if it is random. Mixing of states increases the entropy. In general, the following rules of thumb estimate Shannon entropy.

- If there is certainty about the content of the information or message, the Shannon entropy is zero.

- The higher the uncertainty about the next piece of information content, the higher the Shannon entropy.

In other words, entropy quantifies the amount of information we gain when we measure x. An example of an entropy function known as the *binary entropy function* (BEF) is shown in Figure 4-12. In information theory, BEF is defined as the entropy of a Bernoulli trial with the probability of success p. Mathematically, the Bernoulli trial is modelled as a random variable x that can take on only two values: 0 and 1. The event $x = 1$ is considered a success and the event $x = 0$ is considered a failure, where the two events are mutually exclusive and exhaustive.

If the probability of $x = 1$ is given by $Pr(x = 1) = p$, then probability that $x = 0$ is given by $Pr(x = 0) = (1 - p)$. In his case, the entropy of x is given by

$$S_{SHE}(x) = -p \log_2 p - (1-p)\log_2(1-p)$$

4.16

where the logarithm is usually taken as base 2 for binary operations.

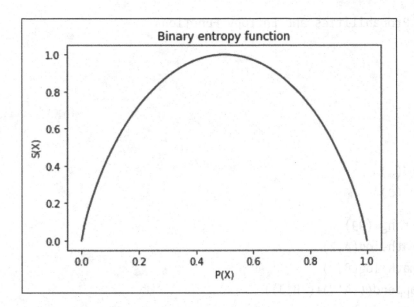

Figure 4-12. *Binary entropy function for Shannon entropy*

As shown in Figure 4-12, the entropy attains its maximum value when $p = 0.5 = \dfrac{1}{2}$. The maximum entropy occurs for the instance with the *least amount of knowledge.* When each of the possible outcomes is equally likely, then the discrete probabilities p_i reflect the least amount of knowledge about the outcome of a measurement. Hence, with n possible outcomes, each of the probabilities is given by

$$p_i = \frac{1}{n} \qquad\qquad 4.17$$

The case where $p = \dfrac{1}{2}$ is the case of the unbiased bit, which is the most common unit of information entropy. Snippets of the simple Python code `BinaryEntropyFunction.` `ipynb` used to generate the plot in Figure 4-12 is shown in Listing 4-3. The probabilities and entropies are defined, and values generated based on the logarithmic functions to give the entropy plot.

Listing 4-3. Libraries and Functions in `BinaryEntropyFunction.ipynb`

```
# Import Libraries
import matplotlib.pyplot as plt
import math
```

```python
# Define Probabilities and Entropy Functions
px0 = []
px1 = []
Sx0=[]
Sx1=[]
A=0.99
B=0.01
px0.append(1)
Sx0.append(0)

for i in range(99):
    V = math.log(A,2)
    J = math.log(B,2)
    Sx0.append((-A*V)+(-B*J))
    px0.append(A)
    px1.append(B)

  # Sx1.append(-A*(math.log(A,2)) -B*(math.log(B,2)))
    A -= 0.01
    B += 0.01
px0.append(0)
Sx0.append(0)

# Print the values and the graph
for i in range(20):
    print()
    print("p(x): ",end="")
    for j in range(5):
        print(str(px0[j*i+i])+",",end="")
print()

for i in range(20):
    print()
    print("S(x): ",end="")
    for j in range(5):
        print(str(Sx0[j*i+i])+",",end="")
plt.plot(px0, Sx0, color='blue')
```

```
plt.xlabel('P(X)')
plt.ylabel('S(X)')
plt.title('Binary entropy function')
plt.show()
```

Mathematically, if $p = 0.3$, then the probability of finding the alternative is $1 - p = 0.7$.

This indicates a scenario where there is knowledge about the state before measuring because one possible value is far more likely than the other. In this case, referring to equation 4.2, we have

$$-p\log_2 p - (1-p)\log_2(1-p) = -0.3\log_2(0.3) - (1-0.3)\log_2(1-0.3) = 0.88$$

On the other hand, if we had if $p = 0.05$, with a probability for an alternative as 95%, then we get

$$-p\log_2 p - (1-p)\log_2(1-p) = 0.29$$

Total uncertainty about the outcome means all possible outcomes are equally likely. The general rule is larger the entropy, the greater the uncertainty of the outcome before measurement.

Von Neumann Entropy

The amount of entropy in a *quantum state* is determined by an analog to the Shannon entropy. It is achieved by using density operators instead of elements of the probability distribution as in equation 4.1. The *quantum information theory* is somewhat similar to its classical counterpart in that the information is carried by a *quantum system*. Hence, there is a difference since qubits carry information in more than one sense.

- Classical bits can be encoded into qubits.

- Unknown qubits can carry information hidden and protected as a direct consequence of the limitations imposed by the no-cloning theorem or distinguishability.

If we recall equation 1.29 in Chapter 1, the density operator of a mixed quantum system is as follows.

$$\rho = \sum_{i=1}^{n} \rho_i p_i = \sum_{i}^{n} p_i |\psi_i\rangle\langle\psi_i| \qquad 4.18$$

where $|\psi_i\rangle$ represent n possible states in an ensemble in a mixed quantum system, and p_i is the probability that a member is in the corresponding state $|\psi_i\rangle$.

Recall that the density operators were a way of describing possible states of a quantum system, or equivalently possible outcomes of measurements, in a way very similar to probability distributions. The entropy of a quantum state with density operator ρ is called the *von Neumann entropy* and is given by

$$S(\rho) = -Tr(\rho \log_2 \rho) \qquad 4.19$$

where Tr depicts the trace of the density operator.

In a mixed system with density states ρ and ϵ, the *relative von Neumann entropy* is given by

$$S(\rho|\epsilon) = Tr(\rho \log_2 \rho) - Tr(\rho \log_2 \epsilon) \qquad 4.20$$

Generally, $S(\rho|\epsilon) \geq 0$. It is only equal if $\rho = \epsilon$.

If the eigenvalues of the density operator are given by λ_i, then the von Neumann entropy is given by

$$S(\rho) = -\sum_{i} \lambda_i \log_2 \lambda_i \qquad 4.21$$

In general, in n dimensions, the entropy of a quantum state follows the inequality.

$$\log_2 n \geq S(\rho) \geq 0 \qquad 4.22$$

A composite state in a quantum system is often a state formed by tensor products of qubit states. If a composite state is separable, then it forms a product state of the form $\rho \otimes \epsilon$. In this case, the entropy is additive and given by

$$S(\rho \otimes \epsilon) = S(\rho) + S(\epsilon) \qquad 4.23$$

Generally, entropy is subadditive. In other words, the reduced density matrices of a composite system are given by the inequality

$$S(\rho_{12}) \leq S(\rho_1) + S(\rho_2)$$ 4.24

Equation 4.10 indicates that to have the most information about an entangled system, we need to consider the *whole* system; that is, $S(\rho)$ is smaller than the combined entropies of the reduced density matrices because knowledge increases when the whole system is considered. The users, in possession of the reduced density matrices ρ_1 and ρ_2, have less knowledge about the state when considering only portions of the system.

Evolution of States

Due to their inherent nature, the state of a system usually evolves with time in quantum physics. In the circuit and gate model of quantum computing, states evolve according to *unitary matrix operations* (see Chapter 1). The evolution of a state vector amounts to performing linear algebra. On quantum annealing platforms, the evolution of a state is modeled using the *Hamiltonian* (see Chapter 1). There are two important kinds of states: mixed states and pure states. *Pure states* have the property that they can always be represented as the "*outer product*" of a state vector with itself. *Mixed states* are generally expressed as a sum of orthogonal state vectors. In pure states, the von Neumann entropy is always zero and is analogous to the deterministic Shannon entropy. In mixed states, the entropy can be represented by a range of values.

Properties of von Neumann entropy can be summarized as follows.

1. $S(\rho) = 0$ if and only if ρ is a pure state.

2. For $\dim(\mathbb{H}) = N$ (the dimension of the Hilbert space),
    $S(\rho) = \log N$ gives the maximum value of $S(\rho)$ for a *maximally mixed state*.

3. $S(\rho)$ is invariant under basis change (i.e., $S(\rho) = S(U\rho U^{\dagger})$ for any unitary matrix U).

4. For a set of numbers $\{\lambda_i\}$ such that $\sum_i \lambda_i = 1$ and $\lambda_i > 0$, and for a corresponding collection of density matrices $\{\rho_i\}$ we have

$$S\left(\sum_i \lambda_i \rho_i\right) > \sum_i \lambda_i S(\rho_i)$$

5. For a set of numbers, $S(\rho)$ satisfies

$$S\left(\sum_i \lambda_i \rho_i\right) \leq \sum_i \lambda_i S(\rho_i) - \sum_i \lambda_i \log \lambda_i$$

6. $S(\rho_M \otimes \rho_N) = S(\rho_M) + S(\rho_N)$ for independent systems M and N.

7. $S(\rho_{123}) + S(\rho_2) \leq S(\rho_{12}) + S(\rho_{23})$

8. $S(\rho_{12}) \leq S(\rho_1) + S(\rho_2)$

A hands-on example where a system evolves from a pure state to a mixed state and then reverts to a pure state again may help with understanding the process. In this process, the von Neumann entropy first increases and then gradually decreases back to zero.

For the first step, we define the spin-up and spin-down state vectors and the pure states given by the outer products. This is shown in Listing 4-4a. The libraries can be found in `qiskitStateEvo.ipynb,` the Jupyter notebook code sample, where *the full code is presented with additional worked-out examples*. We use Qiskit to explore the quantum evolution of states.

To start some coding, we call the `qiskit` library to showcase GHZ states. `Qiskit` is a Python library that creates quantum circuits and runs them either on a simulator or on IBM's quantum devices.

Listing 4-4a. Spin-up, Spin-down Definition at `qiskitStateEvo.ipynb`

```python
import numpy as np
from qiskit import *
import math as m
from scipy import linalg as la
%matplotlib inline

# Spin-up
u = np.matrix([[1],
               [0]])
```

```
# Spin-down
d = np.matrix([[0],
              [1]])
# Pure state |0><0|
P11 = np.dot(u, u.H)
# Pure state |1><1|
P22 = np.dot(d, d.H)
```

The state evolution is plotted using Listing 4-4b in qiskitStateEvo.ipynb.

Listing 4-4b. von Neumann Definiton in qiskitStateEvo.ipynb

```
import matplotlib.pyplot as plt
p = np.arange(0.001, 1., 0.01)

# create list of von Neumann entropies
vn_y = list()
for value in p:
    rho_p = value*P11 + (1-value)*(P22)
    vn_y.append(von_neumann_entropy(rho_p))

# convert list of matrices to list of nos. for plotting
y = list()
for value in vn_y:
    y.append(value.tolist()[0][0])

# plot p vs. S(rho_p)
plt.xlabel('p')
plt.ylabel('von Neumann Entropy')
plt.title('S(rho_p)')
plt.plot(p, y, color='tab:red')
plt.show()
```

The output of Listing 4-4b is shown in Figure 4-13.

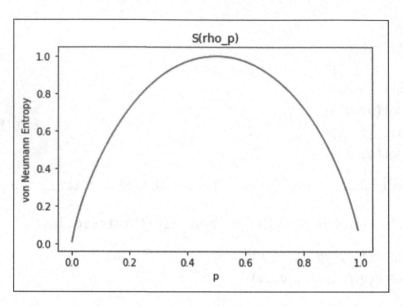

Figure 4-13. *von Neumann entropy*

A comparison between the von Neumann plot of Figure 4-13 and the classical Shannon entropy of Figure 4-12 shows that they are quite similar.

The one-parameter family of density matrices for von Neumann entropy is shown in Listing 4-4c.

Listing 4-4c. One-Parameter Family in `qiskitStateEvo.ipynb`

```
# Define spin-up
u = np.matrix([[1],
               [0]])
# Define spin-down
d = np.matrix([[0],
               [1]])

# Define the pure state |0><0|
P11 = np.dot(u, u.H)

# Define the pure state |0><1|
P12 = np.dot(u, d.H)

# Define the pure state |1><0|
P21 = np.dot(d, u.H)
```

```python
# Define the pure state |1><1|
P22 = np.dot(d, d.H)

# Define the mixed state |0><0| + |1><1|
I = (P11+P22)/2

# Define the pure state |0><0|+|0><1|+|1><0|+|1><1|
D = (P11+P12+P21+P22)/2

p = np.arange(0.001, 1., 0.01)


# Create list of von Neumann entropies
vn_y = list()
for value in p:
    rho = value*I + (1-value)*(D)
    vn_y.append(von_neumann_entropy(rho))

# list of matrices to list of numbers for plotting
y = list()
for value in vn_y:
    y.append(value.tolist()[0][0])

# plot p vs. S(rho_p)
plt.xlabel('p')
plt.ylabel('von Neumann Entropy')
plt.title('S(rho_p)')
plt.plot(p, y, color='tab:blue')
plt.show()
```

Listing 4-4c produces the output shown in Figure 4-14.

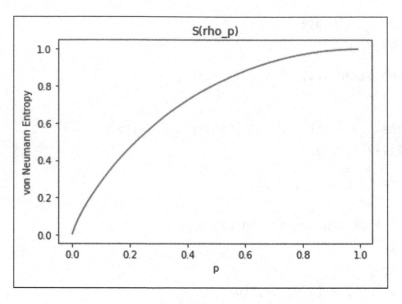

Figure 4-14. *von Neumann entropy for one parameter density matrices*

GHZ State

In quantum information theory, the GHZ state stands for Greenberger–Horne–Zeilinger state, first studied by Daniel Greenberger, Michael Horne, and Anton Zeilinger in 1989 [29]. It is a certain type of entangled quantum state that involves more than two subsystems (e.g., particle states or qubits). The GHZ state is a maximally entangled quantum state. For a system comprised of n subsystems, each of which is two dimensional (i.e., qubits), the GHZ state is given by

$$|GHZ\rangle = \frac{|0\rangle^{\otimes n} + |1\rangle^{\otimes n}}{\sqrt{2}}$$

4.25

The three-qubit GHZ state is the simplest one. It exhibits non-trivial multipartite entanglement.

$$|GHZ\rangle = \frac{|000\rangle + |111\rangle}{\sqrt{2}}$$

A GHZ state is a quantum superposition of all subsystems being in state 0, with all of them being in state 1 where states 0 and 1 of a single subsystem are fully distinguishable.

A GHZ can be prepared using a quantum circuit. Listing 4-4d is a Qiskit code snippet from the Jupyter notebook `qiskitStateEvo.ipynb`.

176

Listing 4-4d. GHZ in `qiskitStateEvo.ipynb`

```
from qiskit import *
import numpy as np
%matplotlib inline

qr = QuantumRegister(3)
cr = ClassicalRegister(3)
GHZ = QuantumCircuit(qr, cr)

GHZ.h(0)
GHZ.cx(0,1)
GHZ.cx(1,2)
GHZ.draw(output='mpl')
```

Listing 4-4d gives the quantum circuit for the GHZ output, as shown in Figure 4-15.

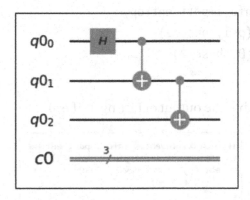

Figure 4-15. *GHZ quantum circuit*

The `Statevector()` function from Qiskit in Listing 4-4e prints the state vector. The state vector is used for obtaining the von Neumann entropy of the corresponding density matrix using the `entropy()` function in Qiskit. The `DensityMatrix()` function from Qiskit is passed to the `GHZ_state` ket vector.

Listing 4-4e. Statevector in `qiskitStateEvo.ipynb`

```
backend = Aer.get_backend('statevector_simulator')
GHZ_state = execute(GHZ,backend).result().get_statevector()
print(np.matrix(GHZ_state).H)
```

```
from qiskit.quantum_info import Statevector
psi = Statevector(GHZ_state)
print(psi)
```

The von Neumann entropy can now be calculated using the Qiskit `entropy()` function. This can then be passed either by the `Statevector` function or the `DensityMatrix` function. In the latter case, the output is effectively zero entropy, whereas in the first, it is exactly zero. This makes sense because the state is a pure state, and the density matrix is computed as $D = |\psi\rangle\langle\psi|$.

Noticeable is the maximum which is at an intermediate point in time, before the final mixed state is achieved. Generally, entropy tends to increase over time. However, in effectively "closed" quantum systems where substantial control can be exercised, the entropy can be reduced to a lower level (see Listing 4-4f).

Listing 4-4f. Entropy in qiskitStateEvo.ipynb

```
from qiskit.quantum_info import entropy
vn_entropy1 = entropy(psi, base=2)
vn_entropy2 = entropy(D, base=2)
print(vn_entropy1)
```

The following shows that the output of Listing 4-4f is 0.

```
from qiskit.quantum_info import entropy
vn_entropy1 = entropy(psi, base=2)
vn_entropy2 = entropy(D, base=2)
print(vn_entropy1)

0

print(vn_entropy2)

1.6017132519074586e-16
```

The entropy of the system that evolved from the state given by the density matrix D of the GHZ-state to the system described by the density matrix ρ can be calculated by Listing 4-4g.

Listing 4-4g. Entropy Calculation of Evolved State in qiskitStateEvo.ipynb

```
from qiskit.quantum_info import random_density_matrix
rho = random_density_matrix(8, rank=None, method='Hilbert-Schmidt',
seed=None)
print(rho)

time = np.arange(0.001, 1., 0.01)

vn_entropy = list()
for t in time:
    vn_entropy.append(entropy(D, base=2)*t + entropy(rho, base=2)*(1-t))

# Create list of von Neumann entropies
vn_y = list()
for value in p:
    M = value*rho + (1-value)*(D)
    vn_y.append(entropy(M))

# plot p vs. S(rho_p)
plt.xlabel('time')
plt.ylabel('von Neumann Entropy')
plt.title('S(rho(t))')
plt.plot(p, vn_y, color='tab:blue')
plt.show()
```

The output plot is shown in Figure 4-16.

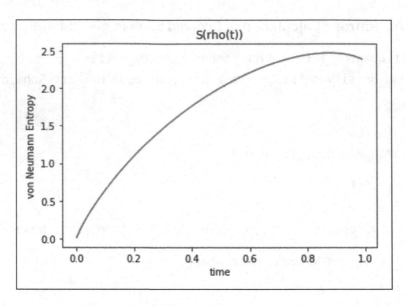

Figure 4-16. *Entropy of the evolved GHZ system*

In Figure 4-16, the maximum is at an intermediate point in time, just before the final mixed state is achieved. Generally, entropy tends to increase over time. However, in effectively closed quantum systems where substantial control can be exercised, the entropy can be reduced to a lower level.

No-Cloning Theorem Revisited

Chapter 1 discussed the *no-cloning theorem* of quantum mechanics, which states that quantum states cannot be directly copied or cloned via unitary transformations such as gate operations. This implies that if $|\psi\rangle$ is a quantum state, and U is some unitary transformation such that $U|\psi 0\rangle = |\psi\psi\rangle$, then it is impossible that this property holds for arbitrary states $|\psi\rangle$. The theory follows from the possibility of linear quantum superpositions of two orthogonal states (such as $|0\rangle$ and $|1\rangle$ in the single-qubit case) and a U such that $U|\psi 0\rangle = |\psi\psi\rangle$ and $U|\phi 0\rangle = |\phi\phi\rangle$. If we consider the superposition $|\varphi\rangle = \frac{1}{\sqrt{2}}(|\psi\rangle + |\phi\rangle)$, then

$$U|\varphi 0\rangle = \frac{1}{\sqrt{2}}(U|\psi 0\rangle + U|\phi 0\rangle) = \frac{1}{\sqrt{2}}(|\psi\psi\rangle + |\phi\phi\rangle)$$

4.26

$$|\varphi\varphi\rangle = \frac{1}{\sqrt{2}}(|\psi\psi\rangle + |\psi\phi\rangle + |\phi\psi\rangle + |\phi\phi\rangle)$$

4.27

From equation 4.26 and 4.27, we see that

$$|\varphi\varphi\rangle \neq U|\varphi 0\rangle \qquad\qquad 4.28$$

In essence, the no-cloning theorem rules that we cannot copy an *unknown quantum state*. However, this does not impose any restrictions on copying a *known quantum state*. *Indeed,* it should be possible to do so; in reality, it says that we can prepare many identical copies of the same known state. The no-cloning theorem has major consequences for quantum communications since it implies that the construction of an amplifier that preserves arbitrary quantum information is not possible. The no-cloning theorem also makes it impossible for hackers to copy information and violate security.

Quantum Teleportation

Quantum teleportation is a process by which a quantum qubit state can be transmitted by sending only two classical bits of information. This option works by distributing entangled quantum particles. These entangled particles, once distributed in a controlled manner between two ends, create virtual channels.

Quantum teleportation, shown in Figure 4-17, is a phenomenon that leverages entangled Bell pairs for an EPR (Einstein-Podolsky-Rosen) source [32].

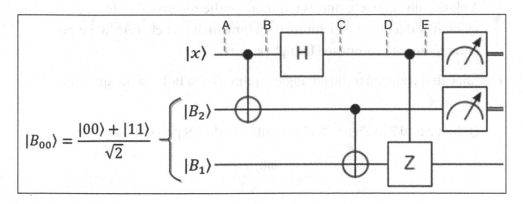

Figure 4-17. *Quantum teleportation with measurement circuit*

The circuit shown in Figure 4-17 is quantum teleportation *with measurement* and has an initial state is $|000\rangle$; the message qubit is converted to a non-trivial state by applying random powers of the X and Y gates. The Alice and Bob qubits are entangled as an EPR pair. A Bell measurement is performed on the message and Alice qubits. Classical information from this measurement then controls two single-qubit operations on the Bob qubit. The process is as follows.

- A sender, Alice, transmits one particle of an entangled pair to the destination while retaining the other particle as the source.

- For an eavesdropper Eve, it is deemed impossible by the *no-cloning theorem* to observe and measure the entangled particle without destroying its state. As soon as she measures a specific state, the act of measurement causes the same state to collapse. In other words, there is no way for Eve to breach security. This property of quantum entanglement renders it a *secure* channel.

- Not only is the security tight for an entangled pair, theoretically, but a pair of particles can also remain entangled independent of distance, even at a cosmic scale. Experimentally, quantum entanglement has been proven for up to an approximate distance of 1200 km between space and earth and 50 km on terrestrial experiments [33].

- A classically open channel is required for the receiver Bob to efficiently decode the quantum data for which an Eckert91 [34] type entanglement is proposed for tight security.

- Alice first transports the entangled qubit $|B_1\rangle$ to Bob via the quantum channel.

  Qubits 2 and 3 in Figure 4-15 are entangled EPR pairs in Bell state.

$$|B_{00}\rangle = \frac{|00\rangle + |11\rangle}{\sqrt{2}}$$

4.29

- $|B_{00}\rangle$ in equation 4.25 has subscripts as 00 to indicate that as per convention, all qubits are initialized at state $|0\rangle$.

- Figure 4-15 shows the three-qubit state in five characteristic points of a quantum circuit. The input three-qubit state at point A is given by

$$|\psi_A\rangle = \frac{|x00\rangle + |x11\rangle}{\sqrt{2}}$$

4.30

- The first qubit is used as a control qubit, and the second qubit is the target qubit for the CNOT (Control-X) gate. Therefore, the state at point B becomes,

$$|\psi_B\rangle = \frac{|xx0\rangle + |x(1-x)1\rangle}{\sqrt{2}}$$

4.31

- The Hadamard gate (H) performs mapping on the first qubit.

$$H|x\rangle = \frac{|0\rangle + (-1)^x |1\rangle}{\sqrt{2}}$$

4.32

- The state at point C is given by

$$|\psi_C\rangle = \frac{|0x0\rangle + (-1)^x |1x0\rangle + |0(1-x)1\rangle + (-1)^x |1(1-x)1\rangle}{\sqrt{2}}$$

4.33

- The second qubit is used as a control qubit. The third is a target qubit in the CNOT gate. The quantum state at point D is given by

$$|\psi_D\rangle = \frac{|0xx\rangle + (-1)^x |1xx\rangle + |0(1-x)x\rangle + (-1)^x |1(1-x)x\rangle}{\sqrt{2}}$$

4.34

- Finally, the state at point E is given by the following. (The first qubit is used for control and the third as the target for the controlled-Z gate.)

$$|\psi_E\rangle = \frac{|0xx\rangle + |1xx\rangle + |0(1-x)x\rangle + |1(1-x)x\rangle}{\sqrt{2}}$$

4.35

$$= \frac{|0x\rangle + |1x\rangle + |0(1-x)\rangle + |1(1-x)\rangle}{\sqrt{2}} \otimes |x\rangle$$

4.36

- As the last step, measurement is performed on the first two qubits, which are destroyed, and the third qubit is delivered to the receiver.

- A comparison between the destination and source qubits offers the conclusion that the correct quantum state was teleported. This analysis does not take into account any errors introduced by the quantum channel.

Note that the teleported state is *not a superposition* state. Hence, any arbitrary state $|B_1\rangle$ can be teleported by the Figure 4-15 scheme. Alice measures her side-channel data and sends it to Bob via the classical channel. Bob reconstructs the quantum data using classically controlled Pauli X (or CNOT) and Z gates in Figure 4-15. This scenario is inherently secure because

- Measurement changes the state of the entangled particles.

- An attacker would require access to both the quantum and classical channels to decode successfully.

This section is a hands-on session on Cirq to see how teleportation works. The coding exercise shows that the final state of Bob's qubit is guaranteed to be in whatever state the message qubit was in originally. This is only possible given that an entangled state was pre-shared between Alice and Bob.

In the following exercise, the message (qubit 0) is prepared in a random state by applying the X and Y gates. Alice is in possession of both the message qubit and qubit 1, which is part of an EPR pair with Bob's qubit 2. Alice now performs a Bell state measurement on her pair of qubits, getting one of four possible results. She then transmits that result, which is equivalent to two classical bits, to Bob. Bob performs unitary operations on his qubit that depends on what he received from Alice.

In Listing 4-5a, continued from our earlier Cirq exercise, the three qubits, Hadamard gates, and measurement is defined to form the teleportation circuit, as commented in the body of code.

Listing 4-5a. Teleporation Circuit Definition `Cirq.ipynb`

```
import random
# Define three qubits: msg = qubit[0], qalice = qubit[1], qbob = qubit[2]
qubit=[0]*(3)
qubit[0] = cirq.NamedQubit('msg')
```

```
qubit[1] = cirq.NamedQubit('qalice')
qubit[2] = cirq.NamedQubit('qbob')
circuit = cirq.Circuit()
# Create a Bell state entangled pair to be shared between Alice and Bob.
circuit.append([cirq.H(qubit[1]), cirq.CNOT(qubit[1], qubit[2])])
# Creates a random state for the Message.
ranX = random.random()
ranY = random.random()
circuit.append([cirq.X(qubit[0])**ranX, cirq.Y(qubit[0])**ranY])

# Unitary operator rotating the two-qubit basis of the Message and Alice's
entangled qubit;
# rotates the Bell state basis to the computational basis:
circuit.append([cirq.CNOT(qubit[0], qubit[1]), cirq.H(qubit[0])])
# Combining now with a measurement in the computational basis,
# we effectively have projected this two-qubit state onto one of the four
states of
# the Bell state basis:
circuit.append(cirq.measure(qubit[0], qubit[1]))
```

The following uses the two classical bits from the Bell measurement to recover.

Listing 4-5b. Recover Original Message in Bob's Entangled Qubit Cirq.ipynb

```
# Use the two classical bits from the Bell measurement to recover the
# original quantum Message on Bob's entangled qubit.
circuit.append([cirq.CNOT(qubit[1], qubit[2]), cirq.CZ(qubit[0],
qubit[2])])
print("Circuit:")
print(circuit)
```

This gives the following teleportation circuit as output, which is analogous to Figure 4-17.

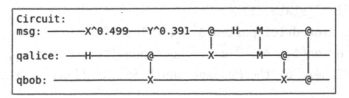

The next step is similar to the previous exercise on Bell states. We run cirq.
Simulator() to define the gates that create the message and outputs the Bloch sphere
location (see Listing 4-5c).

Listing 4-5c. Bloch Sphere Output `Cirq.ipynb`

```
sim = cirq.Simulator()
# Simulation that applies the random X and Y gates to
# create the message.
q0 = cirq.LineQubit(0)
message = sim.simulate(cirq.Circuit([cirq.X(q0)**ranX, cirq.Y(q0)**ranY]))
print("\nBloch Sphere of the Message qubit in the initial state:")
expected = cirq.bloch_vector_from_state_vector(message.final_state,0)
print("x: ", np.around(expected[0], 4), " y: ", np.around(expected[1], 4),
" z: ", np.around(expected[2], 4))
```

This gives the following as output which may change if the code is run several times,
for example.

```
Bloch Sphere of the Message qubit in the initial state:
x:  -0.1992  y:  -0.9796  z:  0.0252



Bloch Sphere of the Message qubit in the initial state:
x:  0.0026  y:  -1.0  z:  0.0009
```

This behavior is a direct result of the probabilistic nature of quantum mechanics.
The last step is to find the Bloch sphere of final state, as shown in Listing 4-5d.

Listing 4-5d. Bloch Sphere of the Final state `Cirq.ipynb`

```
# Records the final state of the simulation.
final_results = sim.simulate(circuit)
print("\nBloch Sphere of Bob's qubit in the final state:")
teleported = cirq.bloch_vector_from_state_vector(
final_results.final_state, 2)
print("x: ", np.around(teleported[0], 4), " y: ",
np.around(teleported[1], 4), " z: ", np.around(teleported[2], 4))
print("\nBloch Sphere of the Message qubit in the final state:")
```

```
message_final = cirq.bloch_vector_from_state_vector(
final_results.final_state, 0)
print("x: ", np.around(message_final[0], 4), " y: ", np.around(message_
final[1], 4), " z: ", np.around(message_final[2], 4))
```

The final output gives the Bloch states as follows.

```
Bloch Sphere of Bob's qubit in the final state:
x:  -0.1992  y:  -0.9796  z:  0.0252

Bloch Sphere of the Message qubit in the final state:
x:  0.0  y:  0.0  z:  -1.0
```

If the code is run several times, it becomes clear that the final state of the Message qubit is always either $|0\rangle$ or $|1\rangle$, which is trivial and a manifestation of the no-cloning theorem. Fans of *Star Trek* may recall that this is why Dr. McCoy objected to using the teleportation device on the Starship Enterprise because the teleportation process appears to destroy the original information and then re-create it instantaneously somewhere else!

A variation on the quantum teleportation circuit with measurement we performed is the same circuit but without measurement. The measurement of qubits 0 and 1 is omitted in the computational basis as shown in Listing 4-5e.

Listing 4-5e. Teleportation Without Measurement

```
# Unitary operator rotating the two- qubit basis of the Message and Alice's
entangled qubit;
# rotates the Bell state basis to the computational basis:
circuit.append([cirq.CNOT(qubit[0], qubit[1]), cirq.H(qubit[0])])
# This time skip the measurement
# circuit.append(cirq.measure(qubit[0], qubit[1]))
# Use the same operations as before to recover the
# original quantum Message on Bob's entangled qubit.
circuit.append([cirq.CNOT(qubit[1], qubit[2]), cirq.CZ(qubit[0],
qubit[2])])
```

The teleportation *without measurement* works just as well as it did *with measurement*. This is a manifestation of the *principle of deferred measurement*, which implies that the operation of measuring a qubit commutes with the operation of using it as a control for a controlled gate operation.

Fundamental of the way teleportation works involves substituting a classical communications channel for part of the swap, which allows Bob's qubit to be arbitrarily far away from Alice's qubit. Currently, the speed of teleportation is limited only by the fact that the classical information cannot be transmitted faster than the speed of light. Bob could choose to measure his qubit before receiving the classical transmission from Alice, since 25% of the time, he already has the correct quantum state without performing any "corrections." When Alice's message finally shows up, Bob learns whether his teleported message was valid.

Quantum teleportation over large distances is not a thought experiment in a lab anymore. Working with Caltech and AT&T, Fermilab is commissioning a high-fidelity, high-rate quantum teleportation system that uses near-infrared photons moving over standard telecom fiber, with superposition of two-time bins as the qubit. Other national labs are developing similar systems (see Joe Lykken lecture notes [35]).

Gate Scheduling

Before we delve into the intricacies of quantum algorithms in Chapter 5, let's address some basics that most quantum algorithms are built upon. This section addresses instruction-level optimization of gate scheduling and associated subjects. A sequence of gate operations on *logical* qubits is referred to as a *schedule*. *Logical* qubits consist of one or more *physical* qubits, can be in superposition, and are deemed to have more coherence time than the physical qubits. *Physical* qubits are actual quantum implementation of qubits which are *physically realized* such as electrons or atoms. As technology stands today, physical qubits suffer from imperfections and challenges of coherence. Limitations of current physically built multi-qubit systems can be overcome by building a logical qubit from several imperfect physical qubits. Logical qubits can be used for programming and can be implemented by a simulator running on a normal CPU on a desktop or laptop to allow the development, testing, and debugging of quantum algorithms.

There are *data dependencies* between the gates of a quantum circuit. The gate *sequence ordering* defines that data dependency between gates. For the circuit in Figure 4-18 with qubits q1, q2 and q3, gate G2 depends on gate G1 if they share a logical qubit.

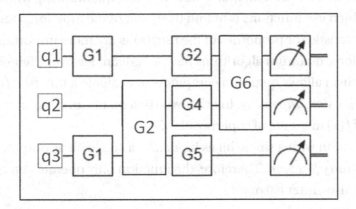

Figure 4-18. *Gate sequencing*

A qubit can be involved in one quantum gate at a time. Since quantum circuits execute from left to right, data dependencies determine the order of the *sequential* and *parallel* execution of gates in the quantum circuit. Two sequential gates, which are back-to-back, have, generally a well-defined ordering constraint, in other words,

$$-\boxed{A}-\boxed{B}- \;\neq\; -\boxed{B}-\boxed{A}-$$

If gates A and B both are arbitrary unitary operators, then swapping the order of their occurrence generally would yield different results. In some special cases, unitary matrices sometimes can be ordered, even though not always equivalent. In these cases, they are said to *commute* with each other.

Generally, two side-by-side *parallel gates* have no ordering constraints. The impact of gate scheduling can be quite significant in gate-based quantum circuits. The implementation of many algorithms depends on the execution of gates in parallel to achieve non-trivial speedups.

Scheduling of gates in quantum algorithms differs from their classical counterparts since gate scheduling enables an additional degree of freedom for the users. Quantum gate scheduling, in comparison to classical instruction scheduling, has been relatively less studied with fewer proposed systematic approaches and, consequentially, a subject matter of current active research trying to find about newer constraints.

Quantum Parallelism and Function Evaluation

The rudimentary basic of the quantum algorithm begins with *quantum parallelism*. Quantum parallelism is a fundamental feature of many quantum algorithms. The focus in many cases which use functions is to find the result of the function. However, in cases where the characteristics or properties of the functions are more important than the function evaluations, quantum algorithms have an advantage over their classical parts. Quantum parallelism allows quantum computers to evaluate a function $f(x)$ for many different values of x *simultaneously*. In other words, a quantum computer can evaluate various values of $f(x)$ in a state of superposition.

If $f(x)$ is a function with a single bit as input and a single bit as output, then $f(x)$ is referred to as a *binary function*. Therefore, the function only operates on either a 0 and a 1 and its result is also either a 0 or a 1.

To define the function, let,

$$f(x) = \begin{cases} 0 & \text{when } x = 0 \\ 1 & \text{when } x = 1 \end{cases}$$

For example, the identity functions $f(x) = 0$ when $x = 0$ and $f(x) = 1$ when $x = 1$ and the constant functions $f(x) = 0$ and $f(x) = 1$. The bit flip functions also belong to the same category: $f(x) = 1$ when $x = 0$ and $f(x) = 0$ when $x = 1$.

The identity and bit flip functions are also known as *balanced* because their output is opposite in half of the input. When $f(0) = f(1)$, then the function is referred to as *constant*, whereas when $f(0)^{-1} = f(1)$, the function is referred to as *balanced*. Note that an operation that states $x \rightarrow f(x)$ is not unitary in general and hence, not suitable for quantum computation.

Generally, given an output $f(x)$ of a function, it is not always possible to invert $f(x)$ to obtain input x. In other words, $f(x)$ must be computed in such a way that it is guaranteed the computation can be undone. To address this, the first task is to model such functions as quantum circuits. However, we cannot simply apply any gate to a qubit because, as per postulates of quantum mechanics, all quantum operations must be *unitary* and *reversible* (refer to Chapter 1).

The reversible unitary transformation is shown in Figure 4-19.

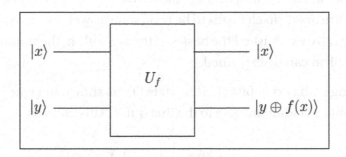

Figure 4-19. *Unitary transformation (Source [35])*

Figure 4-19 shows the unitary transformation where an imaginary unitary transformation U_f acts on two qubits such that

$$U_f|x,y\rangle = |x,y \oplus f(x)\rangle \ \ , x,y \in \{0,1\}$$

4.37

Interpretation of equation 4.37 shows that the first qubit remains unaffected by the action of U_f. Whereas the second qubit is subjected to an *XOR* operation (given by the $\oplus$ operator).

For the initial value of $y = 0$, we get the following from equation 4.37.

$$U_f|x,0\rangle = |x,f(x)\rangle$$

We can prove that U_f is reversible. If we apply U_f once again on its output, we get

$$U_f|x,(y \oplus f(x))\rangle = |x,(y \oplus f(x)) \oplus f(x)\rangle = |x,y\rangle$$

where we have used the modulo 2 property that $f(x) \oplus f(x) = 0$ (Note: The modulo 2 operation states: adding the same bit twice and dividing by two leaves a remainder of zero).

Please note that we are treating U_f as a "black box" or an "oracle" (i.e., the inner details of U_f is not something we concern ourselves with. An oracle describes a "quantum black box," which we assume we can implement and run but cannot "inspect inside" to see its implementation details. Internally, U_f is composed of one or more quantum gates, maybe a circuit even, but we typically don't know which gates or what kind of circuit. By querying the oracle (e.g., by sending input and measuring the output), we can learn more about the properties of the oracle. Because an oracle is composed of quantum gates, the oracle itself needs to be reversible as well.

Now that we have established the basics of the algorithm, the fundamentals of quantum parallelism can be explained.

1. Starting with two qubits at initial state $|0\rangle$, as shown in Figure 4-20, we apply a Hadamard gate to the first qubit. This gives us

$$H|0\rangle = \frac{1}{\sqrt{2}}(|0\rangle + |1\rangle)$$

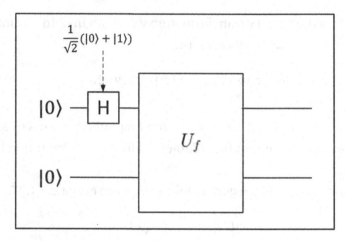

Figure 4-20. *Circuit for quantum parallelism*

2. The second qubit, also at initial state $|0\rangle$, after the unitary transformation gives the final state $|\psi_f\rangle$.

$$|\psi_f\rangle = U_f|H|0\rangle|0\rangle\rangle = U_f(\frac{1}{\sqrt{2}}(|0\rangle + |1\rangle)|0\rangle) = \frac{1}{\sqrt{2}}(U_f|00\rangle + U_f|10\rangle)$$

$$= \frac{1}{\sqrt{2}}(|0, 0 \oplus f(0)\rangle + |1, 0 \oplus f(1)\rangle)$$

It should be clear from inspection of the result that the circuit of Figure 4-19 contains information on $f(0)$ and $f(1)$ simultaneously; in other words, the circuit has produced a superposition state that has information about both $f(0)$ and $f(1)$ in a single step. And we have what is known as *quantum parallelism*—information about both states simultaneously in one operation.

The example of quantum parallelism appears quite cool on the surface. But if we stopped to think about how we get any information, such as measurements out of this system, we see a looming problem: if we measure the state $\sum_{0}^{1} |x| f(x) \rangle$ then we obtain only one value of x and $f(x)$ for the computational basis. After we perform a measurement, the value of the function we obtain is for a random value of x. Hence, this operation by itself does not provide a very useful result as we cannot even choose which value we wish to reveal: $f(0)$ or $f(1)$. This is where Deutsch's algorithm becomes important.

Deutsch's Algorithm

Balanced and constant functions were described in the previous section. However, it is often a challenge to find out whether a function is indeed constant or balanced! To answer that question, David Deutsch, in 1985, provided an excellent proof of concept that, in certain settings, quantum computers are strictly more powerful than classical ones. The Deutsch algorithm [37] can be stated simply by the following framework.

1. Determine the value of the parity of $f(0) \oplus f(1)$, where $\oplus$ denotes addition in modulo 2 or the *XOR* operation; that is, $1 \oplus 1 = 0 \oplus 0 = 0$ and $1 \oplus 0 = 0 \oplus 1 = 1$.

2. Prove that if $f(x)$ is *constant* then the parity is 0, and if $f(x)$ is *balanced*, then the parity is 1.

The Deutsch algorithm tries to resolve which of the given functions is/are constant or balanced. The properties of balanced and constant functions are global properties because to compare the behavior, it is required to compute both the $f(0)$ and $f(1)$ functions (see Figure 4-21).

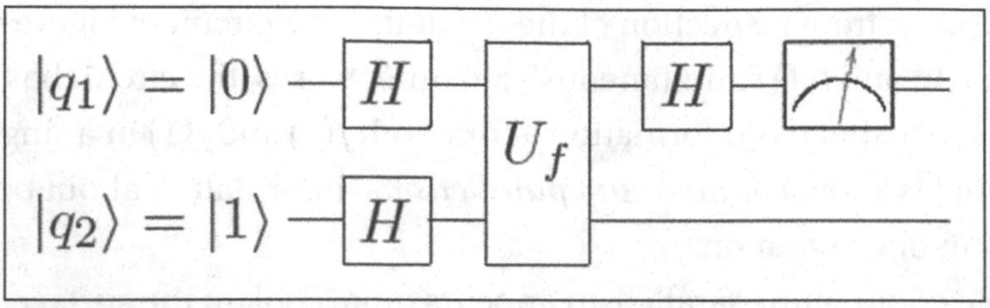

Figure 4-21. *Circuit for the Deutsch algorithm [37] (Source [35])*

The main goal of Deutsch's algorithm is to find the final state $|\psi_f\rangle$ by obtaining information about the global states. The algorithm achieves this by using the superposition state that the system achieves as a result of properties of quantum parallelism. The final state is computed as follows.

$$|\psi_f\rangle = (H \otimes I)U_f(H \otimes H)|01\rangle \qquad 4.38$$

where the initial state of the system is $|\psi_i\rangle = |01\rangle$ and $\otimes$ denotes the tensor product of the states as usual.

The following are the steps through Deutsch's algorithm.

1. Evaluate $|\psi_i\rangle$ by applying Hadamard gates to the input states. By the definition of Hadamard gates, this action produces a product state of superposition. To calculate $|\psi_{i+1}\rangle$, we have the following.

$$|\psi_i\rangle = |01\rangle$$

$$|\psi_{i+1}\rangle = (H \otimes H)|01\rangle = \left[\frac{1}{\sqrt{2}}(|0\rangle + |1\rangle)\right]\left[\frac{1}{\sqrt{2}}(|0\rangle - |1\rangle)\right]$$

$$= \frac{1}{2}(|00\rangle - |01\rangle + |10\rangle - |11\rangle)$$

The operation creates a superposition state that contains all possible values of the combination $(x, f(x))$.

2. In step 2, we apply U_f to $|\psi_{i+1}\rangle$ as follows.

$$|\psi_{i+2}\rangle = U_f|\psi_{i+1}\rangle = \frac{1}{2}(|0\rangle|f(0)\rangle - |0\rangle|1 \oplus f(0)\rangle + |1\rangle|f(1)\rangle - |1\rangle|1 \oplus f(1)\rangle)$$

3. In this step, we break down our analysis into two cases: when $f(x)$ is balanced and when it is constant.

Case 1: $f(x)$ is constant $\Rightarrow f(0) = f(1)$. Therefore, substituting $f(0)$ for $f(1)$ in $|\psi_{i+2}\rangle$, we have

$$|\psi_{i+2}\rangle = \frac{1}{2}\big(|0\rangle|f(0)\rangle - |0\rangle|1 \oplus f(0)\rangle + |1\rangle|f(0)\rangle - |1\rangle|1 \oplus f(0)\rangle\big)$$

$$= \frac{1}{2}\big((|0\rangle + |1\rangle) \otimes |f(0)\rangle - (|0\rangle + |1\rangle) \otimes |1 \oplus f(0)\rangle\big)$$

$$= \frac{1}{2}\big((|0\rangle + |1\rangle) \otimes (|f(0)\rangle - |1 \oplus f(0)\rangle)\big)$$

$$= \frac{1}{\sqrt{2}}|+\rangle \otimes (|f(0)\rangle - |1 \oplus f(0)\rangle)$$

where, $|+\rangle = \frac{1}{\sqrt{2}}(|0\rangle + |1\rangle)$, as usual. This means the qubit is now at the $|+\rangle$ state. Therefore,

$$|\psi_{i+3}\rangle = \frac{1}{\sqrt{2}}|0\rangle \otimes (|f(0)\rangle - |1 \oplus f(0)\rangle)$$

This implies that qubit 1 is in state $|0\rangle$. Therefore, if a measurement is now performed on qubit 1 in the standard basis, the result is a 0 with complete certainty.

Case 2: $f(x)$ is balanced $\Rightarrow f(0) \neq f(1)$ and $f(0) \oplus 1 = f(1)$ and $f(1) \oplus 1 = f(0)$.

Therefore,

$$|\psi_{i+3}\rangle = \frac{1}{2}\big(|0\rangle|f(0)\rangle - |0\rangle|f(1)\rangle + |1\rangle|f(1)\rangle - |1\rangle|f(0)\rangle\big)$$

$$= \frac{1}{2}\big((|0\rangle - |1\rangle) \otimes |f(0)\rangle - (|0\rangle - |1\rangle) \otimes |f(1)\rangle\big)$$

$$= \frac{1}{2}(|0\rangle - |1\rangle) \otimes (|f(0)\rangle - |f(1)\rangle)$$

$$= \frac{1}{\sqrt{2}}|-\rangle \otimes (|f(0)\rangle - |f(1)\rangle)$$

where, $|-\rangle = \frac{1}{\sqrt{2}}(|0\rangle - |1\rangle)$, as usual. This means the qubit is now at $|-\rangle$ state. Therefore,

$$|\psi_{i+3}\rangle = \frac{1}{\sqrt{2}}|1\rangle \otimes \left(|f(0)\rangle - |f(1)\rangle\right)$$

This implies that qubit 1 is in state $|1\rangle$. Therefore, if a measurement is now performed on qubit 1 in the standard basis, the result is a 1 with complete certainty.

In conclusion, Deutsch's algorithm gives 0 as output if f is constant and 1 as ouput if f is balanced. Hence the algorithm uses only *a single query* to determine whether f is balanced or constant. This calculation of the global properties of a function without explicitly evaluating the values of the same function has been possible by *quantum parallelism.*

Deutsch's Algorithm with Cirq

Now that you understand Deutsch's algorithm, it is time to try out an exercise using the code DeutschCirq.ipynb. The program in Listing 4-6a randomly chooses the function $f(x) = 1$ to be evaluated and measure the final state of the first qubit. First, we import the libraries and construct a function to define the circuit analogous to Figure 4-20. Listing 4-6a defines the following three functions.

- main(): generates a function $(0, 1) \rightarrow (0, 1)$
  - calls moracle to map the generated function to the corresponding oracle
  - builds the quantum circuit with the generated oracle
- moracle(): maps the function to a corresponding oracle
- deutsch_circuit(): creates the quantum circuit
  - initializes two qubits
  - applies X gate to qubit 2
  - applies Hadamard gate to each qubit to generate superposition
  - applies the oracle

- ∘ applies Hadamard gate to the first qubit

- ∘ performs measurement

Listing 4-6a. Deutsch's Algorithm First Step `DeutschCirq.ipynb`

```python
import cirq
from cirq import H, X, CNOT, measure
import numpy as np
import random

def main():
    # Choose qubits to use.
    q0, q1 = cirq.LineQubit.range(2)

    # Pick a secret 2-bit function and create a circuit to query the oracle.
    secret_function = [random.randint(0, 1) for _ in range(2)]
    oracle = moracle(q0, q1, secret_function)
    print('Secret function:\nf(x) = <{}>'.format(
        ', '.join(str(e) for e in secret_function)))

    # Embed the oracle into a quantum circuit querying it exactly once.
    circuit = deutsch_circuit(q0, q1, oracle)
    print('Circuit:')
    print(circuit)

    # Simulate the circuit.
    simulator = cirq.Simulator()
    result = simulator.run(circuit)
    print('Result of f(0)⊕f(1):')
    print(result)

def moracle(q0, q1, secret_function):
    """ Gates implementing the secret function f(x)."""

    # coverage: ignore
    if secret_function[0]:
        yield [CNOT(q0, q1), X(q1)]

    if secret_function[1]:
```

197

```
      yield CNOT(q0, q1)
def deutsch_circuit(q0, q1, oracle):
    c = cirq.Circuit()

    # Initialize qubits.
    c.append([X(q1), H(q1), H(q0)])

    # Query oracle.
    c.append(oracle)

    # Measure in X basis.
    c.append([H(q0), measure(q0, key='result')])
    return c
```

Next, the main method is applied in Listing 4-6b. This is the method that runs the program and generates the final output.

Listing 4-6b. Generate Output DeutschCirq.ipynb

```
if __name__ == '__main__':
    main()
```

The code, when run, gives the final output.

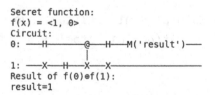

```
Secret function:
f(x) = <1, 0>
Circuit:
0: ───H───────────@───H───M('result')───
                  │
1: ───X───H───X───X──────────────────────
Result of f(0)⊕f(1):
result=1
```

Quantum Computing Systems

This section aims to give a high-level overview of quantum computing (QC) systems and associated key components. Today's QC systems, irrespective of variety (gate-based or annealing), consist of three *main* components common to all platforms: the hardware layer, software layer, and application layer. In gate-model QC, the software layer couples with the hardware layer for reasons of information abstraction and can be further categorized in a virtual layer, a layer for quantum error correction, and a logical layer to address the programming substrate of gate-based universal QC. It is shown in Figure 4-22, referenced from Jones, van Meter et al. [39].

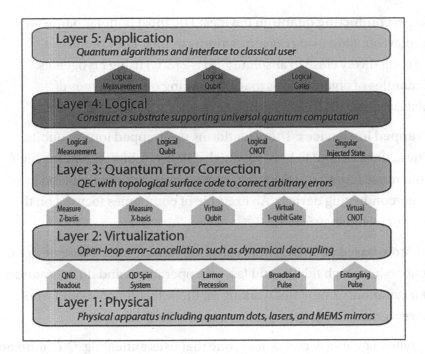

Figure 4-22. *Layered architecture of quantum computing systems (Source [39])*

Today's QC space has various platforms, each of which supports efficiency in certain areas. An application can benefit from quantum computing. We now need to find the right device. The field of quantum computing is a vast one, and as such, there have been many different advancements by different companies manufacturing QC systems. As a result, many different quantum devices were developed, each with its own advantages and limitations. When developing an application, it is important to know which device best serves the purpose of solving a specific problem.

Some of the most popular quantum devices are gate-based quantum computers, where operations are done on qubits using quantum logic gates. You learned about the programming and two types of gate-based properties, such computers in this chapter, namely IBM Q (for Qiskit) and Google Cirq. The logic gates used by this type of QCs are analogous to the logic gates found in classical computers but are more complex and diverse. You saw examples of gates, including Hadamard, Pauli, CNOT, and rotation.

There are many different varieties of implementations of gate-based quantum computing.

- **Superconducting quantum devices**: The most popular hardware implementation using noisy quantum circuits. They require to be cooled very close to absolute zero (–273°C) using cryogenics. Examples of organizations that are working on these types of platforms are IBM, Google, and Amazon.

- **Trapped ion devices**: These platforms use trapped ions as qubits. These devices tend to have very high-fidelity qubits with much higher coherence times but, at present, a lower qubit count compared to superconducting devices. An example of companies focused on this type of technology is IonQ.

- **Photon-based devices**: These use photons as qubits. These devices tend to have high fidelity and fast gate operations and do not require refrigeration. Companies working on these technologies and who have significant progress are Xanadu and PsiQuantum.

Another type of quantum computer is one that uses annealing. Quantum annealers, such as the ones built by D-Wave, are very efficient specifically for solving optimization problems. The classic knapsack problem, the traveling salesman problem, and the Max-Cut problem are very efficiently solved by such platforms. This is done on quantum annealers by first initializing the qubits into superposition. After this, the qubits and the connections between them are slowly tuned. At the end of runtime, the configuration corresponds to the optimal solution of interest. The way D-Wave computers work can be summarized as follows.

- Parameters are turned into voltages, currents, and magnetic fields. The problem is programmed onto the D-Wave QPU (quantum processing unit).

- The qubit spins begin in their superposed states.

- The qubit spins evolve and explore the problem space.

- When the annealing process finishes, the system reaches the ground state or a low excited state of the submitted problem.

- The states of the spins are read, any preprocessing required applied, and the results are handed back to the user.

- This can be done hundreds of times per second.

Some of the prevailing challenges and subjects of intense international research interests in QC are *quantum error correction* (QEC), *coherence,* and *quantum memory.* Some QCs, such as superconducting quantum devices, are prone to noise which can lead to errors. These errors are categorized into three main types: bit flips, phase flips, and readout errors. These errors can be corrected using quantum error correction/ mitigation methods; however, these may require additional qubits and gates in the quantum circuits.

Open source software is becoming crucial in the design and testing of quantum algorithms [38]. Many of the tools are backed by major commercial vendors with the goal to make it easier to develop quantum software. This mirrors how well-funded *open machine learning frameworks* enabled the development of complex models and their execution on equally complex hardware. There is now a wide range of open source software for quantum computing, covering all stages of the quantum toolchain from quantum hardware interfaces through quantum compilers to implementations of quantum algorithms, as well as all quantum computing paradigms, including quantum annealing and discrete and continuous-variable gate-model quantum computing.

A gate-model quantum computer typically has the workflow shown in Figure 4-23.

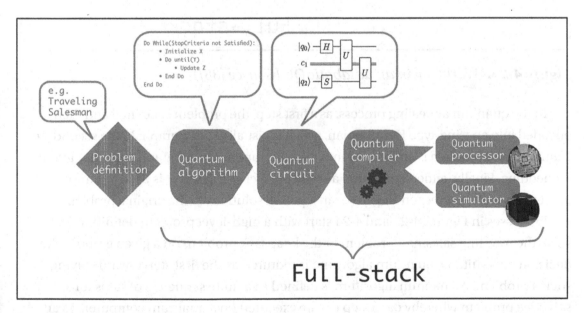

Figure 4-23. *Workflow in a gate-model QC (source [38])*

In a gate model, the problem is defined at a high level. A suitable quantum algorithm is chosen based on the nature of the problem. Next, the quantum algorithm is expressed as a quantum circuit that needs to be compiled to a specific quantum gate set. Finally, the quantum circuit is either executed on a quantum processor or simulated with a quantum computer simulator.

Figure 4-24 shows a typical workflow when using quantum annealing.

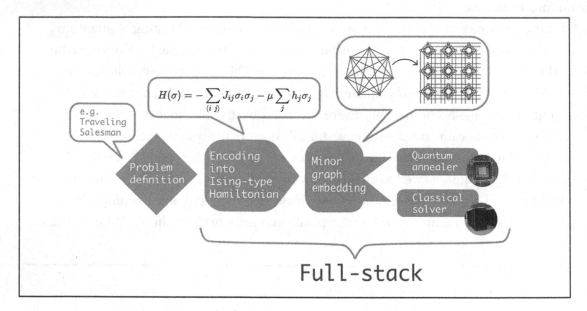

Figure 4-24. *Workflow in an adiabatic QC (Source [38])*

In the quantum annealing process, as a first step, the problem is defined and then encoded into an Ising-type Hamiltonian, which is visualized as a graph. In the second step, Hamiltonian's problem is embedded into the quantum hardware graph via minor graph embedding. Finally, either a quantum annealer or a classical solver is used to sample low-energy states corresponding to (near-)optimal solutions to the original problem.

Processes in Figures 4-23 and 4-24 start with a high-level problem definition (e.g., solve the traveling salesman problem or the knapsack problem on a given graph). The decision on a suitable quantum algorithm is required as the first step toward solving such a problem. A quantum algorithm is defined as a finite sequence of steps for solving a problem whereby each step can be executed on a quantum computer. As an example, in the traveling salesman problem, we face a discrete optimization problem.

Hence, the user can consider an appropriate algorithm (e.g., the Quantum Approximate Optimization Algorithm (QAOA) [40] designed for noisy discrete gate model quantum computers or quantum annealing to find the optimal solution).

Summary

Quantum information refers to data that can be physically stored in a quantum system. Quantum information theory is the study of how such information can be encoded, measured, and manipulated. This chapter went through basic algorithms, quantum parallelism, teleportation, and associated programming in Qiskit and Cirq. The next chapter covers the essentials of quantum machine learning basics, such as the Deutsch-Jozsa algorithm, optimization theory, information encoding, and quantum complexity.

CHAPTER 5

QML Algorithms I

You can't cross the sea merely by standing and staring at the water.

—Rabindranath Tagore

The first quantum computers were built a few years ago. Currently, QC technology has various choices of platforms, each of which exhibits efficiency in certain areas. Applications can benefit from quantum computing. You just need to find the right device. When developing an application, it is important to know which device serves the purpose best for the solution of a specific problem and which algorithms are best supported on the same platform.

Thoughts on boosting classical machine learning with quantum computing have been around for quite a few years. These ideas are nearly not as new or radical as some may think. Indeed, machine learning demonstrations with the D-Wave quantum annealer date back to as early as 2009 [60].

Note In previous chapters, we did hands-on programming related to gates and basic algorithms such as Deutsch with Qiskit and Google's Cirq. In this chapter, we do the same with Rigetti's Forest SDK. In upcoming chapters, we delve into the D-Wave's quantum annealing solutions and Xanadu's photonic-based quantum machine learning library.

S. Ganguly, *Quantum Machine Learning: An Applied Approach*, https://doi.org/10.1007/978-1-4842-7098-1_5

By processing quantum information encoded in a superposition of all possible inputs, a quantum computer can simultaneously compute each output value for every possible input. This is called *quantum parallelism*. The field of quantum computation and associated algorithms was initiated through Deutsch's work on universal quantum computation from 1985 [37], based on earlier ideas of Feynman [41]. This field can be divided into various sections associated with complementary research efforts such as.

- Develop quantum algorithms for computing and communication

- Develop hardware for computing and communication

- Seek solution for robust fault-tolerance

- Develop quantum algorithms for machine learning

- Improve scalability

- Seek proofs of universality

Peter Shor proposed a quantum algorithm to factorize numbers into prime numbers in 1994. His algorithm offered to realize the task in a significantly faster time than classical computing counterparts. As a result, the possibility of realizing a quantum computer became a security issue. Code such as RSA encryption, considered effectively secure based on a mathematically unproven complexity assumption,[1] became suddenly vulnerable; no longer due to the nonexistence of a mathematical proof, but rather because of a new type of computer whose existence is permitted by the laws of physics. However, paradoxically, the solution to the problem of unconditional security was also offered by quantum theory in quantum key distribution (QKD); even a quantum computer cannot render quantum cryptography insecure.

Today's quantum computers are noisy (i.e., their qubits are error-prone at physical levels), and hence, their actual states are not stable (i.e., the states decay over short periods of time). Additionally, the implementations of the gates used to manipulate the qubits are error-prone (i.e., a gate does not manipulate the qubits it operates on exactly), resulting in small deviations from the expected result, and such errors propagate during the execution of an algorithm. The phenomenon that qubits are erroneous is referred to as *decoherence*. The phenomenon that gates are erroneous is referred to as *gate infidelity*.

---

[1] It is assumed that this code is too hard to break by a classical computer. For example, there is no classical algorithm known to factorize numbers in an efficient amount of time to break the RSA encryption.

In addition to decoherence, error, and infidelity, current technologies of building physical quantum computers have limited scaling capabilities where the number of qubits on a device is concerned. Qubits above a certain number on a single device and the engineering methodology to control and connect them, introduce disturbances of the qubits and surrounding systems. Hence, today's quantum technology has scalability limited to (at best) intermediate scale only. The properties of being noisy and of intermediate scale prompted QC Guru John Preskill to coin the term *noisy intermediate-scale quantum* (NISQ) computers [68].

Going by the definitions of quantum parallelism and superposition, the power of quantum computers in universal QC lies in the fact that quantum algorithms should make use of numerous (as many as possible) qubits; otherwise, these algorithms can be simulated on classical computers. However, largely due to the limitations of engineering and practical implementation, most NISQ era quantum algorithms end up with a "low depth."

Depth is one of the *three main complexities* of circuit-based quantum computation. The computation complexity can be specified in terms of the amount of time or space the machine uses to complete the said computation. For the circuit model of computation, one natural measure of complexity is the *number of gates* used in the circuit. A second measure is the *depth* of the circuit. The measures of circuit complexity are illustrated in Figure 5-1.

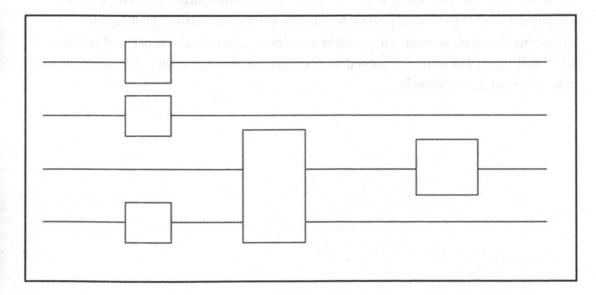

Figure 5-1. *A circuit of depth 3, space, or width 4 with a total of 5 gates*

If we visualize the circuit as being divided into a *sequence of discrete time-slices*, where the application of a single gate requires a single time-slice, the *depth* of a circuit is its total number of time-slices. This is not necessarily the same as the total number of gates in the circuit since gates that act on disjoint bits can often be applied in parallel; for example, a pair of gates could be applied to the bits on two different wires during the same time-slice. The *third measure* of complexity for a circuit is analogous to a Turing machine's space (or width). This is the total number of bits or "wires" in the circuit, sometimes called the *width* or *space* of the circuit.

A low-depth algorithm is also called a *shallow* algorithm. As it stands today, if an algorithm requires a high depth (a deep algorithm), it can make use of a few qubits only, then it can be simulated. This implies that quantum advantage must be shown based on shallow algorithms (i.e., algorithms that use only a few layers of parallel gates).

The low depth of shallow algorithms imposed by the physical qubit limitations of today's NISQ machines automatically mandate the realization of hybrid algorithms (i.e., algorithms that are split into classical parts and shallow quantum parts). The classical parts and quantum parts of a hybrid algorithm are performed in a loop where results are optimized from iteration to iteration [56] [69]. The act of optimization may affect both the quantum parts of a hybrid algorithm and its classical parts [70].

Figure 5-2 shows the general structure of a hybrid algorithm. Its quantum part prepares the state to be manipulated by the unitary transformation representing the algorithm properly. The result produced by the unitary transformation is measured and passed on to the classical part of the workflow. If the measured result is deemed acceptable based on accuracy, it is delivered as the result; else, the measured result is processed again. The output is passed as input to a preprocessing step that controls the state preparation accordingly.

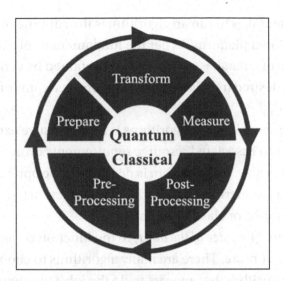

Figure 5-2. *General structure of a quantum algorithm (Source [48])*

The structure of a hybrid algorithm follows the generic structure of most quantum algorithms (i.e., the quantum state processed by the algorithm, such as the unitary transformation) must be prepared, which typically requires classical preprocessing. The output produced by measuring the result of the algorithm proper needs to be processed again (post-processed) (e.g., by assessing or improving its quality). In many cases, the post-processing step does not kickoff another iteration (i.e., it will not pass input to the preprocessing step).

The target QC platform on which the circuit implementing the algorithm should run impacts the overall algorithm because it might support a different gate set than assumed by the algorithm or because the target machine has certain hardware restrictions. For example, a gate-based algorithm with an annealing-based platform as a target quantum machine will not be a successful implementation.

Before we delve into the nitty-gritty details of quantum algorithms and machine learning, it is good to understand more about *quantum computational complexity*.

Quantum Complexity

You had an initial look at computational complexity in Chapter 1. The *complexity* of a problem refers to the classes of problems that involve engaging resources from the perspective of algorithms, computation time, and so on. To estimate the overall computational cost, an analysis of resources is required. This is done by running an algorithm that is needed to solve the problem.

The time a computer takes to run an algorithm is the runtime of that algorithm on that specific computational platform. As per the fundamentals of classical computation, a solution for the target of a database search can be obtained by verifying items in the database until the one desired item is found. This exercise requires a worst-case scenario of time complexity $O(n)$.

The *computational complexity* of algorithms is a quantitative expression of how the algorithm's time, size, and associated resource requirements evolve with the growth in the dataset or problem size. An algorithm is designed to be applied to an unknown function or a set of data to determine properties of the function (such as a period or a minimum), factor a number, or classify the data.

The number of inputs (i.e., *size of the data*) to the function could be anything from a single digit to trillions or more. There are many algorithms to choose from. Try them out until you find the algorithm that appears to do the job. Once we select an algorithm that we think serves our purpose or goal, we need to decide on the response to the following question: how does the algorithm's running time behave with an increase in the number of inputs on which the algorithm operates? This is known as the *sensitivity* of the algorithm to *size*.

As an algorithm's problem size grows, so does its running time. This is known as *time complexity*. The following are varieties of time complexities that are addressed in the next chapters.

- Constant time complexity

- Polynomial time complexity

- Non-polynomial time complexity

If the hardware circuitry can grow as the dataset grows (or utilize a larger number of processors from a very large pool of existing servers), then the time taken for the computation may not be large, and it is possible to trade *time complexity* for *space complexity*.

When speaking about quantum over classical methods, current interest is in relative speed-up of the quantum computing and communication domain. To achieve that, commonly, a hardware black-box type approach finds out the time advantage, if any, offered by the quantum algorithm.

If *space complexity* is considered, then the circuitry in the algorithms grows *linearly* at the worst and *logarithmically* most of the time. The task is to take an expensive *exponential algorithm classically* and find a *polynomial algorithm* using *quantum computation*. Therefore, it is estimated that the linear or logarithmic growth of the hardware is overcome by the favorable time cost and hence, can be ignored.

The field of quantum complexity theory has been developed as an extension to classical complexity theory [44] [45]. It is based on whether quantum computers are in principle able to solve computational problems faster in relation to the runtime complexity.

To compare algorithms, it is customary to use a metric or tool. One such metric is the *asymptotic notation*, sometimes referred to as the "Big O" notation. This notation indicates an estimate of resources needed to perform a given computational task and variational characteristics of the same resources when the problem varies in size. It *does not* give a qualitative indication of how much of the resources are needed but rather an asymptotic tendency. For example, consider some specific algorithm that requires $(25n^2 + n + 21)$ time steps, where n is the number of bits required to define the problem. In this case, it said that the problem has a quadratic time complexity given by $O(n^2)$, because $25n^2$ is by far the largest factor dominating those steps for a large n. In stating the complexity estimation as $O(n^2)$, the constant 25 is ignored because for a large n, when it comes to comparing the algorithm to time complexity of another one, the importance of the constants is trivial.

The asymptotic notation that allows us to compare algorithms for sufficiently large n is an important tool when it comes to the theory of categorizing decision problems. This is referred to as *computational complexity* theory and groups computational problems together depending on how resource-efficient their best-known solution is. These groups are called *complexity classes,* and there are quite a few of them around. In the context of this book, only a few are considered. These classes are all efficient when it comes to spatial resources, which is defined here as classes of problems that have a *spatial complexity* $\leq O(n^p)$. These are solutions that *do not require* more than a polynomial increase of memory as the problem grows. This class is known as PSPACE. *Time efficiency* is defined in the same way: a time-efficient solution requires several time steps proportional to a polynomial in n. When a solution is referred to as *efficient,* that usually means that the solution exhibits *both* time efficiency and spatial efficiency.

A *runtime advantage* in this context is called a *quantum enhancement, quantum advantage*, or simply a *quantum speedup*. Demonstrating an exponential speedup has been controversially referred to as *quantum supremacy*.

Although the holy grail of quantum computing remains finding a provable exponential speedup, the wider definition of quantum advantage in terms of the asymptotic complexity opens a lot more avenues for realistic investigations such as:

- Quantum algorithms often must be compared with classical sampling, which is likewise non-deterministic and has close relations to quantum computing.

- A well-defined computational task

- A plausible quantum algorithm for the problem

- An amount of time/space allowed to any classical competitor

- A complexity-theoretic assumption, and optionally

- A verification method that can efficiently distinguish the quantum algorithm from any classical competitor using the allowed resources

Here *plausible* means *ideally* on near-term hardware and likely includes the need to handle noise and experimental imperfections.

A *complexity class* is a set of decision problems that all have similar complexity in some sense, for example, the ones that can be solved with polynomial time or polynomial space.

The following are some of the main classical complexity classes.

- **P**: Class of problems that can be solved by *classical deterministic* computers using polynomial time.

- **BPP**: Problems that can be solved by *classical randomized* computers using polynomial time with error probability $\leq \frac{1}{3}$ on every input.

- **NP**: Class of problems where the Yes instances can be verified in polynomial time if some prover provides a polynomial-length "witness." Some problems in this class are NP-hard problems where we know algorithms that still run in exponential time but are much faster than brute-force search. An example is the famous traveling salesperson problem (TSP) which we solve hands-on in Chapter 8. There are also NP-complete problems, meaning that any other problem in NP can be reduced to it in polynomial time. *Hence the NP-complete problems are the hardest problems in NP.* An example is the problem of satisfiability: we can verify that a given n-variable Boolean formula is satisfiable if a prover gives us a satisfying assignment. Hence it is in NP, and it can be proven that it is NP-complete. Other examples are integer linear programming, traveling salesperson (a real-life implementation is in Chapter 7), ability to color graphs, and so on.

- **PSPACE**: Problems that can be solved by classical deterministic computers using polynomial space.

Quantum analogs of all the classes can also be defined, an effort that was started by Bernstein and Vazirani [44].

- **EQP**: A class of problems that can be solved *exactly* by quantum computers using polynomial time. This class depends on the set of elementary gates allowed in the computation.

- **BQP**: Problems that can be solved by quantum computers using polynomial time with error probability $\leq \frac{1}{3}$ on every input. This class is accepted as *efficiently solvable by quantum computers.*

- **Quantum NP**: A class of problems where the Yes instances can be verified efficiently if some prover provides a "quantum witness" of a polynomial number of qubits. For every Yes instance, there should be a quantum witness that passes the verification with probability 1, while for No instances, every quantum witness should be rejected with probability 1. This class is also dependent on the elementary gates allowed and with error probability $\leq \frac{1}{3}$ on every input, a class called Quantum Merlin-Arthur (QMA) is obtained.

- **QPSPACE**: Problems that can be solved by quantum computers using polynomial space. It is the same as classical PSPACE.

Quantum Feature Maps

Classical machine learning methods require wrangling their input data in a different space to make it easier to work with or because the new space offers properties convenient for the related model analysis. Support vector machines (SVM) are an example. SVMs classify data using a linear hyperplane. A linear hyperplane works well when the data is already linearly separable in the original space; however, this is unlikely to be true for many datasets. A work-around could be to transform the data into a new space where it is linear through a feature map.

Let φ be a set of input data and ϑ be a feature space. Then a feature map Ω is defined as

$$\Omega : \varphi \rightarrow \vartheta$$

The output data-points $\varphi(x)$ on the output map are called *feature vectors*. ϑ is usually a vector space. The transformation relation is shown in Figure 5-3a. While in the original space of training inputs, data from the two classes (red squares and green circles) are not separable by a simple linear model (left), we can map them to a higher dimensional feature space where a linear model is indeed sufficient to define a separating hyperplane that acts as a decision boundary (right). For more information, see Schuld and Killoran [52].

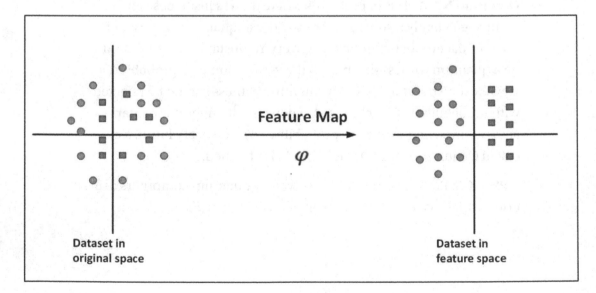

Figure 5-3a. *Quantum feature space transform*

In the context of a *quantum feature map,* the feature vectors are quantum states. ϑ is a member of the Hilbert space. The mapping $\Omega : \varphi \rightarrow \vartheta$ applies a unitary transformation U and transforms $x \rightarrow |\varphi(x)\rangle$. Typically, U, in this case, is a variational circuit, and the transformation happens as depicted in Figure 5-3b.

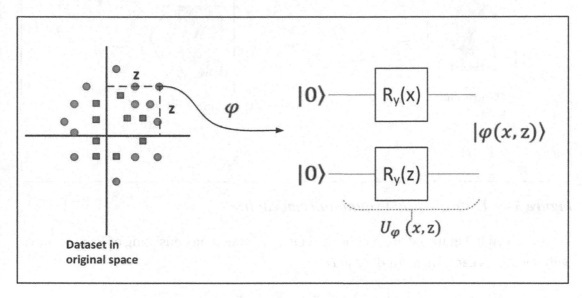

Figure 5-3b. *Quantum feature space transform with variational circuits*

Quantum Embedding

Generally, a quantum algorithm is fed input data that is required for producing a result. This input data must be represented as a quantum state to be understood by and manipulated on a quantum computer. Most applications for quantum machine learning today use quantum feature maps to map classical data to quantum states in a Hilbert space. This is a very important aspect of designing quantum algorithms that directly impact their computational cost. The process is known as *quantum embedding* and involves translating a classical x data point into a set of gate parameters in a quantum circuit, creating a quantum state $|\psi_x\rangle$. More details can be found in [5] and [47].

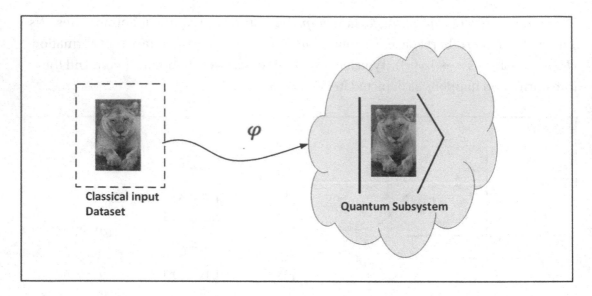

Figure 5-4. *From classical to quantum embedding*

As shown in Figure 5-4, if we consider classical input data consisting of P data points, with R features each, then, for dataset D,

$$D = x^1, x^2, \ldots, x^p, \ldots, x^P$$

where, x^p is a R-dimensional vector, and $p = 1, 2, \ldots, P$. This data can be embedded or encoded into n quantum *subsystems*. These subsystems can be n qubits for discrete-variable (DV) QC or n-qumodes for continuous-variable (CV) quantum computing.

Information Encoding

Information can be encoded into an n-qubit system described by a state in various ways. Generally, the encoding method varies with the nature of the datasets and the problem to be solved. For quantum machine learning and data mining, the importance of encoding is paramount. There are quite a few methods of encoding around. This book mainly focuses on the following: basis encoding, amplitude encoding, tensor product encoding, and Hamiltonian encoding. This chapter looks at the fundamental properties of these encoding methodologies. It expands on them in future chapters, namely in Chapters 6, 7, and 8, where we explore them in relation to quantum machine learning algorithms.

As per current practice, *input preparation* has a classical part called *preprocessing,* which creates a circuit that can be processed on a quantum computer to prepare the quantum state. The latter part, which prepares the quantum state, is referred to as *state preparation.* The process of p*reprocessing* during the classical part, in this case, may constitute a manual task performed by a human or an automatic task performed by a program.

To prepare input for a quantum algorithm as a quantum state, a quantum circuit is defined, which prepares the corresponding state. This circuit can be generated in a classical preprocessing step, as shown in Figure 5-5.

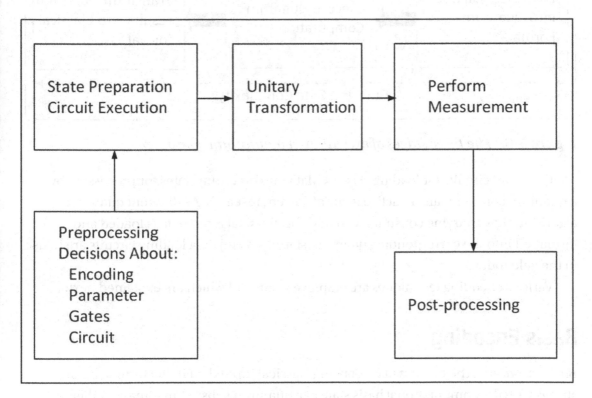

Figure 5-5. *Preparation of classical inputs for a quantum algorithm (Adapted from [55])*

The generated circuit is prepended to the circuit of the algorithm, sent to a quantum device, and executed. The generated circuit is represented in the quantum instruction language of the target device, for example, Cirq, QaSM, or Quil. Hence, to evaluate the executability of a given algorithm on a particular device, the effort and complexity in terms of additional gates and qubits required to prepare the algorithm's input properly must be considered.

Generating the number of gates required to prepare an arbitrary quantum state from classical data is discussed in reference [50]. In this context, efficiency in time and space complexity in encoding classical data into a quantum superposition state suitable to be processed by a quantum algorithm is critical.

The three generic stages of execution in a classical data-driven quantum computing process are shown in Figure 5-6.

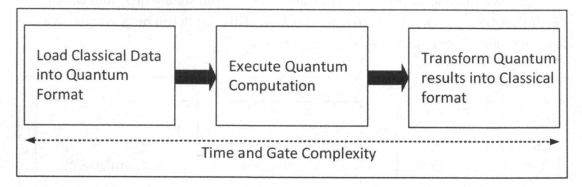

Figure 5-6. *The three stages of execution in a quantum computer*

Quantum circuits for loading classical data into quantum states for processing by a quantum computer are an active focus of current research. As different quantum algorithms have varying constraints on how the classical input data is loaded and formatted into the corresponding quantum states, several data loading circuits are in use in the field today.

Various encoding techniques are employed, some of which are explained next.

Basis Encoding

Basis encoding is the process of associating classical input data in the form of binary strings with the computational basis state of a quantum subsystem. Binary or digital encoding is the representation of data as qubit strings whereas, analog encoding represents data in the amplitudes of a state. If data must be processed by arithmetic computations, a digital encoding is preferable. On the other hand, analog encoding is preferred for mapping data into the large Hilbert space of quantum devices, which is often needed in machine learning algorithms. A brief overview of several encoding schemes like basis encoding, amplitude encoding, or product encoding can be found in [52], while [51] discusses their use in quantum machine learning. Details on generic mechanisms to initialize arbitrary quantum states are presented in reference [51].

Basis encoding is primarily used when real numbers must be arithmetically manipulated in a quantum algorithm. Such an encoding represents real numbers as binary numbers and then transforms them into a quantum state in a computational basis.

In basis encoding, a real number $x \in \mathbb{R}$ is approximated by k decimal places and transformed into the binary representation of this approximation.

$$x \approx \sum_{i=1}^{n} a_i 2^i + \sum_{i=1}^{k} a_{-i} \cdot \frac{1}{2^i}$$

where a_i, $a_{-i} \in 0, 1$, and the sign of the real number is encoded by adding a leading binary number, such as "0" for a "+" and 1 for a −. Hence, the real number is approximated by $(n + k + 2)$ bits and a $(n + k + 2)$-dimensional quantum state. Figure 5-7 shows the input encoding procedure for basis states.

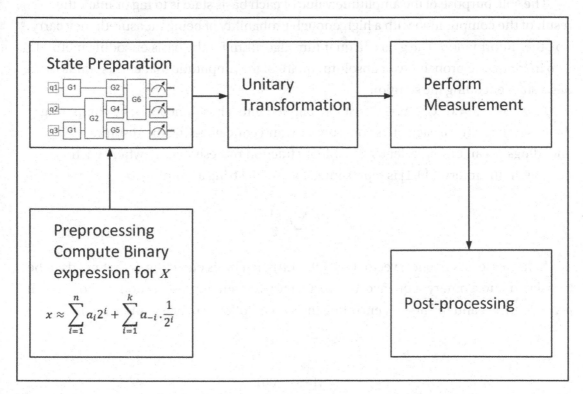

Figure 5-7. *Input preparation for basis encoding (Adapted from [55])*

The following are the steps for basis encoding.

1. In the first step, the preprocessing for the classical data computes
    the binary approximation $a_s a_n \dots, a_0, a_{-1}, a_{-k}$ of the real number,
    where a_s is the bit encoding the sign of the number.

2. The corresponding circuit is generated by applying the gate X^{a_i}
    on qubit i. This circuit creates the quantum state representing the
    real number x.

3. The generated circuit is prepended to the circuit of the algorithm
    proper (represented by the unitary transformation in the figure)
    and executed on a quantum device such as quantum computers
    from IBM or Rigetti.

The sole purpose of the amplitude value of each basis state is to tag or mark the
result of the computation with a high enough probability of being measured; they carry
no other information. The goal of a quantum algorithm for the basis encoding method
is to increase the probability or absolute square of the amplitude that corresponds to the
basis state encoding the solution.

Any vector $x = x_1, x_2, \dots x_n \in \mathbb{R}^n$ can be mapped into a basis encoding by computing
the basis encoding of each of its component x_i and concatenating the resulting
encodings. As an example, let us try a binary fraction representation, where each
number in the interval $[0, 1]$ is represented by a T-bit string according to

$$x = \sum_{i=1}^{T} a_i \frac{1}{2^i}$$

In this case, to encode a vector $(-0.7, 0.1, 1.0)^T$ in a basis encoding, it first needs to be
translated into a binary sequence, where a binary fraction representation is chosen with
precision $T = 4$ and the first bit encoding the sign as follows: 0 for a + and 1 for a -.

$$-0.7 \rightarrow 11011$$

$$0.1 \rightarrow 0\ 0001$$

$$0.1 \rightarrow 01111$$

Hence, the vector $(-0.7, 0.1, 1.0)^T$ translates to a binary sequence of 11011 01001 01111, and encoded as $|x\rangle = |\,11011\ 01001\ 01111\rangle$ in a basis encoding.

When a dataset $D = \{x^1, x^2, ..., x^p\}$ is binary embedded, the representation of D in binary encoding is the uniform superposition of the binary encoded states of the elements of D.

$$|D\rangle = \frac{1}{\sqrt{p}} \sum_{i=1}^{p} |x^i\rangle$$

For an algorithm of time complexity $O(mn)$ to create such a superposition, please see reference [56]. The individual elements of a state representation of a dataset $D = \{x^1, x^2, ..., x^p\} \subseteq \mathbb{R}^n$ of binary encoded states can be accessed by the following quantum random access memory (QRAM) representation.

$$\frac{1}{\sqrt{p}} \sum_{i=1}^{p} |j^i\rangle |x_i^j, ..., x_n^j\rangle$$

For example, for a classical dataset containing two examples $x^1 = 10$ and $x^2 = 01$, the corresponding basis encoding uses two qubits to represent $|x^1\rangle = |\,10\rangle$ and $|x^1\rangle = |\,01\rangle$, giving

$$|D\rangle = \frac{1}{\sqrt{2}} \left[|10\rangle + |01\rangle \right]$$

An architecture for implementing such a QRAM with time complexity $O(\log p)$ is proposed by reference [57]. Data from $\mathbb{R}^n$ can be encoded in its memory cells in complexity $O(\sqrt{n})$.

Amplitude Encoding

The *amplitude encoding* technique involves encoding data into the amplitudes of a quantum state. Although arithmetic manipulation of data by quantum algorithms is not of primary importance, more compact representations of data are useful. Specifically, the large Hilbert space of a quantum device is properly exploited in such encodings.

In this scheme, a normalized classical N-dimensional x data point is represented by the amplitudes of a n-qubit quantum state $|\psi_x\rangle$ as,

$$|\psi_x\rangle = \sum_{i}^{N=2^n} x_i |i\rangle$$

where x_i is the i-th element of x, $|i\rangle$ is the i-th computational basis state and $N = 2^n$. This encoding requires $\log_2 n$ qubits to represent an n-dimensional data point.

For example, if we wish to encode the four-dimensional floating-point array $x = 1.5, 0.0, -5.0, 0.0$ using amplitude encoding, then the first step would be to normalize it in the following way.

$$x_{norm} = \frac{1}{\sqrt{27.25}}[1.5, 0.0, -5.0, 0.0]$$

Then the corresponding amplitude encoding uses two qubits to represent x_{norm} as

$$|\psi_{x_{norm}}\rangle = \frac{1}{\sqrt{27.25}}\left[1.5|00\rangle + 0.0|01\rangle - 5.0|10\rangle + 0.0|11\rangle\right]$$

$$= \frac{1}{\sqrt{27.25}}\left[1.5|00\rangle - 5.0|10\rangle\right]$$

If we consider the previous classical dataset $D = \{x^1, x^2, ..., x^p\}$, then its amplitude encoding can be done if we concatenate all the input examples x^i together into one vector.

$$\alpha = c_{norm} x_1^1, ... x_N^1, ... x_1^2, ..., x_N^2, ..., x_1^P, ..., x_N^P$$

where, c_{norm} is the normalization constant, and the normalization criteria is $|\alpha|^2 = 1$. The input dataset is represented in the computational basis as

$$|D\rangle = \sum_{i=1}^{N=2^n} \alpha_i |j\rangle$$

where, α_i is the element of the amplitude vector α. $|i\rangle$ is the computational basis state. The number of amplitudes to be encoded is given by $N \times P$ and since, a system of n qubits is associated with 2^n amplitudes, amplitude encoding requires $n \geq \log_2 NP$ number of qubits.

During amplitude encoding, the state can be created by a unitary transformation $U = U_1 \otimes U_2 \otimes \ldots \otimes U_k$, where each U_i is either a one-qubit gate or a CNOT, and k is of the order of 4^n for an arbitrary x. In this case, x_i can have different numeric data types (e.g., integer or floating-point).

If x should be prepared from the $|0\rangle$ base state (i.e., $|x\rangle = U|0\rangle$), then $2n$ gates suffice as shown by Shende and Markov [50]. This complexity can be reduced further for x that satisfy certain constraints. For example, Soklakov and Schack [58] give a polynomial algorithm for unit-length states that are specifically bounded (i.e., only a limited set of states can be prepared if exponential complexity in preparing quantum states should be avoided). Schuld et al. [59] present an amplitude encoding for not necessarily unit-length vectors that especially shows how the often-made assumption that n is a power of 2 can be removed by padding.

Even though amplitude encoding requires only $(\log_2 n)$ qubits to encode $x \in R^n$, the preparation of such a state is generally exponentially expensive.

Tensor Product Encoding

The *tensor product encoding*, unlike amplitude encoding, requires n- qubits to represent n-dimensional data but is, in terms of complexity, cheaper to prepare as it requires one rotation on each qubit. This encoding is directly useful for processing data in quantum neural networks and is referred to in this context as angle encoding in reference [5]. Figure 5-8 shows the process for the tensor product encoding.

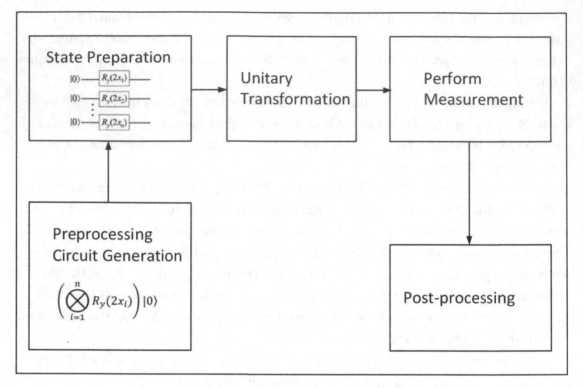

Figure 5-8. *Input preparation for Tensor product or angle encoding (Adapted from [55])*

The process in Figure 5-8 is as follows.

Let $x = x_1, x_2, ..., x_n \in \mathbb{R}^n$. The tensor product encoding on each component x_i is given by

$$|x_i\rangle = \cos x_i |0\rangle + \sin x_i |1\rangle$$

which gives the following complete vector representation

$$|x\rangle = \begin{pmatrix} \cos x_1 \\ \sin x_1 \end{pmatrix} \otimes \begin{pmatrix} \cos x_2 \\ \sin x_2 \end{pmatrix} \otimes ... \otimes \begin{pmatrix} \cos x_n \\ \sin x_n \end{pmatrix} \qquad 5A$$

Pauli matrices and rotation operators were discussed in Chapter 1. The Pauli X, Y and Z matrices when exponentiated, give rise to the rotation operators, which rotate the Bloch about the X, Y and Z axes, by a given angle θ.

$$R_x(\theta) \equiv e^{-i\frac{\theta}{2}X} = \cos\frac{\theta}{2}I - i\sin\frac{\theta}{2}X = \begin{bmatrix} \cos\dfrac{\theta}{2} & -i\sin\dfrac{\theta}{2} \\ -i\sin\dfrac{\theta}{2} & \cos\dfrac{\theta}{2} \end{bmatrix}$$

$$R_y(\theta) \equiv e^{-i\frac{\theta}{2}Y} = \cos\frac{\theta}{2}I - i\sin\frac{\theta}{2}Y = \begin{bmatrix} \cos\dfrac{\theta}{2} & -\sin\dfrac{\theta}{2} \\ \sin\dfrac{\theta}{2} & \cos\dfrac{\theta}{2} \end{bmatrix}$$

$$R_z(\theta) \equiv e^{-i\frac{\theta}{2}Z} = \cos\frac{\theta}{2}I - i\sin\frac{\theta}{2}Z = \begin{bmatrix} e^{-i\theta/2} & 0 \\ 0 & e^{i\theta/2} \end{bmatrix}$$

Apply the unitary $R_y(2x)$ to the vector $|x\rangle$ in Figure 5-6.

$$R_y(2x) = \begin{bmatrix} \cos x & -\sin x \\ \sin x & \cos x \end{bmatrix}$$

It implies the following.

$$R_y(2x_i)|0\rangle = \cos x_i |0\rangle + \sin x_i |1\rangle = x_i$$

The preceding equation, in turn, implies that

$$\left(\bigotimes_{i=1}^{n} R_y(2x_i)\right)|0\rangle = \begin{pmatrix} \cos x_1 \\ \sin x_1 \end{pmatrix} \otimes \begin{pmatrix} \cos x_2 \\ \sin x_2 \end{pmatrix} \otimes \ldots \otimes \begin{pmatrix} \cos x_n \\ \sin x_n \end{pmatrix} = |x\rangle \qquad 5B$$

where we get the original vector $|x\rangle$ back from equation 5A. In equation 5B, $\bigotimes_{i=1}^{n} R_y(2x_i)$ is the tensor product between $R_y(2x_i)$ and state $|0\rangle$ across the entire range of i.

The preprocessing step can generate the circuit of Figure 5-6 such that the state preparation step prepares the quantum representation of x. The circuit can then be generated as a code, prepended to the algorithm proper, send to a quantum computer, which is then executed there.

Hamiltonian Encoding

Quantum many-body systems (addressed later in this chapter) exhibit an extremely diverse range of phases and physical phenomena, which forms the central object of study in many areas of physics and beyond. *Hamiltonian encoding* uses an implicit technique by encoding information in the evolution of a quantum system. The quantum machine learning algorithms in later chapters use Hamiltonian encoding and few other quantum routines. In the remainder of the book, we go through them more rigorously in an example-driven approach.

Some well-known quantum algorithms, such as Deutsch's algorithm, were introduced in Chapter 4. We introduce a few more quantum routines in this chapter.

Deutsch-Jozsa Algorithm

The *Deutsch–Jozsa algorithm* of 1992 [41] inspired the works of Shor and Grover, probably the most well-known quantum algorithms and based on Deutsch's algorithm [37], which David Deutsch and Richard Jozsa extended to a more convincing algorithmic separation into what is now called the *Deutsch-Jozsa algorithm*. Grover's algorithm of 1996 is for searching a database [42] (discussed in Chapter 1 and later in this chapter). What these ideas have in common is that they illustrate the potential of quantum information processing to provide solutions for problems that are defined in purely classical terms and perceived as incapable of being solved efficiently through classical information processing. Similar to what quantum cryptography achieves for classical communication, quantum algorithms offer potentially better ways to perform certain classical computations, even though at intermediate stages, both the communications and computations would rely on quantum resources and processing.

Fundamentally, there are two main categories of quantum algorithms: those based on the *quantum Fourier transform* (QFT) corresponding to general implementations of the hidden subgroup problem and quantum search algorithms. The Shor and Deutsch–Jozsa algorithms belong to the former category, while the latter consists of variations of the Grover algorithm. An example of a class of algorithms that fit in neither of the two categories is *quantum simulation*. In this case, the quantum computation simulates a quantum system, as Feynman originally envisaged it [41].

The notion of simulating a Hamiltonian is the most convenient starting point for defining quantum computation over continuous quantum variables on qumodes. Quantized harmonic oscillators are commonly called *qumodes*. There are two common encodings of quantum information. The commonly known one based upon discrete two-level systems—hence, called *qubits*. The other approach relies on infinite-dimensional quantum systems, such as quantized harmonic oscillators (also called *qumodes*), and these are more reminiscent of classical analog encodings. Schemes based on qubit encodings are commonly referred to as discrete variable (DV) approaches. In contrast, those exploiting infinite dimensional systems and the possibility of preparing and measuring quantum information in terms of variables with a continuous spectrum are called *continuous variable* (CV) schemes.

We looked at quantum parallelism, entanglement, and Deutsch's algorithm in Chapters 1 and 4, along with some related coding examples. Entangled qubits in a state $|\psi_{ent}\rangle$, by definition, is not separable anymore, and the computational basis is given by

$$|\psi_{ent}\rangle = \lambda_1|0\ldots00\rangle + \lambda_2|0\ldots01\rangle + \ldots + \lambda_{2^n}|1\ldots11\rangle \qquad 5.1$$

where, $\lambda_i \in \mathbb{C}$, for $i = 1\ldots2^n$ and $\sum_i^{2^n}|\lambda_i|^2 = 1$. The basis $\{|0\ldots00\rangle\ldots|1\ldots11\rangle\}$ defines the computational basis for the n qubits.

In equation 5.1, the following common convention to represent tensor products is used: $|x\rangle \otimes |y\rangle = |xy\rangle$. The following representation of Dirac vectors in a computational basis is also very useful going forward.

$$|\psi\rangle = \sum_i^{2^n} \lambda_i|i\rangle \qquad 5.2$$

We looked at Deutsch's algorithm from a purely quantum circuit-based point-of-view. If you are more familiar with the fundamentals of quantum mechanics, it may be useful to have an alternate visualization of the same problem via the path in the Feynman diagram shown in Figure 5-9a.

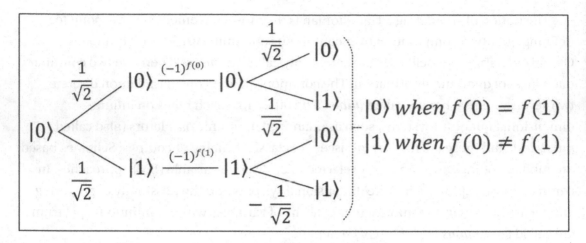

Figure 5-9a. *Deutsch's algorithm: Feynman path diagram*

For the sake of simplicity, let us use the simplified phase oracle model depicted by the circuit in Figure 5-9b.

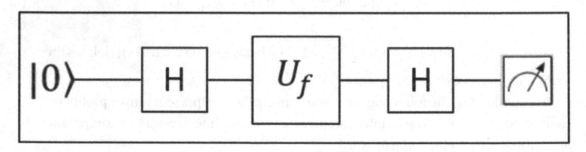

Figure 5-9b. *Deutsch's algorithm: simplified phase Oracle model of Figure 4-20*

The following steps summarize the quantum states for each step of the circuit of Figure 5-2b in terms of Figure 5-2a.

1. Input is $|0\rangle$

2. $|+\rangle = \dfrac{1}{\sqrt{2}}(|0\rangle + |1\rangle)$

3. $\dfrac{1}{\sqrt{2}}\left((-1)^{f(0)}|0\rangle + (-1)^{f(1)}|1\rangle\right) = \dfrac{(-1)^{f(0)}}{\sqrt{2}}\left(|0\rangle + (-1)^{[f(1)-f(0)]}|1\rangle\right)$

$$= \begin{cases} (-1)^{f(0)}|-\rangle & \text{if } f(1) \neq f(0) \\ (-1)^{f(0)}|+\rangle & \text{if } f(1) = f(0) \end{cases}$$

4. $\begin{cases} (-1)^{f(0)}|1\rangle & \text{if } f(1) \neq f(0) \\ (-1)^{f(0)}|0\rangle & \text{if } f(1) = f(0) \end{cases}$

In the expressions $(-1)^{f(0)}$ is a constant factor called the *global phase* of a qubit. Global phases can be ignored from a measurement point of view because the global phase of a quantum state is not detectable. In quantum mechanics, *only energy differences*, not the absolute values of dynamics of a physical system, are important. A classic analogy is the motion of electrons in wires where only the differences in voltages, as opposed to absolute values, are measured. This shows that the result of the final measurement is always 0 if $f(1) = f(0)$ and is always 1 if $f(1) \neq f(0)$.

Deutsch's algorithm can be elegantly explained using what is known as *phase kickback*. The phase kickback scheme considers an input $|x\rangle|-\rangle$, for $x \in \{0, 1\}$ and applies oracle U_f to that input. For the $|\psi\rangle$ state, this gives

$$|\psi\rangle = U_f|x\rangle|-\rangle = \frac{1}{\sqrt{2}}\left(U_f|x\rangle|0\rangle - U_f|x\rangle|1\rangle\right)$$

$$= \frac{1}{\sqrt{2}}\left(|x\rangle|f(x)\rangle - |x\rangle|1 \oplus f(x)\rangle\right)$$

$$= |x\rangle \otimes \frac{1}{\sqrt{2}}\left(|f(x)\rangle - |1 \oplus f(x)\rangle\right) \qquad 5.3$$

Equation 5.3 may be recognized as the general case for Deutsch's algorithm for a single input (see Chapter 4). For equation 5.3, there can be two different scenarios: $f(x) = 1$ or $f(x) = 0$. In case $f(x) = 0$, equation 5.3 gives

$$|\psi\rangle = |x\rangle \otimes \frac{1}{\sqrt{2}}\left(|0\rangle - |1\rangle\right) = |x\rangle|-\rangle$$

This shows that the input $|x\rangle|-\rangle$ remains unchanged by the action of U_f. On the contrary, if $f(x) = 1$, then equation 5.3 gives

$$|\psi\rangle = |x\rangle \otimes \frac{1}{\sqrt{2}}\left(|1\rangle - |0\rangle\right) = -|x\rangle|-\rangle$$

This indicates a phase of (-1) is added to the input $|x\rangle|-\rangle$. These two cases can be generalized with the following expression.

$$|\psi\rangle = U_f|x\rangle|-\rangle = (-1)^{f(x)}|x\rangle|-\rangle \qquad 5.4$$

Equation 5.4 is the *phase kickback* equation.

The solution to Deutsch's algorithm is exact in the sense that it produces the correct answer with an error probability of zero, and it only requires one oracle query, giving it a constant time complexity $O(1)$. The algorithm employs $n + 1$ qubits giving a linear memory consumption $O(n)$.

Deutsch's algorithm works great when $f: \{0, 1\} \to \{0, 1\}$ acts on a *single input* bit which is not very interesting. Of interest is a more general scenario which involves the design and analysis of quantum algorithms for determining properties of functions $f: \{0, 1\}^n \to \{0, 1\}$ which acts on *multiple* input values. This has the potential to offer exponential speedup over any classical deterministic solution.

The *Deutsch-Jozsa algorithm* is an n-bit generalization of Deutsch's algorithm. This algorithm, like Deutsch's algorithm, determines whether a function $f(x)$ is balanced or constant for *multiple* input values. If $f(x)$ is constant then output is the same for all input values of x (i.e., $f(x)$ is always 0 or $f(x)$ is always 1). If $f(x)$ is balanced then $f(x) = 0$ for half of the inputs and $f(x) = 1$ for the other half of the inputs.

During *classical computation*, often in the worst scenario, it becomes necessary to query $(2^{n-1} + 1)$ times, requiring inputs for more than half of the domain $\{0, 1\}^n$ for determining whether $f(x)$ is balanced or constant. When using a randomized algorithm, the exercise may achieve a bounded error ε on the result in $\log \frac{1}{\varepsilon}$ number of queries.

For a quantum solution, a Feynman diagram to facilitate tracking the quantum state is shown in Figure 5-10.

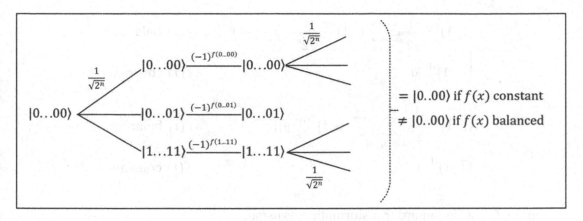

Figure 5-10. *Deutsch-Jozsa algorithm: Feynman path diagram*

The following steps summarize the quantum states for each step of Figure 5-3, which is analogous to the path in the Feynman diagram for Deutsch's algorithm.

1. Input state is $|\psi_0\rangle = |0\rangle^{\otimes n}$

2. $|+\rangle^{\otimes n} = \frac{1}{\sqrt{2}}(|0\rangle + |1\rangle) \dots \frac{1}{\sqrt{2}}(|0\rangle + |1\rangle) = \frac{1}{\sqrt{2^n}}\sum_{i\in\{0,1\}^n}|i\rangle$

3. $\frac{1}{\sqrt{2^n}}\sum_{i\in\{0,1\}^n}(-1)^{f(i)}|i\rangle$

Expansion of step 3 gives step 4.

$$= \begin{cases} \dfrac{1}{\sqrt{2^n}}\sum_{i\in\{0,1\}^n}(-1)^{f(i)}|i\rangle, & f(i) \text{ balanced} \\[3mm] (-1)^{f(i)}\dfrac{1}{\sqrt{2^n}}\sum_{i\in\{0,1\}^n}|i\rangle = (-1)^{f(i)}|+\rangle^{\otimes n}, & f(i) \text{ constant} \end{cases}$$

$$= \begin{cases} (-1)^{f(i)}\dfrac{1}{\sqrt{2^n}}\sum_{i\in\{0,1\}^n}(-1)^{f(i)}H^{\otimes n}|i\rangle, & f(i) \text{ balanced} \\[3mm] (-1)^{f(i)}|0\rangle^{\otimes n}, & f(i) \text{ constant} \end{cases}$$

$$= \begin{cases} (-1)^{f(i)} \dfrac{1}{\sqrt{2^n}} \displaystyle\sum_{i\in\{0,1\}^n} (-1)^{f(i)} \dfrac{1}{\sqrt{2^n}} \displaystyle\sum_{j\in\{0,1\}^n} (-1)^{i\cdot j}|j\rangle, & f(i) \text{ balanced} \\[2em] (-1)^{f(i)} |0\rangle^{\otimes n}, & f(i) \text{ constant} \end{cases}$$

$$= \begin{cases} (-1)^{f(i)} \dfrac{1}{2^n} \displaystyle\sum_{i\in\{0,1\}^n} \displaystyle\sum_{j\in\{0,1\}^n} (-1)^{ij+f(i)}|j\rangle, & f(i) \text{ balanced} \\[2em] (-1)^{f(i)} |0\rangle^{\otimes n}, & f(i) \text{ constant} \end{cases}$$

where, H is a Hadamard transformation as usual.

A circuit similar to Figure 5-11 is used for the Deutsch-Jozsa algorithm for a general quantum solution.

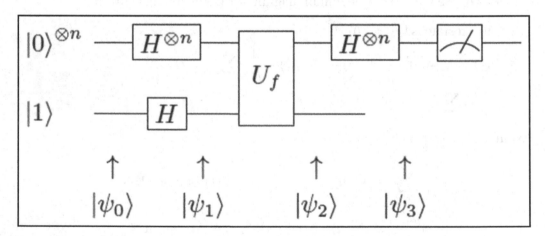

Figure 5-11. *Quantum circuit for Deutsch-Jozsa algorithm*

In Figure 5-11, the general Deutsch-Jozsa algorithm is deduced via the following steps.

1. The input state is $|\psi_0\rangle = |0\rangle^{\otimes n} \otimes |1\rangle$

2. Then, the Hadamard gates are applied, giving

$$|\psi_1\rangle = \frac{1}{\sqrt{2^n}} \sum_{i\in\{0,1\}^n} |i\rangle \left[\frac{1}{\sqrt{2}}(|0\rangle - |1\rangle) \right]$$

3. The state from step 2, after analysis, gives

$$|\psi_2\rangle = \frac{1}{\sqrt{2^n}} \sum_{i \in \{0,1\}^n} (-1)^{f(i)} |i\rangle \left[\frac{1}{\sqrt{2}} (|0\rangle - |1\rangle) \right]$$

4. After applying Hadamard transformation to all the n qubits, we have

$$H^{\otimes n} |i\rangle = \frac{1}{\sqrt{2^n}} \sum_{j \in \{0,1\}^n} (-1)^{i \cdot j} |j\rangle \qquad 5.5$$

The preceding step gives the following.

$$|\psi_3\rangle = \frac{1}{2^n} \sum_{i \in \{0,1\}^n} \sum_{j \in \{0,1\}^n} (-1)^{ij + f(i)} |j\rangle \left[\frac{1}{\sqrt{2}} (|0\rangle - |1\rangle) \right]$$

Finally, the probability of measuring the state $|0\rangle^{\otimes n}$ is given by

$$p\left(|0\rangle^{\otimes n}\right) = \left| \frac{1}{2^n} \sum_i (-1)^{f(i)} \right|^2$$

If $f(i)$=constant, then we get constructive interference, and this gives $p(|0\rangle^{\otimes n}) = 1$. If $f(i)$ is balanced then destructive interference gives us 0 as probability.

In summary, the Deutsch-Jozsa algorithm works because of the following reasons.

In case 1, the oracle is *constant*. In this case, it does not affect the input qubits as far as the global phase is concerned. The H gate is its own inverse, hence

$$H^{\otimes n} \begin{bmatrix} 1 \\ 0 \\ 0 \\ . \\ . \\ . \\ 0 \end{bmatrix} = \frac{1}{\sqrt{2^n}} \begin{bmatrix} 1 \\ 1 \\ 1 \\ . \\ . \\ . \\ 1 \end{bmatrix} \xrightarrow{apply\ U_f} H^{\otimes n} \frac{1}{\sqrt{2^n}} \begin{bmatrix} 1 \\ 1 \\ 1 \\ . \\ . \\ . \\ 1 \end{bmatrix} = \begin{bmatrix} 1 \\ 0 \\ 0 \\ . \\ . \\ . \\ 0 \end{bmatrix} \qquad 5.6$$

233

In case 2, the oracle is *balanced*. In this case, after step 2, the input register is an equal superposition of all the states in the computational basis, and *phase kickback* adds a *negative* phase to exactly half these states.

$$U_f \frac{1}{\sqrt{2^n}} \left(\begin{bmatrix} 1 \\ 1 \\ 1 \\ . \\ . \\ . \\ 1 \end{bmatrix} \right) = \begin{bmatrix} -1 \\ 1 \\ -1 \\ . \\ . \\ . \\ 1 \end{bmatrix} \qquad 5.7$$

This shows that the quantum state after querying the oracle is orthogonal to the quantum state before querying the oracle. As such, applying the H gates gives a quantum state that is orthogonal to the all-zero $|000...0\rangle$ state implying that the al-zero state should never be measured.

Quantum query complexity of *Deutsch-Jozsa algorithm*. The *cost function* in minimizing while solving Deutsch's problem is the number of quantum queries applied to U_f. This is an example of the model of quantum query complexity, which many quantum algorithms were developed with. In the study of quantum query complexity, one is given a black box U_f implementing function f, and asked what the minimum number of required queries to U_f is, to determine a desired property of f. Note that the quantum algorithm computing this property can consist of, for example, 999999999 quantum gates; if it contains only two queries to U_f, then we consider the cost of the algorithm as 2 (i.e., all "non-query" operations are considered free).

Deutsch-Jozsa with Cirq

Chapter 4 included a hands-on introduction to Google's open source quantum computing library called Cirq, most of which were confined to building circuits, gates, and finally, Deustch's algorithm. This chapter ventures into constructing a two-qubit Deutsch-Jozsa algorithm with Cirq. Deutsch-Jozsa algorithm can distinguish constant from balanced functions like these using a single query to the oracle. The goal of this exercise is to write a quantum circuit that can distinguish these.

We start by importing the required libraries as shown in Listing 5-1a for the Deutsch-Jozsa_cirq.ipynb Jupyter notebook downloadable from the book's website.

Listing 5-1a. Libraries for Deutsch-Jozsa_cirq.ipynb

```
import cirq
import numpy as np
import matplotlib.pyplot as plt
```

In discussions about Deutsch's algorithm, you saw that boolean functions for one input bit are either constant or balanced. However, we have also seen from the section on the Deutsch-Jozsa algorithm that for boolean functions from two input bits, not all functions are constant or balanced. There are two constant functions $(x_1, x_2) = 0$, and $f(x_1, x_2) = 1$, and $\binom{4}{2} = 6$ balanced functions. The following code gives the operations for these functions where we take two input qubits and compute the function in the third qubit.

Listing 5-1b. Input States for Deutsch-Jozsa_cirq.ipynb

```
# Define three qubits to use.
q0, q1, q2 = cirq.LineQubit.range(3)

# Define the operations to query each of the two constant functions.
constant = (
    [],
    [cirq.X(q2)]
)

# Define the operations to query each of the six balanced functions.
balanced = (
    [cirq.CNOT(q0, q2)],
    [cirq.CNOT(q1, q2)],
    [cirq.CNOT(q0, q2), cirq.CNOT(q1, q2)],
    [cirq.CNOT(q0, q2), cirq.X(q2)],
```

```
    [cirq.CNOT(q1, q2), cirq.X(q2)],
    [cirq.CNOT(q0, q2), cirq.CNOT(q1, q2), cirq.X(q2)]
)
```

Next, we define a function for a quantum circuit that can distinguish a constant from balanced functions on two bits.

Listing 5-1c. Function Definition for Deutsch-Jozsa_cirq.ipynb

```python
# Define a function

def deJo_circuit(oracle):
    # Phase kickback trick.
    yield cirq.X(q2), cirq.H(q2)

    # Get an equal superposition over input bits.
    yield cirq.H(q0), cirq.H(q1)

    # Query the function.
    yield oracle

    # Use interference to get result, put last qubit into |1>.
    yield cirq.H(q0), cirq.H(q1), cirq.H(q2)

    # Use a final OR gate to put result in final qubit.
    yield cirq.X(q0), cirq.X(q1), cirq.CCX(q0, q1, q2)
    yield cirq.measure(q2)
```

Then we verify our Deutsch-Jozsa circuit by running the following cell, which simulates the circuit for all oracles.

Listing 5-1d. Run Cirq Simulator to Verify Deutsch-Jozsa_cirq.ipynb

```python
simulator = cirq.Simulator()

print("\nResult on constant functions:")
for oracle in constant:
    result = simulator.run(cirq.Circuit(deJo_circuit(oracle)),
            repetitions=10)
    print(result)
```

```
print("\nResult on balanced functions:")
for oracle in balanced:
    result = simulator.run(cirq.Circuit(deJo_circuit(oracle)),
            repetitions=10)
    print(result)
```

The following output shows that the circuit works as expected for different oracles .

```
              Result on constant functions:
              2=0000000000
              2=0000000000

              Result on balanced functions:
              2=1111111111
              2=1111111111
              2=1111111111
              2=1111111111
              2=1111111111
              2=1111111111
```

As with Deutsch's algorithm using a single-bit case, we always measure 0 for constant functions and always measure 1 for balanced functions.

Quantum Phase Estimation

Phase estimation is a quantum routine that helps find the eigenvalues of a unitary matrix *U* applied to a quantum register as a controlled gate. Phase estimation uses QFT to write information encoded in the phase of an amplitude. Typically, this technique is used by *quantum machine learning* (QML) algorithms to evaluate eigenvalues of operators which contain information about a QML training set.

Toward the end of the last section, we noted that the final Hadamard gate in the Deutsch algorithm, and the Deutsch–Jozsa algorithm, was used to get at information encoded in the relative phases of a state. The Hadamard gate is self-inverse (reference equations 5.6 and 5.7) and thus does the opposite; namely, it can *encode information* into the phases.

Let's look at our derivation of Hadamard state of equation 5.5, and then consider an H gate acting on basis state $|x\rangle$ for $x \in \{0, 1\}$, then, for $n = 1$, we have, from equation 5.5,

$$H|x\rangle = \frac{1}{\sqrt{2}} \sum_{y \in \{0,1\}} (-1)^{x.y} |y\rangle \tag{5.8}$$

The Hadamard gate can be thought of as holding *encoded information* about the value of x into the relative phases between the basis states $|0\rangle$ and $|1\rangle$. Since the Hadamard gate is self-inverse, applying it to the state on the right side of equation 5.8 gives us back the state $|x\rangle$ again,

$$H\left[\frac{1}{\sqrt{2}}(|0\rangle + (-1)^x |1\rangle)\right] = |x\rangle \tag{5.9}$$

Hence, in equation 5.9, the Hadamard gate can be thought of as *decoding the information* about the value of x that was encoded in the phases. Likewise, if we consider $H^{\otimes n}$ acting on an n-qubit basis state $|x\rangle$, then we obtain equivalent of equation 5.5, which is

$$H^{\otimes n}|x\rangle = \frac{1}{\sqrt{2^n}} \sum_{y \in \{0,1\}^n} (-1)^{x.y} |y\rangle \tag{5.10}$$

The n-qubit Hadamard transformation can thus be interpreted as possessing encoded information about the value of $|x\rangle$ into the *phases* $(-1)^{x.y}$ of the basis states $|y\rangle$. Note that $(-1)^{x.y}$ are phases of a very particular form. In general, a phase is a complex number of the form $e^{2\pi i \omega}$, where $\omega \in \{0, 1\}$ for any real ω. This allows that $\omega = \frac{1}{2}$ when the phase is -1. Generally, it is not possible for the n-qubit Hadamard transformation to fully access information that is encoded in more general ways than the phase. However, it is possible to write a special routine that allows the Hadamard gate to determine the information encoded in phases.

Let $|\psi_n\rangle$, $n = 1, 2, ..., n$ be eigenvectors of unitary operator U. U is defined in Hilbert space for which $|\psi_n\rangle$ form an orthonormal basis. A unitary operation of controlled U on the eigenvector gives,

$$U|\psi_n\rangle = e^{2\pi i \omega}|\psi_n\rangle \tag{5.11}$$

where $\omega \in \{0, 1\}$ for any real ω. Equation 5.12 holds because *eigenvalues* of the unitary operators are phases.

To visualize equation 5.12, we can consider an initial state given by

$$\frac{1}{\sqrt{2^j}} \sum_{y=0}^{2j-1} e^{2\pi i \omega y} |y\rangle \qquad\qquad 5.12$$

To analyze expression 5.12 further, we need to consider basis encoding.

Considering the j-bit strings to be integers from 0 to $(2j-1)$, $|y\rangle$ refers to the *basis state* labeled by $|y\rangle$, where y is the *binary encoding* of the integer y. Looking at equation 5.12, determining the value, or at least a proper estimate of phase parameter ω, is bound to be of interest to find ψ. This is the *phase estimation problem* [43]. To solve this, the phase estimation routine allows the following transformation

$$U|\psi_n\rangle = \tilde{\omega}|\psi_n\rangle \qquad\qquad 5.13$$

where $\tilde{\omega}$ is an estimation for ω. Equation 5.13 implies that the eigenvalues of the orthonormal $|\psi_n\rangle$ are as follows for ω_j.

$$\lambda_j = e^{2\pi i \omega_j} \cong \tilde{\omega} \qquad\qquad 5.14$$

We need to keep in mind that since is U unitary, λ_j must have magnitude 1. We start by defining ω in binary,

$$\omega = 0.p_1 p_2 p_3 \ldots \xrightarrow{\text{means}} \frac{p_1}{2} + \frac{p_2}{2^2} + \frac{p_3}{3} \ldots \qquad\qquad 5.15$$

Likewise, power-of-two multiples of ω can be written as

$$2^k \omega = p_1 p_2 p_3 \ldots p_k . p_{k+1} p_{k+2} \ldots \qquad\qquad 5.16$$

For the phase part of equation 5.12, we also know that $e^{2\pi i k} = 1$ for any integer k, we have

$$e^{2\pi i 2^k \omega} = e^{2\pi i (p_1 p_2 p_3 \ldots p_k . p_{k+1} p_{k+2} \ldots)}$$

$$= e^{2\pi i (p_1 p_2 p_3 \ldots p_k)} . e^{2\pi i (0.p_{k+1} p_{k+2} \ldots)}$$

$$= e^{2\pi i (0.p_{k+1} p_{k+2} \ldots)} \qquad\qquad 5.17$$

Now, if we have a given input state $\dfrac{1}{\sqrt{2^j}}\displaystyle\sum_{y=0}^{2^{j}-1}e^{2\pi i\omega y}\lvert y\rangle$ and consider a quantum circuit to determine an estimation of ω, then, for a one-qubit input state, where $j = 1$, we have $\omega = 0.\,p_1$ and the following from equation 5.5.

$$\frac{1}{\sqrt{2^1}}\sum_{y=0}^{2\cdot 1-1}e^{2\pi i(0.p_1)y}\lvert y\rangle = \frac{1}{\sqrt{2}}\sum_{y=0}^{1}e^{2\pi i\left(\frac{p_1}{2}\right)y}\lvert y\rangle$$

$$=\frac{1}{\sqrt{2}}\sum_{y=0}^{1}e^{\pi i(p_1)y}\lvert y\rangle$$

$$=\frac{1}{\sqrt{2}}\sum_{y=0}^{1}(-1)^{(p_1)y}\lvert y\rangle$$

$$=\frac{1}{\sqrt{2}}\left(\lvert 0\rangle+(-1)^{p_1}\lvert 1\rangle\right)\qquad\qquad 5.18$$

The circuit to solve phase estimation is shown in Figure 5-12.

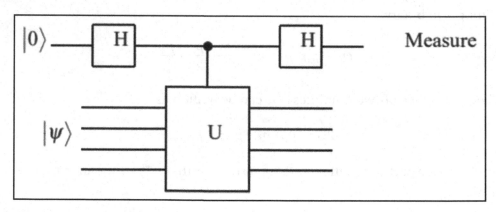

Figure 5-12. *Phase estimation circuit*

In Figure 5-12, after a controlled U transformation, the eigenvector remains unchanged, and eigenvalue λ is included in the phase of the first qubit. A Hadamard transform on the first qubit transforms the information into an amplitude that can be measured. In other words, referring to equations 5.9 and 5.18, to determine the value of p_1 and estimate ω, we have

$$H\left[\frac{1}{\sqrt{2}}(|0\rangle + (-1)^{p_1}|1\rangle)\right] = |p_1\rangle \tag{5.19}$$

Now that you have gone through the phase estimation exercise of a one-qubit input, let us look at a two-qubit state.

$$\frac{1}{\sqrt{2^2}}\sum_{y=0}^{2.2-1}e^{2\pi i(0.p_1p_2)y}|y\rangle = \frac{1}{\sqrt{2}}(|0\rangle + e^{2\pi i(0.p_2)y}|1\rangle)\otimes\frac{1}{\sqrt{2}}(|0\rangle + e^{2\pi i(0.p_1p_2)y}|1\rangle) \tag{5.20}$$

Equation 5.20 shows that p_2 can be determined by applying a Hadamard gate on the first qubit in the same manner as we did in equation 5.19. However, there is still a need to determine p_1, and the second qubit is involved in that process. If $p_2 = 0$, then the second qubit is in the state $\frac{1}{\sqrt{2}}(|0 + e^{2\pi i(0.p_1)y}|1\rangle)$ and p_1 can be determined from this. However, if $p_2 = 1$, then, there is a challenge because the previous straightforward approach will not work. To achieve it, we need to apply a phase rotation operator (reference equation 1.15).

$$R_\theta = \begin{pmatrix} 1 & 0 \\ 0 & e^{i\theta} \end{pmatrix} = \begin{pmatrix} 1 & 0 \\ 0 & e^{2\pi i/2^2} \end{pmatrix} = \begin{pmatrix} 1 & 0 \\ 0 & e^{2\pi i(0.01)} \end{pmatrix}$$

where the 0.01 in the exponent is in base-2. In other words, $0.01 = 2^{-2}$. The inverse of R_θ is given by

$$R_\theta^{-1} = \begin{pmatrix} 1 & 0 \\ 0 & e^{-2\pi i(0.01)} \end{pmatrix} \tag{5.21}$$

Hence, with $p_2 = 1$, we consider the effect of applying a R_θ^{-1}.

$$R_\theta^{-1}\left[\frac{1}{\sqrt{2}}(|0\rangle + e^{2\pi i(0.p_1 1)y}|1\rangle)\right] = \frac{1}{\sqrt{2}}(|0\rangle + e^{2\pi i(0.p_1 1 - 0.01)y}|1\rangle)$$

$$= \frac{1}{\sqrt{2}}(|0\rangle + e^{2\pi i(0.p_1)y}|1\rangle) \tag{5.22}$$

Now, if we apply a Hadamard gate to the result of equation 5.22, then we can determine p_1. If $p_2 = 1$ or $p_2 = 0$ is the factor that determines whether we apply to the Hadamard gate after or before applying the R_θ^{-1} transform to the second qubit. After we applied the Hadamard gate to the first qubit, the state of the first qubit was $|p_2\rangle$.

The approach to phase estimation can be generalized. The same steps taken for one and two-qubit inputs can be used for p_3, p_4 ... and conditionally rotate p_3, p_4... For the analogous circuit on n qubits estimating a phase of form $\omega = 0. \ p_1 p_2 p_3...$, the output of the circuit is given by $|p_n...p_2 p_1\rangle$. Adding some gates to reverse the order of the qubits gives us an $O(n^2)$ efficient circuit, which implements

$$\frac{1}{\sqrt{2^n}} \sum_{y=0}^{2.n-1} e^{2\pi i \frac{p}{2^n} y} |y\rangle \rightarrow |p\rangle \tag{5.23}$$

If we logically label the qubits in equation 5.23 in reverse order, then we get

$$|p\rangle \rightarrow \frac{1}{\sqrt{2^n}} \sum_{y=0}^{2.n-1} e^{2\pi i \frac{p}{2^n} y} |y\rangle \tag{5.24}$$

Equation 5.24 is the unitary transformation realized by applying the phase estimation circuit backward. Applying a circuit "backward" means to replace each gate with its inverse and run the circuit in reverse order. Equation 5.24 resembles classical *discrete Fourier transform* (DFT) used extensively in science and engineering (see Appendix A). A DFT takes a vector X as an input of size N and outputs vector $Y = WX$, where W is the Fourier matrix. QFT is defined as a transformation between two quantum states that are determined using the values of DFT. If $X = \{x_i\}$ and $Y = \{y_i\}$ are vectors such that $Y = WX$, then the QFT is defined as the transformation.

The phase estimation routine uses QFT. The QFT can be interpreted as the inverse of the phase estimation operation and applying the phase estimation circuit backwards gives an efficient circuit for performing the QFT.

A classical DFT process maps a real vector $\bar{\varphi} \in \mathbb{R}^{2^n}$ to another vector $\bar{y} \in \mathbb{R}^{2^n}$ via the following transformation in $O(n2^n)$ steps, where $\mathbb{R}^{2^n}$ are set of real numbers spanning $\{1...2^n\}$.

$$y_k \rightarrow \frac{1}{\sqrt{2^n}} \sum_{j=0}^{2^n} e^{2\pi i \frac{j}{2^n} k} \varphi_j$$

where $k = 1, 2, 3 \ldots 2^n$.

The QFT maps a quantum state with amplitudes encoding the φ_k to a quantum state whose amplitudes encode the y_k.

$$\sum_{j=0}^{2^n} \varphi_j |j\rangle \rightarrow \sum_{k=0}^{2^n} \frac{1}{\sqrt{2^n}} \left(\sum_{j=0}^{2^n} e^{2\pi i \frac{j}{2^n} k} \varphi_j \right) |k\rangle \qquad 5.25$$

where $y_k = \frac{1}{\sqrt{2^n}} \sum_{j=0}^{2^n} e^{2\pi i \frac{j}{2^n} k} \varphi_j$, in $O(2^n)$ steps.

The sum over the j in equation 5.25 vanishes due to the quantum Fourier transform applied to a state that is only in one computational basis state. The basis state $|j\rangle$ corresponds to the classical case in which one φ_j is non-zero, and the rest are zero.

Figure 5-13 shows the circuit representation of an eight-qubit QFT circuit produced with Quirk [6].

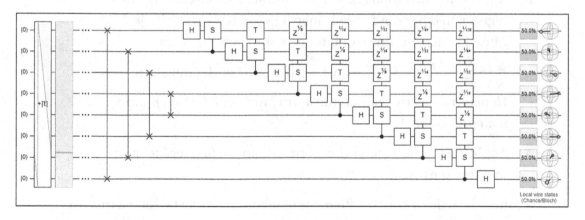

Figure 5-13. *Eight-qubit quantum Fourier transform circuit (Produced with Quirk [6])*

The operation of equation 5.25 requires $n = b + \log\left(2 + \dfrac{1}{2\varepsilon}\right)$ qubits to get the first b bits of the binary string encoding the eigenstate with probability $1 - \varepsilon$. The reverse of QFT is used in the quantum phase estimation procedure.

To get an estimate of ω, via phase estimation technique, the inverse of the unitary QFT U_f^{-1} is used, where U_f is the quantum Fourier transform. The inverse QFT state applied on the basis states $|0\rangle, |1\rangle, ... |m-1\rangle$ is given by

$$U_f^{-1}:|\varphi\rangle = \frac{1}{\sqrt{m}}\sum_{y=0}^{m-1}e^{2\pi i\frac{p}{m}y}|y\rangle$$

5.26

In summary, given a unitary gate U, an eigenvector $|\psi_n\rangle$ of U is given by $U|\psi_n\rangle = \lambda|\psi_n\rangle$. The aim is to find an approximate value (or estimation) of the eigenvalue λ. Since U is unitary, the magnitude of λ must be 1, and we can set $\lambda = e^{2\pi i\omega}$, where $\omega \in \{0,1\}$ and $\omega = 0$. $p_1p_2p_3...$ up to n bits of precision. The aim is to find an estimation of the phase ω.

Then, the phase estimation algorithm is given by the following steps.

1. Initiate with $|0_n\rangle|\psi\rangle$.

2. Apply $H^{\otimes n} \otimes I$ to the first n qubits for $N = 2^n$ which gives

$$\frac{1}{\sqrt{2^n}}\sum_{y=0}^{2n-1}|k\rangle|\psi\rangle$$
.

3. Apply U to the second register for several times given by the first register. In other words, apply map $|k\rangle|\psi\rangle \rightarrow |k\rangle U^k|\psi\rangle = e^{2\pi i\omega k}|k\rangle|\psi\rangle$. After this step, the first n qubits are in the state

$$\frac{1}{\sqrt{2^n}}\sum_{y=0}^{2n-1}e^{2\pi i\omega k}|k\rangle = U_f|2^n\omega\rangle$$, where U_f is the Fourier transform.

Hence, the inverse Fourier transform provides $|2^n\omega\rangle = |p_1...p_2\rangle$ with probability 1.

4. Apply the inverse Fourier transform U_f^{-1} to the first register and measure the result.

If ψ cannot be written exactly with n bits of precision, it can be shown that this procedure can still provide, with high probability, a good n-bit approximation to ψ.

Quantum Programming with Rigetti Forest

Rigetti Computing[2] is a Berkeley, California–based builder of quantum computers powered by their in-house superconducting quantum processors. The company also develops a cloud platform called Quantum Cloud Services (QCS), via which their machines can be integrated into any public, private or hybrid cloud. Rigetti's computing environment is powered by an SDK called Forest, which is a full-stack programming and execution environment for quantum/classical computing.

Forest can develop algorithms for quantum/classical hybrid computing and learn how quantum computers and algorithms work. The algorithms can be simulated on qubits using their Quantum Virtual Machine, or QVM, running in the cloud. It is also possible to interact with *real quantum chips* using simple function calls that execute an active quantum system.

The following are a few terms that you should be aware of (also see documentation on Rigetti [61]).

- **pyQuil**: An open source Python library to help write and run quantum programs. The source code is hosted on GitHub at `http://github.com/rigetti/pyquil`.

- **Quil**: The Quantum Instruction Language standard. Instructions written in Quil can be executed on any implementation of an quantum abstract machine (QAM), such as the quantum virtual machine (QVM) or on a real quantum processing unit (QPU). More details regarding Quil can be found in the whitepaper "A Practical Quantum Instruction Set Architecture" [71].

- **QVM**: The Quantum Virtual Machine is an open source implementation of a quantum abstract machine on classical hardware. The QVM lets you use a regular computer to simulate a small quantum computer and execute Quil programs. QVM can be found on GitHub at `https://github.com/rigetti/qvm` and can be run on standard laptops. We use the QVM for our next exercise.

---

[2]Aspen, Forest, Forest SDK, Pyquil, QCS, Quantum Advantage Prize, Quil, Quilc, Quil-T, QVM, and Rigetti are trademarks or registered trademarks of Rigetti & Co, Inc.

- **QPU**: *Quantum processing unit*. This refers to the physical hardware chip on which quantum programs can be run.

- **Quil Compiler**: The compiler, quilc, compiles Quil written for one QAM to another. According to its supported instruction set architecture, our open source compiler takes arbitrary Quil and compiles it for the given QAM. Find quilc on GitHub at `https://github.com/rigetti/quilc`.

- **Forest SDK**: Rigetti's software development kit, optimized for near-term quantum computers that operate as coprocessors, working in concert with traditional processors to run hybrid quantum-classical algorithms. For references on problems addressable with near-term quantum computers, see "Quantum Computing in the NISQ Era and Beyond" [68].

The Forest SDK is a comprehensive suite of a language that allows the power to define circuits, provides a simulator, and collects quantum algorithms, among other important components. For information on setting it up on various OS, refer to the Forest documentation at `https://pyquil-docs.rigetti.com/en/stable/` [61]. The summary of the set-up on Ubuntu Linux 18.04 LTS is given next. It is recommended that all work in a Linux environment is done in a virtual environment.

Installing the QVM

To use the Rigetti Forest SDK to its full capacity, we need pyQuil, the QVM, and the Quil Compiler. At the time of writing this book, pyQuil 2.0 is Rigetti's library for generating and executing Quil programs on the Rigetti Forest platform. Before installation, it is recommended that a Python 3.6+ virtual environment be activated. Then, pyQuil is installed using `pip` (for Ubuntu, you may need the `sudo` command).

```
$ pip install pyquil
```

or

```
$ sudo pip install pyquil
```

For any existing pyQuil environment, it is always good to run an upgrade.

```
pip install --upgrade pyquil
```

In the next step, the QVM and the Quil Compiler need to be downloaded as follows. The Forest 2.0 Downloadable SDK Preview contains the following.

- The Rigetti Quantum Virtual Machine (qvm) allows high-performance simulation of Quil programs

- The Rigetti Quil Compiler (quilc) allows compilation and optimization of Quil programs to native gate sets

The QVM and the compiler are packed as program binaries that are accessed through the command line. Both provide support for direct command-line interaction, as well as a *server mode*. The server mode is *required* for use with pyQuil.

The Forest SDK can be downloaded at `https://qcs.rigetti.com/sdk-downloads`. The same download site offers links for installations in Windows, macOS, Linux (.deb), Linux (.rpm), and Linux (bare-bones).

Installing the QVM and Compiler on Linux

1. Download the distribution by clicking the appropriate link on the SDK download page (for Ubuntu, download the package for Debian). `https://qcs.rigetti.com/sdk-downloads`

2. Unpack the `tarball` and change to that directory.

   ```
   tar -xf forest-sdk-linux-deb.tar.bz2
   cd forest-sdk-2.0rc2-linux-deb
   ```

3. From the same directory, run the following command.

   ```
   sudo ./forest-sdk-2.0rc2-linux-deb.run
   ```

4. Upon successful installation, you should be able to run the following two commands.

   ```
   qvm --version
   quilc --version
   ```

5. After the SDK is successfully installed, open a terminal window, change to the same virtual environment where the SDK is installed, and start the QVM in server mode with the qvm -S command. This should start the QVM server, as shown in Figure 5-14, where the virtual environment is called qml.

```
Python 3.7.9
(qml) sgangoly@ubuntu:~$ qvm -S
*****************************
* Welcome to the Rigetti QVM *
*****************************
Copyright (c) 2016-2019 Rigetti Computing.

This is a part of the Forest SDK. By using this program
you agree to the End User License Agreement (EULA) supplied
with this program. If you did not receive the EULA, please
contact <support@rigetti.com>.

(Configured with 10240 MiB of workspace and 2 workers.)
(Gates parallelize at 19 qubits.)
(There are 3 kernels and they are used with up to 29 qubits.)
(Features enabled: none)

<134>1 2021-02-02T19:35:24Z ubuntu qvm 14917 - - Compilation mode disabled.
<135>1 2021-02-02T19:35:24Z ubuntu qvm 14917 - - Selected simulation method: pur
e-state
<135>1 2021-02-02T19:35:24Z ubuntu qvm 14917 - - Starting server on port 5000.
<134>1 2021-02-02T19:35:25Z ubuntu qvm 14917 LOG0001 - This is the latest versio
n of the SDK.
```

Figure 5-14. *Starting QVM server inside of a virtual environment*

6. Keeping the terminal for the QVM running, open a new terminal, change to the same virtual environment, and start Quilc in server mode with the quilc -S command.

The quilc -S run is shown in Figure 5-15 inside of the qml virtual environment.

```
(qml) sgangoly@ubuntu:~$ quilc -S
+-----------------+
|  W E L C O M E  |
|    T O   T H E  |
|  R I G E T T I  |
|      Q U I L    |
|  C O M P I L E R|
+-----------------+
Copyright (c) 2016-2020 Rigetti Computing.

This is a part of the Forest SDK. By using this program
you agree to the End User License Agreement (EULA) supplied
with this program. If you did not receive the EULA, please
contact <support@rigetti.com>.

<134>1 2021-02-02T19:35:43Z ubuntu quilc 14958 LOG0001 - Launching quilc.
<134>1 2021-02-02T19:35:43Z ubuntu quilc 14958 - - Spawning server at (tcp://*:5
555) .

<133>1 2021-02-02T19:35:43Z ubuntu quilc 14958 LOG0001 - An update is available
to the SDK. You have version 1.22.0. Version 1.23.0 is available from https://qc
s.rigetti.com/sdk-downloads
```

Figure 5-15. *Starting Quilc server inside a virtual environment*

7. To uninstall Forest SDK, enter the following.

```
sudo apt remove forest-sdk
```

At this point, you should have at least *two terminal windows open and running* in the background: one for qvm and one for quilc.

If you do not have a Jupyter notebook or other Python libraries installed in the same virtual environment, do so with the following command.

```
$ python3 -m pip install -U jupyter matplotlib numpy pandas scipy
scikit-learn
```

The next step is to start the Jupyter notebook. Open a new third terminal window, change to the same virtual environment, and start Jupyter notebook by typing the following.

```
$ jupyter notebook
```

This should open a Jupyter notebook session in your browser. Keep this terminal window open. So, at the end of this process, three different terminal windows are running from the *same virtual environment*: one for qvm, one for quilc, and one for the Jupyter notebook.

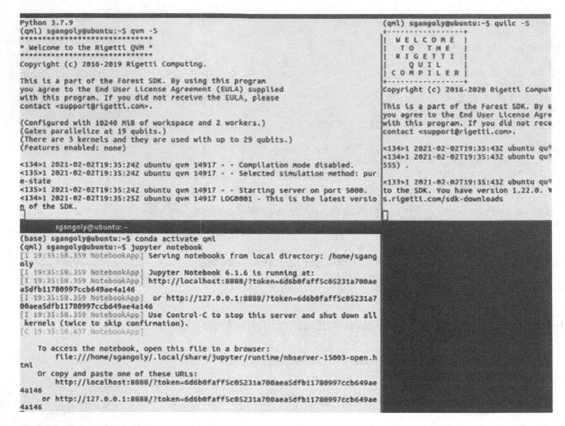

Figure 5-16. *QVM, Quilc, and Jupyter notebook running inside of the same virtual environment*

The final state of the terminals for QVM, Quilc, and Jupyter notebook should resemble the screenshot shown in Figure 5-16.

pyQuil [72] provides a function to ensure that a local `qvm` and `quilc` are currently running in the environment. To make sure both are available, we need to execute `from pyquil.api import local_forest_runtime` and then use `local_forest_runtime()`. This starts `qvm` and `quilc` instances using subprocesses if they have not already been started, as in Listing 5-2.

Listing 5-2. pyQuil Code to Verify qvm and quilc Are Available

```
from pyquil import get_qc, Program
from pyquil.gates import CNOT, Z
from pyquil.api import local_forest_runtime

prog = Program(Z(0), CNOT(0, 1))
```

```
with local_forest_runtime():
    qvm = get_qc('9q-square-qvm')
    results = qvm.run_and_measure(prog, trials=10)
```

The Program object allows us to build up a Quil program. get_qc() connects us to a QuantumComputer object, which specifies what our program should run on (see https://pyquil-docs.rigetti.com/en/stable/qvm.html#qvm). The (*) imports *all* gates from the pyquil.gates module, which allows us to add operations to our program.

Listing 5-3a gives the code to introduce some operations in pyQuil using QVM and builds up to a description of a three-qubit *quantum Fourier transform* (QFT). Initially, we import some libraries and verify that QVM and Quilc servers are reachable from a Jupyter notebook. If this step does not give any errors, and then we are good to go.

Listing 5-3a. Libraries for Quantum Fourier Transform (QFT) qft_forest.ipynb

```
from pyquil import Program, get_qc
from pyquil.gates import *
import cmath
import matplotlib.pyplot as plt
import numpy as np
from pyquil.api import ForestConnection
from pyquil.api import WavefunctionSimulator
from pyquil.api import QVMConnection
from qutip import Bloch
from tempfile import mkdtemp
qvm = QVMConnection()

# Verify QVM and Quilc are available

from pyquil import get_qc, Program
from pyquil.gates import CNOT, Z
from pyquil.api import local_forest_runtime

prog = Program(Z(0), CNOT(0, 1))

with local_forest_runtime():
    qvm = get_qc('9q-square-qvm')
    results = qvm.run_and_measure(prog, trials=10)
```

For the first test, let's generate a state.

Listing 5-3b. Libraries for Quantum Fourier Transform (QFT) qft_forest. ipynb

```
# Test
qvm = QVMConnection()
hello_qubit = Program()
print(qvm.wavefunction(hello_qubit))
```

This gives the following output.

$$(1+0j)|0>$$

As a second test, let's first construct a Bell state and run the program. Qubit 0 is initialized into a superposition state with the Hadamard H gate. Then an entangled state is created between qubits 0 and 1 with the CNOT gate. And then run the program.

Listing 5-3c. Test with Bell State qft_forest.ipynb

```
# construct a Bell State program
p = Program(H(0), CNOT(0, 1))
# run the program on a QVM
qc = get_qc('9q-square-qvm')
result = qc.run_and_measure(p, trials=10)
print(result[0])
print(result[1])
```

The code for Bell state gives the following output.

```
[0 0 0 0 1 0 1 1 0 1]
[0 0 0 0 1 0 1 1 0 1]
```

QC is the simulated quantum computer. The run_and_measure, tells the QVM to run the program, perform a measurement that collapses the state, and return the results. The trials refer to the number of times the program is run.

The call to run_and_measure makes a request to the two servers (qvm and quilc) that were started up. The Quilc server instance is called to compile the Quil program into native Quil, and the QVM server instance is called to simulate and return measurement results of the program ten times. If you open the terminal windows where your servers are running, you should see output printed to the console regarding the requests we made.

In the next step, the Quantum Fourier Transform is done. The task is to apply the QFT on the amplitudes of the states that correspond to the sequence for which we want to compute the DFT (discrete Fourier transform). For this, we need three qubits to transform an 8-bit sequence.

The required sequence to be computed is 01000000, so the initial state is $|001\rangle$. For a bit-string with more than one 1, an equal superposition over all the selected states is needed; for example, 01100000 would be an equal superposition of $|001\rangle$ and $|010\rangle$. One X gate applied to the zeroth qubit sets up the $|001\rangle$ state. It can be verified that this works by computing its wave function with the Wavefunction Simulator. We need to add some "dummy" qubits because otherwise, a wave function would return a two-element vector for only qubit 0. This is shown in Listing 5-3d.

Listing 5-3d. State Preparation for Quantum Fourier Transform (QFT) qft_ forest.ipynb

```
# Quantum Fourier Transform

state_prep = Program(X(0))
qft = Program()
qft += H(0)
qft += CPHASE(π/2, 1, 0)
qft += H(1)
qft += CPHASE(π/4, 2, 0)
qft += CPHASE(π/2, 2, 1)
qft += H(2);

# Add dummy qubits
from pyquil.api import WavefunctionSimulator
add_dummy_qubits = Program(I(1), I(2))  # The identity gate I has no affect
wf_sim = WavefunctionSimulator()
wavefunction = wf_sim.wavefunction(state_prep + add_dummy_qubits)
print(wavefunction)
```

This gives the following output.

(1+0j)|001>

Next, it is defined as a function, qft1, to make a three-qubit QFT quantum program. The algorithm was described earlier in this chapter. The implementation consists of Hadamard and CPHASE gates, with a SWAP gate for bit reversal correction. The qft1 function doesn't compute the QFT, but rather it makes a quantum program to compute the QFT on qubits q0, q1, and q2.

Listing 5-3e. State Preparation for QFT qft_forest.ipynb

```
# 3 qubit QFT Program: DFT of [0, 1, 0, 0, 0, 0, 0, 0], using pyQuil:
from math import pi

def qft1(q0, q1, q2):
    p = Program()
    p += [SWAP(q0, q2),
          H(q0),
          CPHASE(-pi / 2.0, q0, q1),
          H(q1),
          CPHASE(-pi / 4.0, q0, q2),
          CPHASE(-pi / 2.0, q1, q2),
          H(q2)]
    return p

print(qft1(0, 1, 2))
```

The code snippet returns the following as output.

```
SWAP 0 2
H 0
CPHASE(-pi/2) 0 1
H 1
CPHASE(-pi/4) 0 2
CPHASE(-pi/2) 1 2
H 2
```

Next, we run the QFT.

Listing 5-3f. Execute Quantum Fourier Transform (QFT) qft_forest.ipynb

```
compute_qft_prog = state_prep + qft1(0, 1, 2)
wavefunction = wf_sim.wavefunction(compute_qft_prog)
print(wavefunction.amplitudes)
```

This returns the following output.

```
[ 3.53553391e-01+0.j        2.50000000e-01-0.25j
  2.16489014e-17-0.35355339j -2.50000000e-01-0.25j
 -3.53553391e-01+0.j        -2.50000000e-01+0.25j
 -2.16489014e-17+0.35355339j  2.50000000e-01+0.25j
```

In the last step, we run Listing 5-3g to verify the QFT runs with an inverse fast Fourier transform (FFT).

Listing 5-3g. Verify QFT with Inverse FFT qft_forest.ipynb

```
# Verify qft by inverse FFT

from numpy.fft import ifft
ifft(wavefunction.amplitudes, norm="ortho")
```

This produces the following array as an output if we ignore the terms that are on the order of 1e-17, gives [0, 1, 0, 0, 0, 0, 0, 0], which was our input.

```
array([ 0.00000000e+00+0.00000000e+00j,  1.00000000e+00+5.45603965e-17j,
        0.00000000e+00+0.00000000e+00j,  0.00000000e+00-1.53080850e-17j,
        0.00000000e+00+0.00000000e+00j, -7.85046229e-17-2.39442265e-17j,
        0.00000000e+00+0.00000000e+00j,  0.00000000e+00-1.53080850e-17j])
```

Going forward, we use more examples from Rigetti Forest, especially when we look at variational algorithms such as Max-Cut in future chapters.

Measurement and Mixed States

Measurement was addressed in Chapter 1. This section looks at it from a hands-on programming perspective. Measurement is a central concept in quantum mechanics. One way to think about it is as a sample from a probability distribution. As a random variable with many outcomes, each outcome is produced with a certain probability. A system remains in its superposed state unless it is measured—since generally, acts of measurements destroy the quantum state.

Measurement is the core connection between quantum systems and our existing classical worlds. The quantum state in nature cannot be directly observed—only statistics about it can be recorded after performing measurements on their states. There is a subtle reality beyond the apparent notion of a boundary between quantum and classical worlds: a quantum system interacts with its surrounding environment unless it is completely and perfectly isolated from it. This interaction causes the existence of mixed states. The mixed states, within limits, recover classical probabilities to give an interpretation of the quantum states. Quantum measurement is done by having a closed quantum system interact in a controlled way with an external system from which the quantum system's state under measurement can be recovered.

In quantum mechanics, the rudimentary basics of measurements of states include *bra* and *ket*. As a refresher, a *bra* is denoted by $\langle\psi|$ for some quantum state $|\psi\rangle$. Together they form the *bra-ket* or the Dirac notation. A *bra* is the conjugate transpose of a *ket*, and the other way around. This also means that a bra is a row vector. The code for this in Qiskit is shown in Listing 5-4a for `MeasMixedSt_qiskit.ipynb` available in this book's code repository.

Listing 5-4a. Bra and Ket in Qiskit `MeasMixedSt_qiskit.ipynb`

```
import numpy as np
zero_ket = np.array([[1], [0]])
print("|0> ket:\n", zero_ket)
print("<0| bra:\n", zero_ket.T.conj())
```

This gives the following output.

```
|0> ket:
  [[1]
   [0]]
<0| bra:
  [[1 0]]
```

We could also try out a ket and a bra, which is a matrix; essentially, the outer product of the two vectors: $|0\rangle\langle0|$, as shown in code Listing 5-4b.

Listing 5-4b. Bra and Ket in qiskit `MeasMixedSt_qiskit.ipynb`

```
zero_ket.dot(zero_ket.T.conj())
```

That, in turn gives a matrix array as an output.

```
array([[1, 0],
       [0, 0]])
```

This output looks familiar because it is a projection of the first element of the canonical basis (see Chapter 1). In general, that $|\psi\rangle\langle\psi|$ is going to be a projector to $|\psi\rangle$. If we take some other quantum state $|\phi\rangle$ and apply the matrix $|\psi\rangle\langle\psi|$ on it, we get $|\psi\rangle\langle\psi|\phi\rangle$. The two rightmost terms are a bra and a ket, which indicates that it is a dot product and is an overlap between $|\phi\rangle$ and $|\psi\rangle$. Since this is a scalar, it scales the leftmost term, which is the ket $|\psi\rangle$, so in effect, $|\phi\rangle$ got projected on this vector.

A measurement in quantum mechanics generally produces an operator-driven random variable. The theory of measurements is rich and an active area of research which means there are still many unknowns and questions yet to be answered. However, most quantum computers that we have today only implement one very specific measurement, which makes things a lot simpler. This measurement is in a canonical basis. In other words, the measurement contains two projections, $|0\rangle\langle0|$ and $|1\rangle\langle1|$, and this measurement can be applied to any of the qubits of the quantum computer.

Chapter 1 explained how applying a projection on a vector works. To get a scalar value of that, we need to add a bra to the left. For instance, for some state $|\psi\rangle$, we get a scalar for $\langle\psi|0\rangle\langle0|\psi\rangle$. This is called the *expectation value* of the operator $|0\rangle\langle0|$.

To put the theory in context, the projection $|0\rangle\langle0|$ can be made on the superposition state $\frac{1}{\sqrt{2}}\big(|0\rangle-|1\rangle\big)$, the Hadamard transform, which can be represented as the column vector $\frac{1}{\sqrt{2}}\begin{bmatrix}1\\1\end{bmatrix}$. This is shown in Listing 5-4c.

Listing 5-4c. Projection `MeasMixedSt_qiskit.ipynb`

```
ψ = np.array([[1], [1]])/np.sqrt(2)
Π_0 = zero_ket.dot(zero_ket.T.conj())
ψ.T.conj().dot(Π_0.dot(ψ))
```

which gives the following output

$$array([[0.5]])$$

That is exactly one half, the *square* of the *absolute value* of the probability amplitude corresponding to state $|0\rangle$ in the superposition. The following is the mathematical formalism of a claim made in Chapter 1: given a state $|\psi\rangle = \alpha_0|0\rangle + \alpha_1|1\rangle$, we get an output i with probability $|\alpha_i|^2$, $i = 0, 1$. This is a manifestation of the *Born rule*. The Born rule provides a link between the mathematical formalism of quantum theory and experiment. As such, is one of the most important theories related to almost all predictions of quantum physics. In the history of science, on a par with the Heisenberg uncertainty relations, the Born rule is often seen as a turning point where indeterminism entered fundamental physics. This is exactly what is implemented in the quantum simulator. It is the measurement in the simulator that is described here.

For an equal superposition with the Hadamard gate (using the Jupyter notebook for quantum circuits in Chapter 4), we need to apply the measurement and observe the statistics, as shown in Listing 5-4d.

Listing 5-4d. Measurment Statistics `MeasMixedSt_qiskit.ipynb`

```
from qiskit import QuantumCircuit, ClassicalRegister, QuantumRegister
from qiskit import execute
from qiskit import BasicAer
from qiskit.tools.visualization import plot_histogram
backend = BasicAer.get_backend('qasm_simulator')
q = QuantumRegister(1)
c = ClassicalRegister(1)
circuit = QuantumCircuit(q, c)
circuit.h(q[0])
circuit.measure(q, c)
job = execute(circuit, backend, shots=100)
plot_histogram(job.result().get_counts(circuit))
```

The output is shown in the histogram in Figure 5-17a. If it is run a few times, it is a *random output*. Figure 5-17b is an example of a second run of Listing 5-4d.

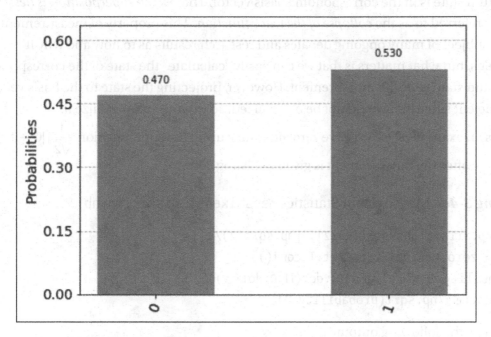

Figure 5-17a. *Output of measurement statistics*

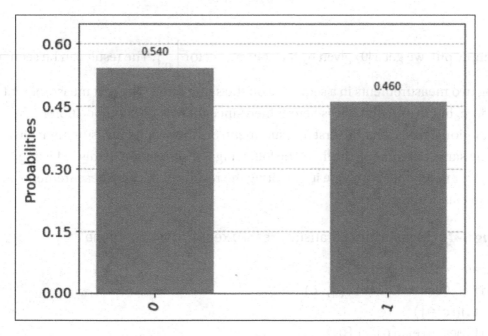

Figure 5-17b. *Random nature of output of measurement statistics*

The outputs depict that roughly half of the outcomes are approximately 0. However, even though the measurement has a random outcome, once it is performed, the quantum state is in the corresponding basis vector. That is, the *superposition is destroyed.* This is referred to as the *collapse of the wave function.* This property of measurement is the subject of many ongoing debates and research results as to how and why it happens, but what matters is that we can easily "calculate" the state of the corresponding quantum state after the measurement. However, projecting the state to the basis vector is insufficient since that would not be normalized, so we must renormalize it.

As an example, if we observe zero after measuring the superposition $\frac{1}{\sqrt{2}}\big(|0\rangle + |1\rangle\big)$, the state after the measurement is given in Listing 5-4e.

Listing 5-4e. Measurment Statistics `MeasMixedSt_qiskit.ipynb`

```
ψ = np.array([[np.sqrt(2)/2], [np.sqrt(2)/2]])
Π_0 = zero_ket.dot(zero_ket.T.conj())
probability_0 = ψ.T.conj().dot(Π_0.dot(ψ))
Π_0.dot(ψ)/np.sqrt(probability_0)
```

It has the following output.

```
array([[1.],
       [0.]])
```

Simply put, we get a $|0\rangle$ given by the column vector $\begin{bmatrix} 1 \\ 0 \end{bmatrix}$. The result can be seen by putting two measurements in a sequence on the same qubit. The first measurement is random, but the second one is determined since there is no superposition in the computational basis after the first measurement. So, the second measurement always gives the same outcome as the first. The following is a simulation of this in Listing 5-4f of our Jupyter notebook. We create it by writing the results of the two measurements in two different classical registers.

Listing 5-4f. Measurment Statistics `MeasMixedSt_qiskit.ipynb`

```
c = ClassicalRegister(2)
circuit = QuantumCircuit(q, c)
circuit.h(q[0])
circuit.measure(q[0], c[0])
```

```
circuit.measure(q[0], c[1])
job = execute(circuit, backend, shots=100)
job.result().get_counts(circuit)
```

The following shows the output.

$$\{'11': 57, '00': 43\}$$

As you can see, there is no output such as 01 or 10.

Now that you have seen measurement behavior in single qubits, it is helpful to measure multi-qubit systems. Most quantum computers implement local measurements, which means that each qubit is measured separately. So, if we have a two-qubit system where the first qubit is in the equal superposition and the second one is in $|0\rangle$, that is, we have the state $\frac{1}{\sqrt{2}}\big(|00\rangle + |01\rangle\big)$, we observe "0 and 0" or "0 and 1" as outcomes of the measurements on the two qubits.

Listing 5-4g. Measurment of Multi-Qubit Systems `MeasMixedSt_qiskit.ipynb`

```
q = QuantumRegister(2)
c = ClassicalRegister(2)
circuit = QuantumCircuit(q, c)
circuit.h(q[0])
circuit.measure(q, c)
job = execute(circuit, backend, shots=100)
plot_histogram(job.result().get_counts(circuit))
```

This gives the histogram as statistical output, as shown in Figure 5-18.

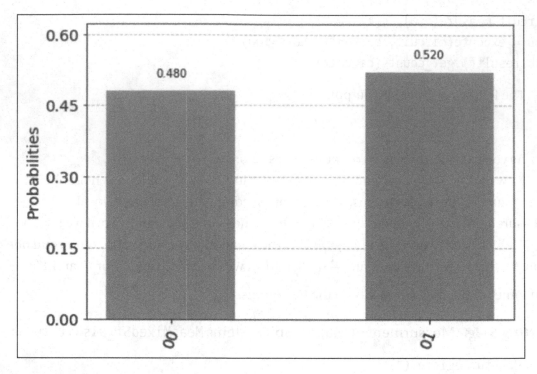

Figure 5-18. *Measurement statistics of multi-qubit systems*

Next, let's make measurements on an entangled state and obtain the statistics after 100 such measurements.

Listing 5-4h. Measurement of Entangled State MeasMixedSt_qiskit.ipynb

```
q = QuantumRegister(2)
c = ClassicalRegister(2)
circuit = QuantumCircuit(q, c)
circuit.h(q[0])
circuit.cx(q[0], q[1])
circuit.measure(q, c)
job = execute(circuit, backend, shots=100)
plot_histogram(job.result().get_counts(circuit))
```

Listing 5-4h gives the histogram shown in Figure 5-19 as statistical output.

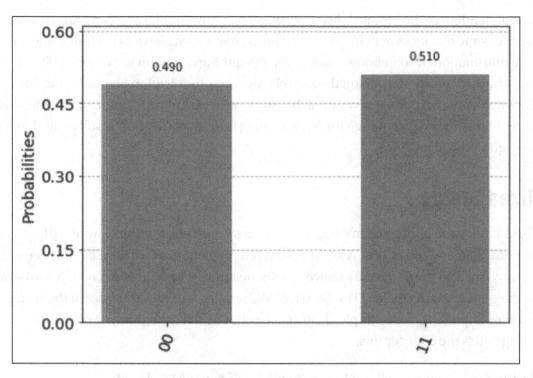

Figure 5-19. *Measurement statistics of entangled systems*

As output, we only observe 00 and 11. Since the state is $\frac{1}{\sqrt{2}}\big(|00\rangle + |01\rangle\big)$, this should not be surprising. Yet, there is something remarkable going on here. At the end of the last code section, there were the same statistics but *measurements on the same qubit*. We have *two spatially separate qubits* exhibiting the same behavior, which is a *very strong form of correlation*. This means that if we measure one qubit and get, say, 0 as the outcome, we know with certainty that if we measured the other qubit, we would also get a 0, even though the *second measurement* is *also a random variable.*

To appreciate this better, imagine that we are tossing two unbiased coins. If we observe heads on one, there is absolutely nothing that we can say about what the other one might be other than a prediction that holds with a probability of 0.5. If we use biased coins, then we might improve our prediction accuracy. However, still, we can never say with certainty what the other coin is based on the outcome we observed on one coin.

There is no activation or instantaneous "faster than the speed of light" signaling involved between the entanglement qubits even though entanglement phenomena are independent of the physical distance between the two qubits and can happen even across cosmic distances of light-years. The *measurement was local to the qubit,* and so is the information. If somebody else is doing the measurement on the other qubit, we would have to inform the person through classical communication channels that we happen to know what the outcome will be. So, while we certainly cannot violate the theory of relativity with entanglement, this strong form of correlation is still central to many quantum algorithms.

Mixed States

Mixed states were addressed in Chapter 1. A *density matrix* is another way of writing a quantum state instead of kets. A ket and a bra is a projection, as we have already seen. But they can also form a density matrix. So, for instance, we could write $\rho=|\psi\rangle\langle\psi|$, where ρ is the density matrix for $|\psi\rangle$. The Born rule still applies, but now the *trace* of the result becomes important. For example, $Tr[|0\rangle\langle0|\rho|]$ would be the probability of seeing a 0. Listing 5-4i is the code for this.

Listing 5-4i. Trace in Mixed States `MeasMixedSt_qiskit.ipynb`

```
ψ = np.array([[1], [1]])/np.sqrt(2)
ρ = ψ.dot(ψ.T.conj())
Π_0 = zero_ket.dot(zero_ket.T.conj())
np.trace(Π_0.dot(ρ))
```

Which gives the output

$$0.4999999999999999$$

In other words, get one half again. The renormalization after a measurement happens similarly, as follows.

$$\frac{|0\rangle\langle0|\rho|0\rangle\langle0|}{Tr\big[|0\rangle\langle0|\rho|\big]}$$

This is shown in Listing 5-4j.

Listing 5-4j. Renormalization MeasMixedSt_qiskit.ipynb

```
probability_0 = np.trace(Π_0.dot(ρ))
Π_0.dot(ρ).dot(Π_0)/probability_0
```

With the output

```
array([[1., 0.],
       [0., 0.]])
```

Even though every state has been created as a ket and a bra, there are other states called mixed states: classical probability distributions over pure states. Formally, a mixed state is written as $\sum \rho_i |\psi_i\rangle\langle\psi_i|$, where $\sum \rho_i = 1$, $\rho_i \geq 0$. We can, as an example, compare the density matrix of the equal superposition $\frac{1}{\sqrt{2}}(|0\rangle + |1\rangle)$ and the mixed state $0.5(|0\rangle\langle 0| + |1\rangle\langle 1|)$, as shown in Listing 5-4k.

Listing 5-4k. Mixed State MeasMixedSt_qiskit.ipynb

```
zero_ket = np.array([[1], [0]])
one_ket = np.array([[0], [1]])
ψ = (zero_ket + one_ket)/np.sqrt(2)
print("Density matrix of the equal superposition")
print(ψ.dot(ψ.T.conj()))
print("Density matrix of the equally mixed state of |0><0| and |1><1|")
print((zero_ket.dot(zero_ket.T.conj())+one_ket.dot(one_ket.T.conj()))/2)
```

The output is as follows.

```
Density matrix of the equal superposition
[[0.5 0.5]
 [0.5 0.5]]
Density matrix of the equally mixed state of |0><0| and |1><1|
[[0.5 0. ]
 [0.  0.5]]
```

The *off-diagonal* elements called *coherence* have disappeared in the second case. The presence of coherence indicates that the state is quantum. The smaller these values, the closer the quantum state is to a classical probability distribution.

The second density matrix has *only diagonal* elements, and they are equal; this is the equivalent way of writing a *uniform distribution*. The uniform distribution has *maximum entropy*, and hence, a density matrix with this structure is called a *maximally mixed state*. In other words, we are perfectly ignorant of which elements of the canonical basis constitute the state.

Ideally, a quantum state would be perfectly isolated from the environment. However, in reality, quantum computers cannot yet achieve a high degree of isolation. Hence, coherences are slowly lost to the environment. This process is called *decoherence*. The speed at which decoherence happens determines the *length of the quantum algorithms* we can run on the quantum computer: if it happens fast, we have time to apply a handful of gates or do any other form of calculation and quickly measure the results.

Density matrices are more expressive than state vectors, as state vectors can only represent pure states. But, even a system in a mixed state, it can be seen as part of a larger system in a pure state. This process of *converting a mixed state* into a pure state of an enlarged system is called *purification*. A mixed state of an n-qubit system can be purified by adding n more qubits and working with the $2n$-qubit system. Once purified, the joint system of $2n$ qubits is in a pure state while the first n qubits are still in the original mixed state.

Open and Closed Quantum Systems

Let us start by tabulating the closed quantum systems insights that you have acquired.

- Closed systems are fundamental, and any system under consideration is represented as a closed system.

- In standard quantum mechanics, one considers closed systems (i.e., systems that are perfectly isolated from their environment). The resulting dynamics are unitary.

- This is an idealization as physical systems cannot be perfectly isolated from their environment (apart, perhaps, from the whole universe).

In a closed quantum system, the dynamics of the system are completely independent of those of the environment. Hence, the observables associated with this system at its creation, for example, its total energy, remain conserved for the duration of the system's evolution. Such systems, though convenient for gaining insight and understanding, are far from being realistic. Figure 5-20 shows the diagrammatic representation of a closed quantum system.

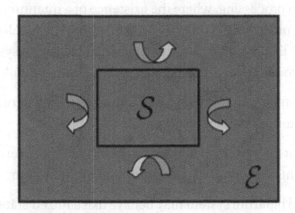

Figure 5-20. *Closed quantum system*

However, since every quantum system interacts with its environment, the idealization of an isolated, closed quantum system obeying perfectly unitary quantum dynamics is an idealization. In reality, every system is open, meaning that it is coupled to an external environment. Figure 5-21 shows an open quantum system.

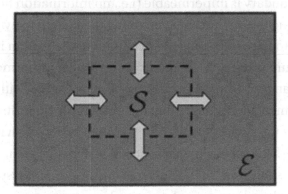

Figure 5-21. *Open quantum system*

The theory of *open quantum systems* is the base for nearly all modern research in quantum mechanics and its applications. At times, the effects of these open systems are small, but they can rarely be neglected. This is particularly relevant in the field of quantum information processing, where the existence of a quantum advantage over classical information processing is often derived first from the idealized, closed system perspective and must then be re-scrutinized in the realistic, open system setting. With these thoughts, we proceed to investigate open systems with the following views in mind.

- Open systems are fundamental, and any system under consideration is represented as an open system.

- The fact that the system of interest interacts with the external environment is essential to our description of the system.

A pure state of the bipartite system may behave like a mixed state when we observe subsystem A alone in a mixed environment. What if we want to know the dynamics of A only? Can we describe its evolution even if we don't have full knowledge of subsystem B? If we assume that the state of the bipartite system undergoes unitary evolution, how do we describe A alone?

The difference between a closed and an open system is in the assumed characteristics of the boundary separating the system from the environment. For the closed system, the boundary is impermeable (i.e., no information flow is allowed between the system and the environment enabling the system to evolve in isolation). Hence, the full Hamiltonian describing the dynamics of the system itself.

In an open quantum system, the boundary is no longer impermeable, and thus, the system is free to exchange information, be that phase or energy, with its environment. Hence, there is no framework to ensure that the observable quantities remain conserved throughout the system's evolution. An open system provides a much better approximation of a realistic scenario than a closed system. As such, anything contained in a universe (i.e., any subsystem of the universe must be an open system). Any system which is considered a closed system is done so only as an approximation.

Quantum states evolve, and their evolution is described by unitary matrices. This leads to some interesting properties in quantum computing. Unitary evolution is true for a closed system; that is, a quantum system perfectly isolated from the environment. This is not the case in the quantum computers we have today, which are open quantum systems that evolve differently due to uncontrolled interactions with the environment. We go through some Jupyter notebook examples in this section for an insight into this process. An in-depth theoretical discussion of open quantum systems and associated dynamics is beyond the scope of this book. I recommend the book by Rivas and Huelga [62] and the MIT open courseware [63].

In this notebook, we take a glimpse at both types of evolution. An operator is considered unitary if it produces an identity matrix as output when multiplied by its own adjoint. A unitary operator has its own inverse equal to its adjoint. The inverse of an operator U is denoted by U^{-1}. This satisfies the following relationship.

$$UU^{-1} = I$$

where I is the identity operator.

For U to be unitary,

$$UU^\dagger = U^\dagger U = I$$

and

$$U^{-1} = U^\dagger$$

The NOT operation is performed by the X gate in a quantum computer. While the generic discussion on gates only occurs in a subsequent notebook, we can study the properties of the X gate. Its matrix representation is $X = \begin{pmatrix} 0 & 1 \\ 1 & 0 \end{pmatrix}$. Listing 5-5a for the Jupyter notebook code `ClosedOpenSystems.ipynb` verifies if it is indeed unitary.

Listing 5-5a. Evolution of Unitary Gates `ClosedOpenSystems.ipynb`

```
import numpy as np
X = np.array([[0, 1], [1, 0]])
print("XX^dagger")
print(X.dot(X.T.conj()))
print("X^daggerX")
print(X.T.conj().dot(X))
```

Listing 5-5a gives the following as output.

```
XX^dagger
[[1 0]
 [0 1]]
X^daggerX
[[1 0]
 [0 1]]
```

Hence, X appears to be a legitimate unitary operation. The unitary nature ensures that the l_2 norm is preserved; that is, quantum states are mapped to quantum states. Listing 5-5b shows the code to find the norm evolution.

Listing 5-5b. Evolution of Norm `ClosedOpenSystems.ipynb`

```python
print("The norm of the state |0> before X is applied")
zero_ket = np.array([[1], [0]])
print(np.linalg.norm(zero_ket))
print("The norm of the state after X is applied")
print(np.linalg.norm(X.dot(zero_ket)))
```

The following output confirms that the norm is preserved.

```
The norm of the state |0> before applying X
1.0
The norm of the state after applying X
1.0
```

Since the unitary operation is a matrix, it is *linear*. Measurements are also represented by matrices. The two observations imply that everything a quantum computer implements are linear. If some form of *nonlinearity* is desired, then that must involve some classical intervention.

Another consequence of unitary operations is reversibility (i.e., any unitary operation can be reversed). Hence, quantum computing libraries often provide a function to reverse entire circuits. The X gate needs to be applied again to be reversed as it is a conjugate transpose of itself, which means $X^2 = 1$.

Listing 5-5c. Evolution of norm ClosedOpenSystems.ipynb

```python
import numpy as np
from qiskit import QuantumCircuit, ClassicalRegister, QuantumRegister
from qiskit import execute
from qiskit import BasicAer
from qiskit.tools.visualization import circuit_drawer
np.set_printoptions(precision=3, suppress=True)
backend_statevector = BasicAer.get_backend('statevector_simulator')
q = QuantumRegister(1)
c = ClassicalRegister(1)
circuit = QuantumCircuit(q, c)
circuit.x(q[0])
circuit.x(q[0])
job = execute(circuit, backend_statevector)
print(job.result().get_statevector(circuit))
```

The output is

$$[1.+0.j \quad 0.+0.j]$$

This is exactly 0 as expected.

As we have stressed before, actual quantum systems are very rarely, if ever, closed. They constantly interact with their environment in a way that is very difficult, sometimes impossible to control as technology stands today, which, in turn, causes them to lose coherence. This is true for current and near-term quantum computers too. This means that their actual time evolution is not described by a unitary matrix but by some other operator known as a *completely positive trace-preserving map*.

Quantum computing libraries often offer a variety of noise models that simulate and mimic different types of interaction. Increasing the strength of the interaction with the environment leads to *faster* decoherence. The timescale for decoherence is often called the T_2 time. Among a couple of other parameters, T_2 time is critically important for the number of gates or the duration of the quantum computation we can perform.

An easy way of studying the effects of decoherence is mixing a pure state with the maximally mixed state $\frac{1}{2^d}$, where d is the number of qubits. This way, we do not have to specify noise models or any other map modeling decoherence. For instance, we can mix the $|\phi+\rangle$ state with the maximally mixed state.

271

Listing 5-5d. Maximally Mixed State `ClosedOpenSystems.ipynb`

```python
def mixed_state(pure_state, visibility):
    density_matrix = pure_state.dot(pure_state.T.conj())
    maximally_mixed_state = np.eye(4)/2**2
    return visibility*density_matrix + (1-visibility)*maximally_mixed_state

φ = np.array([[1],[0],[0],[1]])/np.sqrt(2)
print("Maximum visibility is a pure state:")
print(mixed_state(φ, 1.0))
print("The state is still entangled with visibility 0.8:")
print(mixed_state(φ, 0.8))
print("Entanglement is lost by 0.6:")
print(mixed_state(φ, 0.6))
print("Hardly any coherence remains by 0.2:")
print(mixed_state(φ, 0.2))
```

Listing 5-5d gives the following output.

```
Maximum visibility is a pure state:
[[0.5 0.  0.  0.5]
 [0.  0.  0.  0. ]
 [0.  0.  0.  0. ]
 [0.5 0.  0.  0.5]]
The state is still entangled with visibility 0.8:
[[0.45 0.   0.   0.4 ]
 [0.   0.05 0.   0.  ]
 [0.   0.   0.05 0.  ]
 [0.4  0.   0.   0.45]]
Entanglement is lost by 0.6:
[[0.4 0.  0.  0.3]
 [0.  0.1 0.  0. ]
 [0.  0.  0.1 0. ]
 [0.3 0.  0.  0.4]]
Barely any coherence remains by 0.2:
[[0.3 0.  0.  0.1]
 [0.  0.2 0.  0. ]
 [0.  0.  0.2 0. ]
 [0.1 0.  0.  0.3]]
```

One way to look at a quantum state in an open system is through its equilibrium processes. For example, a steaming cup of tea left alone loses heat due to interacting with the environment, eventually reaching the environment's temperature. This process includes energy exchange. A quantum state exhibits the same behavior.

The equilibrium state is called the *thermal state,* which has a very specific structure, and we revisit it in the future. The *energy* of the statistical samples taken from a thermal state follows a Boltzmann distribution [64] given by

$$P(\epsilon_i) = \frac{1}{Q} e^{-\epsilon_i/kT} = \frac{e^{-\epsilon_i/kT}}{\sum_{j=1}^{M} e^{-\epsilon_i/kT}}$$

where $P(\epsilon_i)$ is the probability of state i. ϵ_i is the energy of state i. k is the Boltzmann constant. T is the temperature of the system. M is the number of all states accessible to the system of interest. Q is the normalization denominator. The higher the temperature, the closer we are to the uniform distribution. In the infinite temperature limit, it recovers the uniform distribution. At high temperatures, all energy levels have an equal probability. In contrast, at zero temperature, the entire probability mass is concentrated on the lowest energy level— the ground state energy. To get a sense of this process, Listing 5-5e plots the Boltzmann distribution with different temperatures.

Listing 5-5e. Boltzmann Distribution `ClosedOpenSystems.ipynb`

```
import matplotlib.pyplot as plt
temperatures = [.5, 5, 2000]
energies = np.linspace(0, 20, 100)
fig, ax = plt.subplots()
for i, T in enumerate(temperatures):
    probabilities = np.exp(-energies/T)
    Z = probabilities.sum()
    probabilities /= Z
    ax.plot(energies, probabilities, linewidth=3, label = "$T_" +
    str(i+1)+"$")
ax.set_xlim(0, 20)
ax.set_ylim(0, 1.2*probabilities.max())
ax.set_xticks([])
ax.set_yticks([])
```

```
ax.set_xlabel('Energy')
ax.set_ylabel('Probability')
ax.legend()
```

This gives the plot in Figure 5-22 as output.

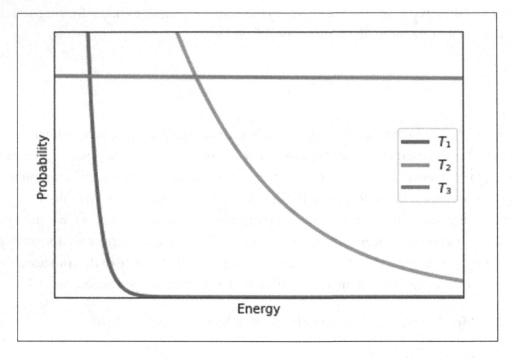

Figure 5-22. *Boltzmann distribution of energy*

The plot shows that T1 < T2 < T3. T1 is a low temperature and expectedly peaks sharply at low energy levels. In contrast, T3 is a very high temperature, and the probability distribution is almost completely flat.

These exercises aim to increase your understanding of how the influences of measurement from an external observer trigger new types of many-body phenomena with no analogs in closed systems. As a recap, the time evolution of an isolated quantum system is described by a single Hermitian operator (i.e., the Hamiltonian). In contrast, under continuous observation, the dynamics become intrinsically non-unitary due to the measurement backaction and are characterized not only by the Hamiltonian but also by a measurement process.

Quantum Principal Component Analysis

Machine learning and analysis exercises usually have many features, some of which are correlated. For example, consider housing prices, which are a function of many features of a house, such as the number of bedrooms and bathrooms, the square footage, the lot size, the date of construction, and the location. Often, you want to limit the number of features to the most important (i.e., features that capture the largest variance in the data). For example, if someone is only considering houses on a particular street, then the location may not be important but the square footage may capture a large variance.

Determining which features capture the largest variance is known as *principal component analysis* (PCA) [65]. Mathematically, PCA involves taking the raw data (e.g., the feature vectors for various houses) and computing the covariance matrix.

The quantum algorithm for performing qPCA (quantum PCA) presented in reference [66] uses the density matrix representation. The algorithm discussed there have four main steps.

1. Encode covariance matrix Σ in a quantum density matrix ρ.

2. Prepare many copies of ρ.

3. Perform exponential *SWAP* operation on each copy and a target system.

4. Perform quantum phase estimation to determine the eigenvalues.

To implement this qPCA algorithm on a noisy simulator, refer to reference [67]. The authors have also made the code in Cirq available on GitHub at `https://github.com/rmlarose/vqsd`. Figure 5-23 shows the steps involved in implementing a quantum algorithm for qPCA on a noisy quantum simulator. The authors call the algorithm *variational quantum state diagonalization* or VQSD.

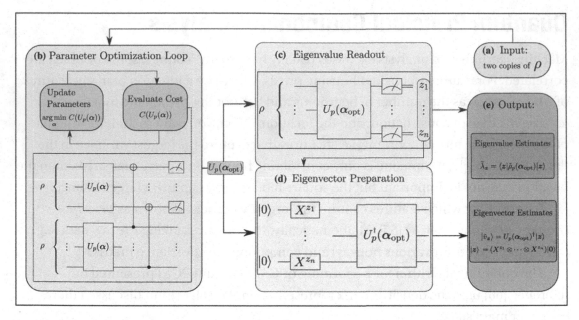

Figure 5-23. *qPCA/VQSD algorithm implementation (Source [67])*

The following explains the steps in the VQSD algorithm, as shown in Figure 5-23.

(a) Inputs are two copies of quantum state ρ. These states are sent to the parameter optimization loop, (b) where a hybrid quantum-classical variational algorithm (to be explained in later chapters) approximates the diagonalizing unitary $U_p\alpha_{opt}$. Here, p is a hyperparameter that dictates the quality of the solution. This optimal unitary is sent to the eigenvalue readout circuit, (c) to obtain bit-strings z, the frequencies of which provide estimates of the eigenvalues of ρ. Along with the optimal unitary $U_p\alpha_{opt}$, these bit-strings are sent to the eigenvector preparation circuit, (d) to prepare the eigenstates of ρ on a quantum computer. Both the eigenvalues and eigenvectors are the outputs (e) of the VQSD algorithm.

Summary

This chapter ventured into the first part of quantum algorithms, building up from where we finished in the last chapter. It looked at some important algorithms and methods such as the Deutsch-Jozsa algorithm, quantum computing complexity, techniques of encoding (basis, tensor product, amplitude, and Hamiltonian), and the process of classical data input into quantum algorithms. We also ventured into measurements and mixed states, quantum Fourier transform, and one of its important applications, namely phase estimation. We had our first look at Rigetti's Forest SDK and QVM and utilized Qiskit and Cirq in other exercises. The next chapter continues into advanced quantum algorithms and associated code.

CHAPTER 6

QML Algorithms II

Let your life lightly dance on the edges of Time like dew on the tip of a leaf.

—Rabindranath Tagore

Quantum machine learning provides unprecedented scope in computing the techniques done in classical machine learning on a quantum computer. The entanglement and superposition of the basic qubit states promise to provide an edge over the performance and scope of classical machine learning.

By now, you should have some insights into how classical-quantum systems work across their interfaces. Currently, there are two promising ways to utilize quantum computers for machine learning tasks.

- The first way constructs a "quantum version" of the classical machine learning models so that the quantum algorithms produce results analogous to the said classical model.

- The second way is to create quantum machine learning models specific to the physical building blocks and fundamental principles of a quantum computing platform. In this case, the model performance is deemed to be of higher importance.

Note This chapter addresses several algorithms in addition to the ones already covered that are important for quantum machine learning: more aspects of measurement such as Schmidt decompositions, SWAP Test, linear models, kernel methods, an overview of k-means clustering, and k-medians algorithms. It also shows kernel methods in both Qiskit and Rigetti's Forest for the Iris dataset that we encountered in Chapters 2 and 3 during our tour of classical machine learning.

© Santanu Ganguly 2021
S. Ganguly, *Quantum Machine Learning: An Applied Approach*, https://doi.org/10.1007/978-1-4842-7098-1_6

This chapter looks at several other algorithms related to some very important properties of quantum computing and quantum machine learning: more measurement aspects, more algorithms, and tricks for data encoding in linear and nonlinear space and associated predictions leading us on a natural path encompassing quantum algorithmic *inference*. The process of computing a prediction for a model is known as *inference*. You have already seen some quantum algorithms. More are explored in this chapter as a deeper inroads toward quantum inference to discover how to compute a prediction if given an input using a quantum system.

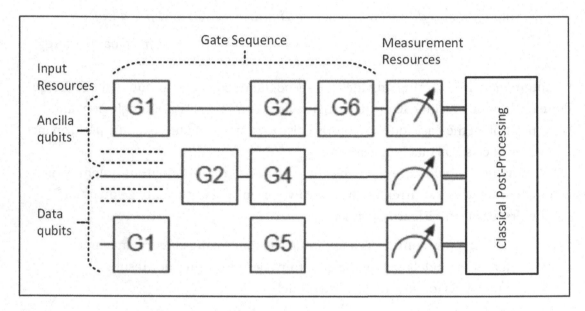

Figure 6-1. *Quantum algorithm development process*

Figure 6-1 shows an approach to developing and optimizing quantum algorithms. An algorithm is optimized for a given set of resources, including ancilla and data qubits and measurement resources. The algorithm is determined by the quantum gate sequence and classical post-processing of the measurement results. To find the algorithm that performs $x \rightarrow f(x)$, a cost function (Chapter 5) that quantifies the difference between the *desired* output and the *actual* output is minimized for a set of training data inputs. In cases where the training data is sufficiently general, the algorithm minimizing the cost should be a general algorithm that computes $f(x)$ for any x.

This chapter looks at these principles and performing measurements, which are greatly important in all quantum computing systems. Toward this path, the most important concept is that of *Schmidt decomposition*.

Schmidt Decomposition

We have looked at entanglement and associated properties, including teleportation. We have also seen the properties of systems comprising of pure and mixed states. However, entanglement does not always refer to pure or mixed states, and often it is a challenge to practically detect the presence of entanglement in a system. From the perspective of detection of entanglement in bipartite composite systems and related measurement of the same, Schmidt decomposition is of great importance.

Unlike in classical physics, particles can become coupled or entangled, making the *composite system* not only unequal but *greater than* the *sum of the components*. For a composite system, equation 4.24 gave the reduced density matrices by the inequality, as follows.

$$S(\rho_{12}) \le S(\rho_1) + S(\rho_2)$$

Composite quantum systems are made up of *two or more distinct physical systems*. The state-space of such a composite system is given by the tensor product of the states of the component physical systems.

Chapter 4 discussed the Shannon entropy and the von Neumann entropy, which coincide only for an ensemble formed from mutually orthogonal pure quantum states. Hence, if an encoded message is sent in a set of orthogonal qubit states where each of those is pure, it can be described by an overall tensor-product state. In this case, the transmission would be equivalent to sending the same information as a set of classical bits because each qubit is perfectly distinguishable once the encoding basis has been determined.

A set of states from a known basis can be cloned with perfect fidelity, in contrast to the general realistic case, which is imperfect and constrained. If unknown quantum states were perfectly distinguishable, they could be perfectly cloned, and the "no-cloning theorem" would cease to be important.

State decompositions take place with the correlations associated with entanglement. As an example, in a bipartite quantum composite system, the mutual quantum information between two subsystems is described by states φ_1 and φ_2 of a composite system described by a joint state φ_{12} is given by

$$I(A:B) \equiv S(1) + S(2) - S(12)$$

$$= S(\varphi_1) + S(\varphi_2) - S(\varphi_{12}) \qquad 6.1$$

The state-space of a composite system is the tensor product of the states of the component physical systems. For example, in pure bipartite states, A and B, the total Hilbert space $\mathcal{H}$ (see Chapter 1) of the composite system becomes

$$|\psi\rangle_{AB} = |\psi\rangle_A \otimes |\psi\rangle_B \in \mathcal{H} = \mathcal{H}_A \otimes \mathcal{H}_B \qquad 6.2$$

where, $\otimes$ signifies the tensor product. $|\psi\rangle_{AB}$ is the state vector for the composite system. If $\mathcal{H}_A$ has dimension m and $\mathcal{H}_B$ has dimension n, this general vector is a superposition of mn basis vectors.

The definition for the composite state vector $|\psi\rangle_{AB}$ indicates that there is at least one orthonormal basis for $\mathcal{H}$ given by $e_i \otimes v_i$, where e_i and v_i are the orthonormal bases for $\mathcal{H}_A$ and $\mathcal{H}_B$ respectively, when the bipartite state can be represented as

$$|\psi\rangle_{AB} = \sum_i^r a_i |e_i\rangle \otimes |v_i\rangle = \sum_i^r a_i |e_i\rangle |v_i\rangle \qquad 6.3$$

where, r is the *Schmidt rank,* and a_i is the *Schmidt basis.* In this representation of Schmidt decomposition, the summation index runs up only to the smaller of the corresponding two Hilbert space dimensions. It is often convenient to take the amplitudes a_i, also called *Schmidt coefficients,* to be real numbers by absorbing any phases into the definitions of $\{e_i\}$ and $\{v_i\}$. As such, $a_i \geq 0$ and $\sum a_i^2 = 1$. Schmidt decomposition in multipartite systems is limited; it is available with certainty only in bipartite states.

As an example of the Schmidt decomposition process, we can consider a vector in a 4- dimensional Hilbert space $\mathcal{H}_A \otimes \mathcal{H}_B$ where each of $\mathcal{H}_A$ and $\mathcal{H}_B$ has dimension 2. This can be given by

$$|\psi\rangle_{AB} = |11\rangle$$

For our example, this state is already written in terms of Schmidt bases as for each of $\mathcal{H}_A$ and $\mathcal{H}_B$, the computational bases are Schmidt bases. Hence, we have the following for the Schmid coefficients.

$$a_0 = 0 \text{ and } a_1 = 1$$

The following are for the computational bases.

$$\{|\phi_{A0}\rangle = |0\rangle, \qquad |\phi_{A1}\rangle = |1\rangle$$

$$\{|\phi_{B0}\rangle = |0\rangle, \qquad |\phi_{B1}\rangle = |1\rangle$$

The bases $|e_i\rangle$ and $|v_i\rangle$ be represented in terms of two unitary matrices $U, V \in U(2^n)$ and a_i's can be represented in terms of a single n-qubit state. If the n-qubit state is given by $Q |00...0\rangle \in U(2^n)$, then, the product in the Schmidt decomposition may be given by a quantum circuit combining U, V and Q with n numbers of CNOT gates, as shown in Figure 6-2 for $n = 6$.

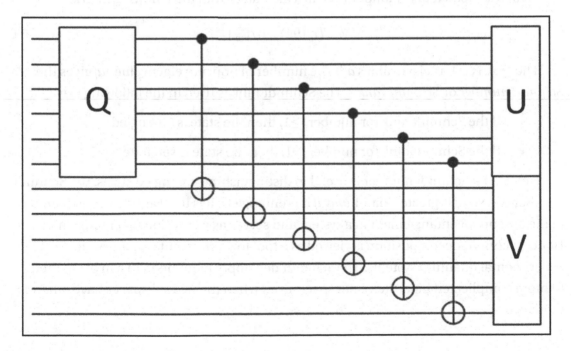

Figure 6-2. *Schmidt decomposition*

The Schmidt decomposition resembles singular value decomposition, where density matrix representation is useful for describing individual subsystems of the composite quantum system. The density matrix of the composite system is given by the following tensor product.

$$\rho_{AB} = \rho_A \otimes \rho_B \qquad 6.4$$

where the components ρ_A and ρ_B are from the composite vector ρ_{AB}.

A subsystem of a composite quantum system is described by a *reduced density matrix*. The reduced density matrix for a subsystem A is defined in terms of a partial trace as

$$\rho_A = Tr_B(\rho_{AB}) \qquad 6.5$$

where Tr_B is a partial trace operator over system B. The probability amplitudes belonging to system B vanish due to the partial trace because of the density operator (refer to Chapter 1), $Tr(\rho) = 1$ for any density matrix. This procedure is also called *tracing out*. Post the partial trace operation, only the amplitudes belonging to the system A remain.

Schmidt coefficients or amplitudes are calculated from the following matrix.

$$\varphi = Tr_B\left(|\psi\rangle_{AB}\langle\psi|_{AB}\right) \qquad 6.6$$

The matrix φ has eigenvalues a_i^2. The number of nonzero eigenvalues a_i gives the *Schmidt number* or *Schmidt rank r*. The Schmidt rank is used in the following way.

- If the Schmidt rank (or number) ≥ 1, then the state is entangled

- If the Schmidt rank (or number) $=1$, then the state is separate

Hence, the Schmidt number is useful in distinguishing entangled states. the Schmidt number of a state is greater than 1 *only if* it is entangled. It is used as a (coarse) quantifier of the amount of entanglement in a system and serves as a criterion for entanglement. Hence, Schmidt decomposition is viewed as a tool to confirm entanglement in real-life experimental quantum systems. The Schmidt decomposition theorem can be applied to more complicated bipartite vector space, even when the two subspaces have different dimensions.

Historically, the original proposal of entanglement concentration included the "Schmidt projection" [74]. The Schmidt projection method requires at least two non-maximally entangled pairs and becomes efficient for large numbers of pairs. The method is based on a collective measurement of the "Hamming weight" [75] of all qubits at one side, projecting all pairs onto a subspace spanned by states having a common Schmidt coefficient. The measurement result is then classically communicated to the other side. Alternatively, the same collective measurement is performed at the other side, which would yield the same result and make classical communication dispensable. This method also works for dimensions $d > 2$ [74]. In the asymptotic limit, turning the total state vector of n *non-maximally* entangled input pairs into that of m *maximally* entangled output pairs can be described via the majorization criterion for deterministic entanglement transformations [3], giving $m < n$ in an entanglement concentration.

In general, an important sign of entanglement is the violation of inequalities imposed by local realistic theories [76]. Any pure two-party state is entangled if and only if, for the chosen observables, it yields a violation of such inequalities.

Note that the reduced states remain pure if and only if the state vector of the composite state is of the tensor product form $|\psi\rangle_{AB} = |\psi\rangle_A \otimes |\psi\rangle_B$, in other words, when the Schmidt number equals 1. Computational examples of Schmidt decomposition can be found in Appendix A.

Quantum Metrology

Precision measurements are important across all fields of science—more so in quantum mechanics as the scale that needs measuring is challengingly small and requires great sensitivity. For example, optical phase measurements can measure distance, position, displacement, acceleration, and optical path length. The field of quantum information technology needs interfaces to perform high-resolution and highly sensitive measurements of physical parameters. It is very important that these measurement interfaces do not interact with the quantum systems under scrutiny, or any such interaction can be distinguished by the process. These measurement criteria are governed by a quantum theory that describes the physical systems and exploits features, such as quantum entanglement and quantum squeezing. The phenomena of quantum entanglement enable a precision much higher than would otherwise be possible.

Quantum metrology uses quanta or individual packets of energy for setting the standards that define units of measurement. Quantum mechanics sets the ultimate limit on the accuracy of any measurement. Quantum metrology, therefore, uses quantum effects to enhance precision beyond that possible through classical approaches.

Quantum metrology is the field concerned with extracting information from a quantum system. Since this is done primarily using measurements, the states, and observables matter as well. The enhancement due to the use of entangled states is quadratic and connects to our previous discussions of Grover and Grover-like algorithms. A complete discussion of the experimental and theoretical results on quantum metrology is reviewed by Vittorio Giovannetti, Seth Lloyd, and Lorenzo Maccone [89].

A hands-on exercise helps you get better insight into entanglement-based measurements that power quantum metrology . Measurements involving light and its properties often involve the *Mach-Zehnder interferometer* [90]. In physics, the Mach-Zehnder interferometer is a device used to determine the relative phase shift variations between two collimated beams derived by splitting the light from a single source. The interferometer can measure phase shifts between the two beams caused by a sample or a change in the length of one of the paths. The apparatus is named after the physicists Ludwig Mach and Ludwig Zehnder, published in an 1892 article [90].

The Mach–Zehnder interferometer is frequently used in aerodynamics, plasma physics, and heat transfer to measure pressure, density, and temperature changes in gases and is shown in Figure 6-3.

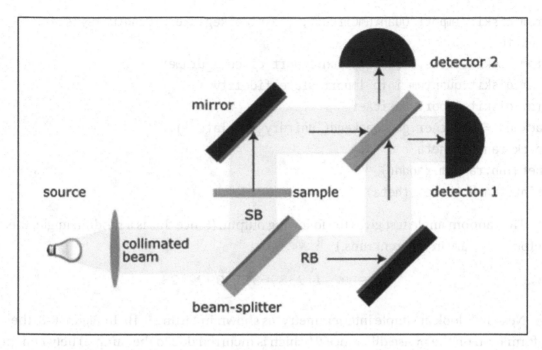

Figure 6-3. *The Mach-Zehnder interferometer (Source [91][1])*

Let's start by importing the required libraries from Qiskit, as shown in Listing 6-1a, for the entanglementM.ipynb Jupyter notebook downloadable from the book's website and test that our library imports have worked by printing a random angle. For this exercise, we use a simulator.

Listing 6-1a. Libraries for entanglementM.ipynb

```
import matplotlib.pyplot as plt
%matplotlib inline
import numpy as np
from math import pi
```

---

[1]Daniel Mader was the original uploader (at en.wikipedia). Created with Inkscape, October 6, 2005, on a SuSE 9.3 box. https://commons.wikimedia.org/w/index.php?curid=2151148.

```
from qiskit import QuantumCircuit, ClassicalRegister, QuantumRegister,
execute
from qiskit.tools.visualization import circuit_drawer
from qiskit.quantum_info import state_fidelity
from qiskit import BasicAer
backend = BasicAer.get_backend('unitary_simulator')
#pick random theta
theta=np.random.randn()
print("rand theta",theta)
```

The random angle test gives the following output. (Since this is a random angle, the output may vary in different runs.)

$$\text{theta } -1.2389631590922523$$

Next, let's look at simple interferometry, as shown in Listing 6-1b. In Figure 6-3, the information on the phase difference θ (which is incurred due to the sample) between the two optical paths of the interferometer can be extracted by monitoring the two output beams by measuring their intensity, which is the photon number. In the absence of any phase difference, all photons exit the apparatus as output at *detector 1*. On the other hand, if $\theta = \pi$ radians, 100% of photons exit at the output *detector 2*. If the value of θ is somewhere between 0 and π (i.e., if $0 < \theta < \pi$), then a fraction $\cos^2 \dfrac{\theta}{2}$ of the photons exit at the output at detector 1 and a fraction $\sin^2 \dfrac{\theta}{2}$ at detector 2.

To compute the circuit, we leverage the rotation operator (Chapter 1, equation 1.26), giving a single-qubit rotation through angle θ (radians) around the Z axis.

$$RZ_\theta \equiv \begin{pmatrix} e^{-i\theta/2} & 0 \\ 0 & e^{i\theta/2} \end{pmatrix}$$

Listing 6-1b. Simple Interferometry entanglementM.ipynb

```
q = QuantumRegister(1)
c = ClassicalRegister(1)
qc = QuantumCircuit(q,c)

qc.h(q[0])
```

```
qc.rz(theta,q[0])
qc.h(q[0])
qc.measure(q,c)
qc.draw()
```

This gives the following circuit as output. Note that the value of `rz` in the output may from run to run vary as θ is initiated at random (`theta=np.random.randn`).

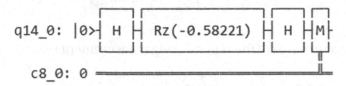

In the next step, we calculate the probability of zero (i.e., $\cos^2\dfrac{\theta}{2}$ and associated errors). This is done by estimating the following.

$$\sum_{x=i}^{N}\frac{x_i}{N}$$
6.7

where $x_i = 0$ *or* 1 depending on the output of the i-th measurement. Here, an x_i is a stochastic[2] variable. For stochastic variables, the *variance* (see Chapter 2) associated with their *average* is the *average of the variances* as defined by the *central limit theorem* [91].

The error given during the measurement of $\cos^2\dfrac{\theta}{2}$ is given by

$$\varepsilon = \Delta\sum_{x=i}^{N}\frac{x_i}{N} \equiv \sqrt{\sum_{x=i}^{N}\frac{\Delta^2 x_i}{N}} = \frac{\Delta x}{\sqrt{N}}$$
6.8

where all the Δx_i spreads are equal to Δx because they are part of the same experiment; hence, Δx_i is the spread of the i-th measurement. The code for this calculation is given in Listing 6-1c. It uses the `qasm` simulator.

---

[2]In probability theory and related fields, a stochastic or random process is a mathematical object usually defined as a family of random variables.

Listing 6-1c. Probability entanglementM.ipynb

```
backend = BasicAer.get_backend('qasm_simulator')
job = execute(qc, backend, shots=1024)
job.result().get_counts(qc)
```

This gives the following output.

$$\{'0': 674, '1': 350\}$$

It shows the occurrences of the states during measurement of $x_i = 0$ *or* 1.

Entanglement Measurement

In the final step, we start with the GHZ state to demonstrate entanglement measurement. A three-qubit GHZ state is the simplest one, which exhibits a non-trivial multipartite entanglement, given by

$$|GHZ\rangle = \frac{|000\rangle + |111\rangle}{\sqrt{2}}$$

A GHZ state is a quantum superposition of all subsystems being in state 0, with all of them being in state 1, where states 0 and 1 of a single subsystem are fully distinguishable. In Listing 6-1d, the initial state is $\frac{|000\rangle + |111\rangle}{\sqrt{2}}$ and the final state is given by

$$|\psi_{Final}\rangle = \frac{e^{\frac{-iN\theta}{2}}|000\rangle + e^{\frac{iN\theta}{2}}|111\rangle}{\sqrt{2}}$$

In our example, we have $N = 3$. Hence, the probability of measurement is now given by

$$p(0) = \cos^2\frac{N\theta}{2}$$

Here, we initiate θ with a value of 1.5 in radians, again for 1024 shots.

Listing 6-1d. Probability entanglementM.ipynb

```
theta = 1.5
q = QuantumRegister(3)
c = ClassicalRegister(1)
qc = QuantumCircuit(q,c)

qc.h(q[0])
qc.cx(q[0],q[1])
qc.cx(q[0],q[2])
qc.rz(theta,q[0])
qc.rz(theta,q[1])
qc.rz(theta,q[2])
qc.cx(q[0],q[2])
qc.cx(q[0],q[1])
qc.h(q[0])
qc.measure(q[0],c)
qc.draw()

backend = BasicAer.get_backend('qasm_simulator')
job = execute(qc, backend, shots=1024)
job.result().get_counts(qc)
```

The output shows the probabilities with their occurrences (again, this may change from run to run due to the nature of random variables).

$$\{'1': 614, '0': 410\}$$

As an exercise, you are encouraged to try the following using this code as a template.

- Compute the variance of the sample for different values of N

- Test with different seed values for theta

Linear Models

In Chapters 2 and 3, you saw classical linear models in the form of perceptrons and linear regression. A linear model [Schuld et al. ref 79] can be described by a function in time $t = \{0, T\}$ (i.e., time 0 to T) which maps N-dimensional inputs given by x_T, where x is given by $(x_1, ..., x_N)^T$ to M-dimensional outputs given by $y_T = (y_1, ..., y_M)^T$. The linear function with a set of weights given by $\mathbf{w}$ can be expressed as

$$f(x; w) = \varphi\left(\mathbf{w}^T . \mathbf{x}\right)$$

$$= b_0 + w_1 x_1 + ... + w_N x_N$$

where $\mathbf{w}^T$ is the matrix representation of the weights multiplied by the input vector $\mathbf{x}$, ϕ is a function dependent on the input and the weights. In this equation, $\mathbf{w}^T$ is a parameter-defined vector that needs to be fitted or learned from the data. As N-dimensional inputs are mapped to M-dimensional outputs, it is possible for the function $f(x, w)$ to be multi-dimensional. A linear regression function (Chapter 2) is also a linear model and, in this case, can have multidimensions.

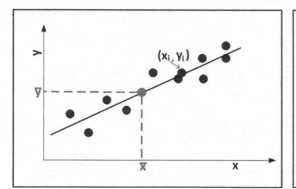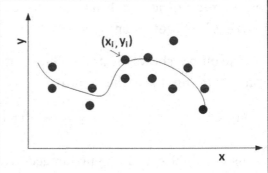

Figure 6-4. *Illustration of linear regression given training data points (x_i, y_i) and the nonlinear map to fit the same data points*

Figure 6-4 shows on the *left* an illustration of linear regression, given training data points (x_i, y_i). The task in machine learning is not to determine the fit itself (in contrast to statistical analysis) but to have an educated guess at the class label $\bar{y}$ of a new input $\bar{x}$ based on the fitted function. On the right, Figure 6-4 illustrates that the inputs can be transferred by a nonlinear map $x \rightarrow \bar{x}$ to a higher dimensional space, thereby increasing the power of linear regression to fit nonlinear functions.

In neural networks with perceptrons, linear functions may also appear as linearized layers. With perceptrons (refer to *bias absorption* in Chapter 3), it is customary to include the bias b_0 when defining $\mathbf{w}^T \equiv (b_0, w_1, ..., w_N)^T$.

Let's look at ways of implementing different linear models to interpret quantum evolution as a general linear model.

Generalized Linear Models

Generalized linear models (GLM) are a generalization and an extension of linear regression. Linear regression assumes a Gaussian distribution of the variable. GLMs do not impose such a limitation. They use a link function to model the relationship between the dependent and linear combinations of independent variables [80, 81]. The $h(x)$ link function is given by

$$h(x) = b_0 + \beta_1 x_1 + ... + \beta_N x_N$$

where β is a vector of predictor weights. In this format, the $h(x)$ link function achieves the important property of transforming the distribution of the dependent or outcome variable to a normal distribution to fit into a linear model, yielding a generalized linear model [80].

GLMs have various statistical extensions, including generalized estimating equations for longitudinal data modeling, generalized linear mixed models for longitudinal data with random effects, and generalized additive models. However, many if not most problems in industry are not easily modeled accurately enough by the exponential family; in addition, linear regression, in general, has many assumptions that may not be met in real-life scientific and industry data. Machine learning algorithms provide alternative ways to minimize the error between predicted values and actual values of a test set. One way to achieve this is the sum of squared error, which is typically used by optimization algorithms. Many supervised learning algorithms are extensions of generalized linear models and have link functions built into the algorithm to model different outcome distributions; examples include boosted regression and Bayesian model averaging. Methods like deep learning and classical neural networks attempt to solve a similar problem through a series of general mappings leading to a potentially novel link function.

Today, the packages available in Python, R, or SAS provide various options for the link function for both GLMs and many of these machine learning models. However, such models struggle when the dataset distribution has many zeros or outliers, and it has now become common that none of the link functions provided in these packages is accepted as the optimum choice for a given dataset. In addition, the computational cost to solve for the right parameters is very high, particularly for boosted regression, topology-based regression methods, and deep learning models [82, 83]. All these factors can contribute to continuous deformations of known distributions that create very large sets of new possible link functions. What is desirable are algorithms that superpose all possible distributions and collapse to fit a dataset.

Quantum gates are a solution to these limitations. The non-Gaussian transformation gates, which perform the task natively and in a computationally efficient way. Exploiting the geometric relationships between distributions through a superposition of states collapsed to the ideal link presents an optimal solution to the problem [84].

An intuitive way to represent a linear model characterized by scalar values in quantum computing format is to encode inputs x into amplitude $|\phi_x\rangle$ and weights w into amplitude $|\phi_y\rangle$. The inner product of the two gives

$$\langle \phi_x | \phi_y \rangle = f(x; w) = \varphi(\mathbf{w}^{\mathrm{T}}.\mathbf{x}) == b_0 + w_1 x_1 + \ldots + w_N x_N \qquad 6.9$$

Measurement of the inner product between two quantum states is an important part of quantum information theory. Superposition between quantum wires in some quantum circuits can be used to perform such measurements. One of the most well-known circuits of the inner product interference is the *swap test*. A method that has been applied widely to quantum machine learning exercises [85] [86].

Swap Test

The quantum circuit for the swap test was first proposed by Buhrman et al. [87], as shown in Figure 6-5a. The significance of this circuit is that by measuring the probability that the first qubit is in state $|1\rangle$ or $|0\rangle$, the square of the inner product of qubit states $|\langle \phi_x | \phi_y \rangle|^2$ can be obtained. This measurement can then be fed into the probability of measuring the state of an ancilla qubit.

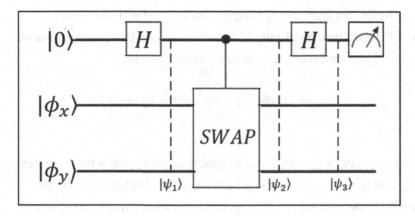

Figure 6-5a. *Swap test circuit*

The SWAP test circuit is shown in Figure 6-5a. The Swap Test consists of an ancilla qubit, two qubit registers in states $|\phi_x\rangle$ and $|\phi_y\rangle$, and three quantum gates: two Hadamard gates and a controlled-*SWAP* gate, also known as a *Fredkin gate* (see Figure 6-5b). The controlled SWAP test has widespread usage in quantum machine learning and has been modified to carry out the task of detection and quantifying entanglement in quantum systems [88].

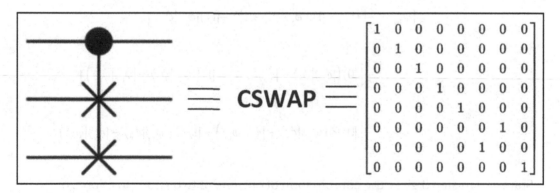

Figure 6-5b. *C-SWAP or Fredkin gate*

Initially, two quantum states $|\phi_x\rangle$ and $|\phi_y\rangle$ (Figure 6-5a) are prepared along with an ancilla qubit in state $|0\rangle$. The ancilla qubit is superposed, and the $|\phi_x\rangle$ and $|\phi_y\rangle$ states are swapped.

1. The Swap Test starts by applying a Hadamard transform to an ancilla qubit that is initially at state $|0\rangle$ and the two quantum states $|\phi_x\rangle$ and $|\phi_y\rangle$ and this leads to the first output state.

$$|\psi_1\rangle = H|0\rangle|\phi_x\rangle|\phi_y\rangle = \frac{1}{\sqrt{2}}(|0\rangle + |1\rangle)|\phi_x\rangle|\phi_y\rangle$$

2. Next, a C-SWAP gate or Fredkin gate is applied to the two registers $|\phi_x\rangle$ and $|\phi_y\rangle$. The registers $|\phi_x\rangle$ and $|\phi_y\rangle$ are conditioned for the ancilla to be in state $|1\rangle$. Hence, this operation swaps the states of the two registers: $|\phi_x\rangle |\phi_y\rangle \longmapsto |\phi_y\rangle |\phi_x\rangle$. At this time, the overall output state is given by

$$|\psi_2\rangle = \frac{1}{\sqrt{2}}\left(|0\rangle|\phi_x\rangle|\phi_y\rangle + |1\rangle|\phi_y\rangle|\phi_x\rangle\right)$$

3. Then, another Hadamard gate is applied to the ancilla qubit resulting in the following state.

$$|\psi_3\rangle = \frac{1}{2}\left[(|0\rangle + |1\rangle)|\phi_x\rangle|\phi_y\rangle + (|0\rangle - |1\rangle)|\phi_y\rangle|\phi_x\rangle\right]$$

$$= \frac{1}{2}\left[|0\rangle\left(|\phi_x\rangle|\phi_y\rangle + |\phi_y\rangle|\phi_x\rangle\right) + |1\rangle\left(|\phi_x\rangle|\phi_y\rangle + |\phi_y\rangle|\phi_x\rangle\right)\right]$$

$$= \frac{1}{2}\left[|0\rangle \otimes \left(|\phi_x\rangle|\phi_y\rangle + |\phi_y\rangle|\phi_x\rangle\right) + |1\rangle \otimes \left(|\phi_x\rangle|\phi_y\rangle - |\phi_y\rangle|\phi_x\rangle\right)\right]$$

State $|\psi_3\rangle$ in step 3 gives the combination of outputs of two superpositions: one containing the sum of the original and swapped states of the two registers and another containing the difference between the two. It is to be noted that if $|\phi_x\rangle = |\phi_y\rangle$, then the swap test always gives a zero.

The outcome of the probability being 1 of measuring the ancilla qubit in state 0 is the *acceptance probability* given by

$$p_{accept} = |\langle 0|\psi_3\rangle|^2 = \frac{1}{2}\left[1 - |\langle\phi_x|\phi_y\rangle|^2\right]$$

6.10

This, in turn, gives the value of the inner product as

$$\left|\langle\phi_x|\phi_y\rangle\right|^2 = 1 - 2p_{accept}$$

or,

$$\left|\langle\phi_x|\phi_y\rangle\right| = \sqrt{1 - 2p_{accept}} \qquad\qquad 6.11$$

Hence, if $\left|\langle\phi_x|\phi_y\rangle\right|$ is near 0, then there is a probability close to 1/2 that we observe a 1. If $|\phi_x\rangle = |\phi_y\rangle$ then we get a probability of 0 to observe a 1.

The overall complexity of evaluating a single dot product is given by $O(\epsilon^{-1}\log N)$.

Linear models can be used in various forms in quantum machine learning, such as interference circuits, unitary linear models, linear models as basis encoding, and amplitude encoding. Detailed treatments of this are beyond the scope of this book; however, interested readers are referred to the book by Schuld et al. [5].

One of the notable limitations of linear models is they do not give very meaningful results to nonlinear data, in other words, data that cannot be separated in linear space. Hence, nonlinear functionalities in models are often desired, as discussed in Chapters 2 and 3 when we looked at classical cases. Nonlinearity can be introduced in cases such as basis encoding and amplitude encoding. Nonlinearities are introduced as nonlinear activation functions ϕ that map an input x to an output $\phi(x)$. We revisit this concept when we encounter implementations of such nonlinear activation functions.

To apply and harvest the benefits of algorithms and models, one of the most important actions is to perform high-precision, highly sensitive measurements, which leads naturally toward *quantum metrology*.

Kernel Methods

There are quite a few methods to implement linear models. The treatment of non-linear data throws up challenges for linear models. Hence, nonlinearities require functions that support such data characteristics. Chapter 5 discussed quantum feature space and feature maps. In Chapter 2, you encountered the nearest neighbor method where the squared-distance method was used as a weight of contribution of a training point toward predicting a new input; here, the measure of similarity is called a *kernel*. The *kernel method* offers many options in classical machine learning, as you saw in the exercises in Chapter 2. Linear models have linear decision boundaries (intersecting hyperplanes). In contrast, nonlinear kernel models (polynomial or Gaussian RBF) have more flexible nonlinear decision boundaries with shapes that depend on the kind of kernel and its parameters. The kernel method also offers parallels to quantum computing–based learning processes.

Kernel methods [93] are very common in machine learning. They were particularly common before deep learning became dominant in the domain of machine learning. In this context, the kernel of the state means the space spanned by vectors corresponding to zero eigenvalues of the state. An example of this is the support vector machine (SVM) discussed in Chapter 1. The SVM is a supervised learning algorithm that learns a given independent and identically distributed set of training instances and draws a decision boundary between two classes of data points by mapping the data into a feature space where it becomes linearly separable. You saw in Chapter 2 that a classical SVM model uses a kernel function, such as sigmoid and tanh, which is defined on the domain of the original input data. Leveraging the principles of kernel methods in classical machine learning, quantum systems can perform computations in an intractably large Hilbert space through the efficient manipulation of data input.

Kernel methods use a *kernel function* $\kappa(x_i, x_j)$ to measure the distance between training data points in the input space and leverage that measurement to compute the distance between each training input and any new input that needs to be classified. The kernel function typically uses a nearest-neighbor type of logic (see Chapter 2) and chooses the class of the "nearer" training data point when a decision for the prediction is to be taken. Such a kernel corresponds to an *inner product* of data points mapped to a higher dimensional feature space. By using kernel functions, nonlinearity is addressed, allowing embedding data to a higher dimensional space where it becomes linearly separable. Figure 6-6 [52] shows a mapping between an original input space and a feature space. On the left is the training input space, where the data from two classes—blue squares and red circles—are not separable in a simple linear model. They can be mapped to the space on the right, however, which shows a higher dimensional feature space (additional *n* dimensions are indicated by the gray arrows). A linear model is sufficient to define a separating hyperplane acts as a decision boundary.

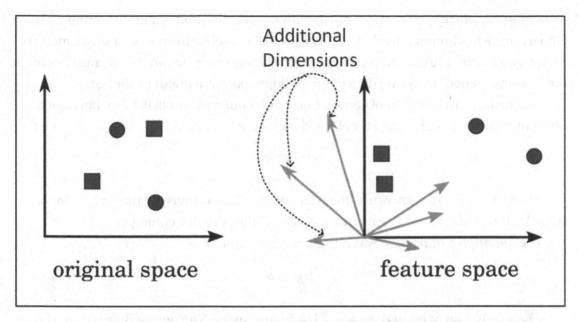

Figure 6-6. *Kernel methods allow embedding of data into a higher-dimensional feature space in which it becomes easier to analyze. (Adapted from source [52])*

The way data-driven models are used is by defining similarities between training datasets and the input new data space; inputs with characteristics similar to the training data points are assumed to have similar outputs as the training set. To factor in the importance of certain input training data points, weights can be assigned to those data points. This is analogous to the nearest neighbor methods (using squared-distance) in classical machine learning discussed in Chapter 2.

Schuld et al. [52] proposed a novel perspective on quantum machine learning in 2018. They proposed a way to map input data into the multi-dimensional Hilbert space (refer to Chapter 1) to perform data analysis.

Hilbert space is a vector space, and it is very, very large. For any vector space over a finite field, the total number of possible vectors is exponentially large in the vector space dimension. A quantum system of n qubits assigns an amplitude to each of the possible classical bitstring of basis states. These amplitudes can be arbitrary complex numbers, hence require *an infinite amount of information* to specify the state of *even a single qubit*. For the ease of counting, let us assume that these amplitudes are discretized to take on d possible values. Since each of the $2n$ complex amplitudes can take on any of the d possible values, the total number of possible quantum states is d^{2n} and the number of bits required to completely specify a state is the *log* base *two* of that number, or $2^n \times \log_2 2d$, which grows exponentially with n. For a 50-qubit quantum computer, the Hilbert space has approximately 1,125,899,907,000,000-dimensions; for a single mode of a continuous-variable quantum computer, the Hilbert space has an infinite number of dimensions. Hence, analyzing data in such a huge space is a major challenge.

Leveraging the definition of kernels from Schölkopf and Smola [94], we have, for a non-empty set $\varphi \in \{x_1, x_2, ..., x_N\}$ for $N \geq 2$, the kernel κ is a map.

$$\kappa \longmapsto \varphi \times \varphi$$

where, $K_{ij} = \kappa(x_i, x_j)$ is known as the *Gram matrix*. The kernels are positive definite, $\kappa(x_i, x_j) \geq 0$ and $\kappa(x_i, x_j) = \kappa(x_i, x_j)^*$, where $\kappa(x_i, x_j)^*$ is the complex conjugate.

The *quantum feature map* was introduced in Chapter 5 as

$$\Omega : \varphi \rightarrow \vartheta$$

where, φ is a set of input data, and ϑ is a feature space. The output data points $\varphi(x)$ on the output map are called *feature vectors*. ϑ is usually a vector space. A kernel κ maps from the input set to complex-valued functions on the input set to create the feature map Ω and a kernel can always be used to construct Ω. A feature space, which is a vector space with an inner product, can be constructed from this feature map.

As a quick recap, the kernel function acts as an inner product on a higher dimensional space and encompasses some $\Omega()$ mapping from the original space of the data points to this space, the kernel function defined by $\kappa(x_i, x_j) = \{\varphi(x_i), \varphi(x_j)\}$. This approach theoretically enables us to map points (which were not linearly separable in the original space) to a higher dimensional space where they can eventually become linearly separable. The $\kappa()$ function may map to an infinite dimensional space and need not be specified. This is defined only when the kernel function $\kappa(x_i, x_j) \geq 0$ and $\kappa(x_i, x_j) = \kappa(x_i, x_j)^*$.

Some kernel-based learning algorithms are instance-based (i.e., the final model retains some or all the training instances and plays a role in the actual prediction). Hence support vector machines (SVMs) are good examples as they are training instances that define the boundary between two classes. Table 6-1 lists some important kernel functions.

Table 6-1. *Important Kernel Functions*

Kernel Name	Kernel Function
Linear	x_i, x_j
Polynomial	$[(x_i, x_j) + a]^\gamma$
Radial basis function	$e^{-\alpha\|x_i - x_j\|^2}$

The choice of kernel and the parameters of the kernel are usually arbitrary. They require either trial and error on the dataset or hyperparameter optimization to help choose the right combination. Quantum computers, due to their quantum nature, naturally give rise to certain kernels. It is worth examining at least one important variant in how it is constructed.

There are two main ways to approach a quantum machine learning process: modify a machine learning algorithm until it contains subroutines with quantum variants or reverse the process and look at a new learning method given a quantum system and its associated constraints. For example, interference is a natural thing to do in quantum physics. So, let us consider the following process: we have the training vectors encoded in some register. This register is entangled with state $|0\rangle$ in the superposition of an ancilla qubit. The ancilla's state $|1\rangle$ of the superposition is entangled with another register that contains a test vector. Applying the Hadamard on the ancilla interferes with the test and training instances. Measuring and post-selecting on the ancilla give rise to a kernel [95].

Kernel Method with Qiskit

We start with Listing 6-2a for the `kernel_qiskit.ipynb` Jupyter notebook downloadable from the book's website. We use Qiskit for this part and follow-up with an example in Rigetti Forest. First, let's import some libraries, call up the `qasm` simulator, and initialize some parameters.

Listing 6-2a. Libraries for and Initialization `kernel-qiskit.ipynb`

```
from qiskit import ClassicalRegister, QuantumRegister, QuantumCircuit
from qiskit import execute
from qiskit import BasicAer

q = QuantumRegister(4)
c = ClassicalRegister(4)
backend = BasicAer.get_backend('qasm_simulator')
```

State Preparation

Preparing a state in a particular encoding can give rise to interesting kernels. Some kernel-based learning algorithms are instance-based (i.e., the final model retains some or all the training instances and plays a role in the actual prediction). Hence support vector machines (SVMs) are good examples as they are training instances that define the boundary between two classes.

To construct an instance-based classifier, we calculate a kernel between all training instances and a test example. In this case, no actual learning takes place and each prediction includes the entire training set. Hence, state preparation is critical to this protocol. The training instances are encoded in a superposition in a register, and the test instances encoded in another register. Let's proceed with this example with the following training instances of the *Iris dataset* [3][13], the same one used in Chapters 2 and 3.

To demonstrate, it is convenient to choose two examples that are relatively easier to handle: we take one example from class 0 and one example from class 1 as *training* instances as follows.

$$\Phi_{TR} = \left\{ \begin{bmatrix} 0 \\ 1 \end{bmatrix}, 0, \begin{bmatrix} 0.78861006 \\ 0.61489363 \end{bmatrix}, 1 \right\}$$

---

[3]Iris dataset: `https://archive.ics.uci.edu/ml/datasets/iris`, reference [13].

The following are the two test instances.

$$\Phi_{Test} = \left\{ \begin{bmatrix} -0.549 \\ 0.836 \end{bmatrix}, \begin{bmatrix} 0.053 \\ 0.999 \end{bmatrix} \right\}$$

Armed with this, we can now define our training and test instances, as shown in Listing 6-2b.

Listing 6-2b. Training and Test Set Definition `kernel-qiskit.ipynb`

```
training_set = [[0, 1], [0.78861006, 0.61489363]]
labels = [0, 1]
test_set = [[-0.549, 0.836], [0.053 , 0.999]]
```

In this case, we try amplitude encoding as referred to in Chapter 5. This means that the second training vector is encoded as $0.78861006\,|0\rangle + 0.61489363\,|1\rangle$. Preparing these vectors needs a rotation, and we need to specify the corresponding angles. The first element of the training set does not need an angle defined since it is the $|1\rangle$ state. Listing 6-2c shows this.

Listing 6-2c. Angle Definition `kernel-qiskit.ipynb`

```
test_angles = [4.30417579487669/2, 3.0357101997648965/2]
training_angle = 1.3245021469658966/4
```

Define the state preparation function, as shown in Listing 6-2d.

Listing 6-2d. State Preparation Function `kernel-qiskit.ipynb`

```
def prepare_state(q, c, angles):
    ancilla_qubit = q[0]
    index_qubit = q[1]
    data_qubit = q[2]
    class_qubit = q[3]
    circuit = QuantumCircuit(q, c)
    # Ancilla and index qubits into uniform superposition
    circuit.h(ancilla_qubit)
    circuit.h(index_qubit)
```

```
# Prepare the test vector
circuit.cx(ancilla_qubit, data_qubit)
circuit.u3(-angles[0], 0, 0, data_qubit)
circuit.cx(ancilla_qubit, data_qubit)
circuit.u3(angles[0], 0, 0, data_qubit)
# Flip the ancilla qubit > this moves the input
# vector to the |0> state of the ancilla
circuit.x(ancilla_qubit)
circuit.barrier()

# Prepare the first training vector
# [0,1] -> class 0
# We can prepare this with a Toffoli
circuit.ccx(ancilla_qubit, index_qubit, data_qubit)
# Flip index qubit ->
moves the first training vector to the
# |0> state of the index qubit
circuit.x(index_qubit)
circuit.barrier()

# Prepare the second training vector
# [0.78861, 0.61489] -> class 1

circuit.ccx(ancilla_qubit, index_qubit, data_qubit)
circuit.cx(index_qubit, data_qubit)
circuit.u3(angles[1], 0, 0, data_qubit)
circuit.cx(index_qubit, data_qubit)
circuit.u3(-angles[1], 0, 0, data_qubit)
circuit.ccx(ancilla_qubit, index_qubit, data_qubit)
circuit.cx(index_qubit, data_qubit)
circuit.u3(-angles[1], 0, 0, data_qubit)
circuit.cx(index_qubit, data_qubit)
circuit.u3(angles[1], 0, 0, data_qubit)
circuit.barrier()
```

```
# Flip the class label for training vector #2
circuit.cx(index_qubit, class_qubit)
circuit.barrier()
return circuit
```

Next, we generate the circuit for preparing state with the first test instance, as shown in Listing 6-2e.

Listing 6-2e. Circuit Generation `kernel-qiskit.ipynb`

```
from qiskit.tools.visualization import circuit_drawer
angles = [test_angles[0], training_angle]
state_preparation_0 = prepare_state(q, c, angles)
circuit_drawer(state_preparation_0)
```

The circuit generation code produces the circuit shown in Figure 6-7.

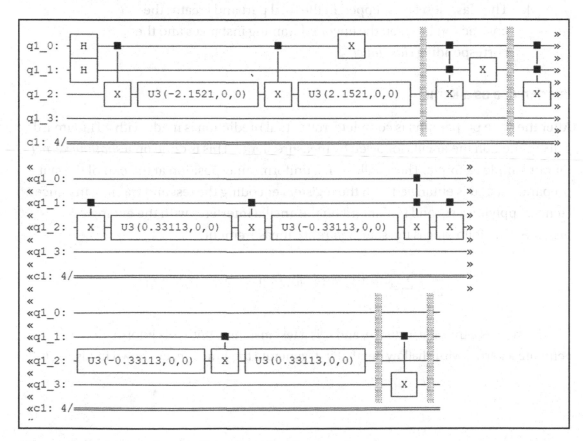

Figure 6-7. *Circuit for the kernel method example*

In Figure 6-7, the vertical lines define a segmentation of the state preparation and act as barriers to ensure that all gates are finished by that point.

1. The *test instance* is prepared within the first barrier. A uniform superposition is formed between the ancilla and index qubits, which are registers 0 and 1, respectively. The test instance is entangled with the ground state of the ancilla.

2. The *first training* instance is created between the first and second barriers. For this, the state $|1\rangle$ is prepared and entangled with the excited state of the ancilla and the ground state of the index qubit with a Toffoli gate and a Pauli X gate. The Toffoli gate is also called the *controlled-controlled-not gate*, describing its action.

3. The *second training* instance is prepared and entangled with the excited state of the ancilla and the index qubit in the third section.

4. The class qubit gets flipped in the final part and creates the connection between the encoded training instances and the corresponding *class label*.

Interference as a Kernel

After the state preparation is complete, the actual prediction is made with a Hadamard gate applied on the ancilla followed by measurements. This is done on a *shallow kernel* for our simple example. The ancilla is in a uniform superposition at the end of the state preparation and is entangled with the registers encoding the test and training instances. Hence, applying a second Hadamard on the ancilla interferes with the entangled registers. The following is the state *before* the measurement.

$$\frac{1}{2\sqrt{2}}\sum_{i=1}^{1}\Big[|0\rangle|i\rangle\big(|x_t\rangle+|x_i\rangle\big)|y_i\rangle+|1\rangle|i\rangle\big(|x_t\rangle-|x_i\rangle\big)|y_i\rangle\Big]$$

where, x_i is a training instance, and x_t is a test instance. With this information, we generate a kernel with shallow depth. First, generate the function, as shown in Listing 6-2f.

Listing 6-2f. Function for Kernel with Shallow Depth `kernel-qiskit.ipynb`

```
def interfere_data_and_test_instances(circuit, q, c, angles):
    circuit.h(q[0])
    circuit.barrier()
    circuit.measure(q, c)
    return circuit
```

A measurement of the ancilla at this point gives the probability of observing 0 as

$$\frac{1}{4N}\sum_{i=0}^{1}|x_t + x_i|^2$$

This creates a kernel given by Listing 6-2g.

Listing 6-2g. Kernel Generation `kernel-qiskit.ipynb`

```
import matplotlib.pyplot as plt
import numpy as np
%matplotlib inline
x = np.linspace(-2, 2, 100)
plt.xlim(-2, 2)
plt.ylim(0, 1.1)
plt.plot(x, 1-x**2/4)
```

Listing 6-2g gives the following output, which is the kernel that performs the classification.

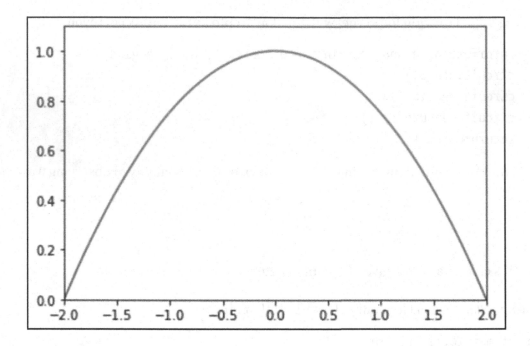

The next step is to perform the post-selection on observing 0 on the measurement on the ancilla and calculate the probabilities of the test instance belonging to either class. To do this, we define a function to perform the task as in Listing 6-2h.

Listing 6-2h. Lambda Function kernel-qiskit.ipynb

```
def postselect(result_counts):
    total_samples = sum(result_counts.values())
    # lambda function: retrieves only results where the
ancilla is in the |0> state
    post_select = lambda counts: [(state, occurences) for state, occurences
    in counts.items() if state[-1] == '0']
    # perform the postselection
    postselection = dict(post_select(result_counts))
    postselected_samples = sum(postselection.values())

    print(f'Ancilla post-selection probability was found to be
    {postselected_samples/total_samples}')
    retrieve_class = lambda binary_class: [occurences for state, occurences
    in postselection.items() if state[0] == str(binary_class)]
    prob_class0 = sum(retrieve_class(0))/postselected_samples
```

```
prob_class1 = sum(retrieve_class(1))/postselected_samples

print('Probability for class 0 is', prob_class0)
print('Probability for class 1 is', prob_class1)
```

This gives the results for the first and second instances of post-selection probability, as shown in Listing 6-2i.

Listing 6-2i. First Instance `kernel-qiskit.ipynb`

```
circuit_0 = interfere_data_and_test_instances(state_preparation_0, q, c,
angles)
job = execute(circuit_0, backend)
result = job.result()
postselect(result.get_counts(circuit_0))
```

The following is the output for the first instance.

```
Ancilla post-selection probability was found to be 0.7392578125
Probability for class 0 is 0.6605019815059445
Probability for class 1 is 0.3394980184940555
```

Listing 6-2j shows the code for the second instance of post-selection probability.

Listing 6-2j. Second Instance `kernel-qiskit.ipynb`

```
angles = [test_angles[1], training_angle]
state_preparation_1 = prepare_state(q, c, angles)
circuit_1 = interfere_data_and_test_instances(state_preparation_1, q, c,
angles)
job = execute(circuit_1, backend)
result = job.result()
postselect(result.get_counts(circuit_1))
```

The following is the output for the second instance.

```
Ancilla post-selection probability was found to be 0.9140625
Probability for class 0 is 0.5288461538461539
Probability for class 1 is 0.47115384615384615
```

Kernel Method with Rigetti Forest

You have seen examples of the routines run in Qiskit for inferences out of the kernel methods. You also had a glimpse at Rigetti's Forest leveraging qvm and quilc in Chapter 5. In that context, it would be interesting to see how similar kernel method–generated inferences can be produced using Rigetti Forest.

To start, let's open three terminals in our Forest virtual environment exactly as we did in Chapter 5. We start QVM in server mode, Quilc in server mode, and Jupyter notebook with the following commands, each in its separate terminal.

- QVM: qvm -S

- Quilc: quilc -S

- Jupyter notebook: jupyter notebook. Then choose Python 3 to create a new notebook instance.

As usual, we first import the required libraries, as shown in Listing 6-3a, for kernel-qiskit.ipynb.

Listing 6-3a. Import Libraries kernel_Inference_forest.ipynb

```
from pyquil import Program, get_qc
from pyquil.gates import *
import cmath
import matplotlib.pyplot as plt
import numpy as np
from pyquil.api import ForestConnection
from pyquil.api import WavefunctionSimulator
from pyquil.api import QVMConnection
from tempfile import mkdtemp
import itertools
import socket
import subprocess
qvm = QVMConnection()
```

State Preparation

To start, we prepare a state for encoding and work with the following dataset of two vectors.

$$\Phi_{TR} = \left\{ \begin{bmatrix} 0 \\ 1 \end{bmatrix}, 0, \begin{bmatrix} \sqrt{2}/2 \\ \sqrt{2}/2 \end{bmatrix}, 1 \right\}$$

The following are the two test instances.

$$\Phi_{Test} = \begin{bmatrix} 1 \\ 0 \end{bmatrix}$$

As before, our objective is to build an interference circuit for the dataset using four qubits: ancilla, index, data, and class.

- The ancilla and index qubits are put in a uniform superposition.

- The test instance is entangled with the ground state of the ancilla.

- An identity gate is applied to the class qubit.

- The solution is placed in an object called a *circuit*.

The part of the code with circuit preparation is shown in Listing 6-3b for kernel_Inference_forest.ipynb.

Listing 6-3b. State Preparation kernel_Inference_forest.ipynb

```
## define a function for amplitudes

def get_amplitudes(circuit):
    wf_sim = WavefunctionSimulator()
    wavefunction = wf_sim.wavefunction(circuit)
    amplitudes = wavefunction.amplitudes
    return amplitudes

ancilla_qubit = 0
index_qubit = 1
```

```
data_qubit = 2
class_qubit = 3
training_set = [[0, 1], [np.sqrt(2)/2, np.sqrt(2)/2]]
labels = [0, 1]
test_set = [[1, 0]]
test_angles = [2*np.arccos(test_set[0][0])/2]
training_angle = (2*np.arccos(training_set[1][0]))/4
angles = [test_angles[0], training_angle]
circuit = Program()

# Create uniform superpositions of the ancilla and index qubits
circuit += H(ancilla_qubit)
circuit += H(index_qubit)

# Entangle the test instance with ground state of ancilla
circuit += CNOT(ancilla_qubit, data_qubit)
circuit += X(ancilla_qubit)
# Apply Identity to Class state
circuit += I(class_qubit)
print('Input = \n', get_amplitudes(circuit))
```

Listing 6-3b prints the input.

```
Input =
 [0. +0.j 0.5+0.j 0. +0.j 0.5+0.j 0.5+0.j 0. +0.j 0.5+0.j 0. +0.j 0. +0.j
 0. +0.j 0. +0.j 0. +0.j 0. +0.j 0. +0.j 0. +0.j 0. +0.j]
```

Then the circuit is extended to prepare the first training instance and entangle it with the excited state of the ancilla and ground state of the index qubit, as shown in Listing 6-3c.

Listing 6-3c. First instance Entanglement kernel_Inference_forest.ipynb

```
circuit += CCNOT(ancilla_qubit, index_qubit, data_qubit)
circuit += X(index_qubit)
print('First training instance \n', get_amplitudes(circuit))
```

For this step, we have the training instance as output.

First training instance
```
[0. +0.j 0. +0.j 0. +0.j 0.5+0.j 0.5+0.j 0.5+0.j 0.5+0.j 0. +0.j 0. +0.j
 0. +0.j 0. +0.j 0. +0.j 0. +0.j 0. +0.j 0. +0.j 0. +0.j]
```

In the next step, the circuit is extended to prepare the second training instance and entangle it with the excited state of the ancilla and the index qubit, as shown in Listing 6-3d.

Listing 6-3d. First Instance Entanglement kernel_Inference_forest.ipynb

```
circuit += CCNOT(ancilla_qubit, index_qubit, data_qubit)
circuit += CNOT(index_qubit, data_qubit)
circuit += H(data_qubit)

circuit += CCNOT(ancilla_qubit, index_qubit, data_qubit)
circuit += CNOT(index_qubit, data_qubit)
circuit += H(data_qubit)

circuit += CCNOT(ancilla_qubit, index_qubit, data_qubit)
circuit += CNOT(index_qubit, data_qubit)
circuit += H(data_qubit)

circuit += CCNOT(ancilla_qubit, index_qubit, data_qubit)
circuit += CNOT(index_qubit, data_qubit)
circuit += H(data_qubit)

print('Second training instance \n', get_amplitudes(circuit))
```

This step gives the second instance as follows.

Second training instance
```
[ 0. +0.j  0. +0.j  0. +0.j  0.5+0.j  0.5+0.j  0.5+0.j -0.5+0.j  0. +0.j
  0. +0.j  0. +0.j  0. +0.j  0. +0.j  0. +0.j  0. +0.j  0. +0.j  0. +0.j]
```

We finish the state preparation circuit by flipping the class qubit conditioned on the index qubit and run a verification code that should print "No" if the bit *is flipped* and "Yes" if it is *not*. We run this code first with Listing 6-3e turned on and then negated with a hash (#) symbol.

Listing 6-3e. First instance Entanglement `kernel_Inference_forest.ipynb`

```
circuit += CNOT(index_qubit, class_qubit)
#circuit += CNOT(index_qubit, class_qubit)
```

In the first case, the bit flip is turned on. No is the output, as expected. In the second case, the bit flip is negated. Yes is the output.

Interference as a Kernel

After the last step, the measurement is given by

$$\frac{1}{2\sqrt{2}}\sum_{i=1}^{1}\left[|0\rangle|i\rangle\big(|x_t\rangle+|x_i\rangle\big)|y_i\rangle+|1\rangle|i\rangle\big(|x_t\rangle-|x_i\rangle\big)|y_i\rangle\right]$$

where, x_i is a training instance and x_t is a test instance. Here, we apply the Hadamard gate on the ancilla to apply the interference, as shown in Listing 6-3f.

Listing 6-3f. First Instance Entanglement `kernel_Inference_forest.ipynb`

```
circuit += H(ancilla_qubit)
print('Circuit \n', get_amplitudes(circuit))
```

The last step gives the following output.

```
Circuit
[ 0.          +0.j  0.        +0.j  0.          +0.j  0.          +0.j
  0.70710678+0.j  0.        +0.j  0.          +0.j  0.          +0.j
  0.          +0.j  0.        +0.j  0.35355339+0.j -0.35355339+0.j
  0.          +0.j  0.        +0.j -0.35355339+0.j -0.35355339+0.j]
```

Note   The values of the circuit parameters vary run to run due to the probabilistic nature of the calculations.

In summary, the kernel trick allows us to construct a variety of models formulated in terms of a kernel function by replacing it with another kernel function. We look at state preparation and information encoding as a quantum feature map.

Quantum *k*-Means Clustering

As a recap from Chapter 2, k-means [98] is an *unsupervised clustering* algorithm that groups together *n* observations into *k* clusters, making sure intracluster variance is minimized. Given a dataset $D \in \{x_1, x_2, ..., x_N\}$ with *N* observations, the objective is to cluster this dataset into *k* partitions. We want to assign a set of points grouped near each other in one cluster and another set of points grouped into another cluster. The *k* centroids are initialized randomly. The objective is to find sets $S \in \{s_1, s_2, ..., s_K\}$ such that the *intra-set* variance is *minimum* and the *inter-set* variance is *maximum*.

The *quantum version of the k-means algorithm* is an unsupervised algorithm that aims to classify data to *k* clusters based on an unlabeled set of training vectors. The training vectors are reassigned to the nearest centroid in each iteration, and then a new centroid is calculated, averaging the vectors belonging to the current cluster. It provides an exponential speed-up for very high dimensional input vectors utilizing the fact that only log*N* qubits are required to load *N*-dimensional input vectors using *amplitude encoding*. The time complexity O(NM) of the classical algorithm using Lloyd's version of the algorithm is linearly dependent on several *N* features and training examples M [99]. The most resource-consuming operation for a k-means algorithm is the calculation of a distance between vectors. We are expecting to gain speed-up by retrieving the distance using a quantum computer.

There are a few versions of the Quantum k-means algorithm. One version of the quantum k-means algorithm utilizes three different quantum subroutines to perform k-means clustering: Swap test, distance calculation (DistCalc), and Grover's optimization (GroverOptim).

As seen before, the s*wap test* subroutine measures the overlap between two quantum states $\langle x|y \rangle$ based on the measurement probability of the control qubit in state $|0\rangle$. The overlap is a measure of similarity between two states. If the probability of the control qubit in state $|0\rangle$ is 0.5, it means the $|x\rangle$ and $|y\rangle$ states are orthogonal; whereas if the probability is 1, then the states are identical. The subroutine should be repeated several times to obtain a good estimator of probability.

The advantage of using the swap test is that the states $|x\rangle$ and $|y\rangle$ can be unknown before the procedure is performed. A simple measurement is required on the control qubit, which has two eigenstates. The time complexity is *negligible* as the procedure does not depend on the number of qubits representing the input states. From equation 6.10, the probability of control qubit being in state $|0\rangle$ is given by

$$p(0) = |\langle 0|\psi\rangle|^2 = \frac{1}{2}\left[1 - |\langle x|y\rangle|^2\right] \qquad 6.12$$

The distance calculation (DistCalc) subroutine follows the Swap Test and presents an algorithm to retrieve the Euclidean distance $|x - y|^2$ between two real valued vectors x and y. The algorithm was described by Lloyd, Mohseni, and Rebentrost [100].

Let's discuss the representation of classical data into a quantum state. The classical information in vector x is encoded as [101]

$$|x|^{-1} x \mapsto |x\rangle = \sum_{i=1}^{N} |x|^{-1} x_i |i\rangle$$

In the DistCalc subroutine, two states are prepared, then a subroutine swap test is applied, and the procedure is repeated to obtain an acceptable estimate of probability. Providing that the states are already prepared, the subroutine swap test does not depend on the size of a feature vector. The preparation of states in subroutine DistCalc is proven to have $O(\log N)$ time complexity [100] as the classical information is encoded in $n = \log_2 N$ qubits, and the time complexity is expected to be proportional. The classical algorithms require $O(N)$ to calculate Euclidean distance between two vectors; hence, there is an exponential speed-up.

After the swap test and DistCalc algorithm, the last step is to run Grover's optimizer, which chooses the closest centroid of the cluster. This completes an overview of the quantum k-means algorithm.

Quantum k-Medians Algorithm

The k-medians algorithm is similar to the k-means algorithm; however, instead of setting new cluster centroids by calculating mean, a median is evaluated. The k-means algorithm has a disadvantage. In some cases, the calculated mean may lie outside the desired region, whereas cluster centroid calculated as a median always belongs to a training vector set. It has even further implications for quantum algorithms, as the calculation of the mean is executed on a classical computer. In contrast, the calculation of median can be implemented using the MedianCalc subroutine [102].

The quantum version of the k-medians algorithm requires the same quantum subroutines as the quantum k-means algorithm: Swap Test, DistCalc, and GroverOptim plus an additional MedianCalc, which also uses GroverOptim. Usage of the GroverOptim quantum minimization algorithm is optional in a k-means algorithm; whereas for k-medians, it may be used to choose the closest centroids and must be used to calculate median.

Summary

This chapter looked at the Schmidt decomposition, linear models, kernel methods, Swap tests, entanglement measurements, in addition to some introduction to unsupervised and supervised quantum models. It also looked at coding with Rigetti's Forest SDK, QVM, and Qiskit. The next chapter looks at advanced quantum machine learning algorithms and methods, such as VQE, QUBO, QAOA, HHL, and qSVM with relevant applications and associated code in D-Wave, Forest, Qiskit, and Cirq.

CHAPTER 7

QML Techniques

The butterfly counts not months but moments and has time enough.

—Rabindranath Tagore

In our exploration of various quantum algorithms and platforms, we have experienced the interesting point that the quantum computer is probabilistic, meaning that it can return multiple answers. Some of these might be the answer that you are looking for, and some might not. At first, this sounds like a bad thing, as a computer that returns a different answer when you ask it the same question does not breed confidence! However, in a quantum computer, returning multiple answers can give important information about the confidence level of the computer.

If we showed the computer an image of an apple and asked it to label the same image 100 times, and it gave the answer *apple* 100 times, then the computer is confident that the image is an apple. However, if it returns the answer apple 50 times and raspberry 50 times, then the computer is uncertain about the image we are showing it. And if we had shown it an image with apples *and* raspberries in, it would be perfectly correct! This uncertainty can be very powerful when designing systems that can make complex decisions and learn about the world.

Note   This chapter focuses on advanced algorithms and techniques of quantum computing and machine learning. It introduces D-Wave's annealing platform and its optimization power. VQE, QAOA, and QUBO are covered from a theoretical perspective supported by code in D-Wave, pyQuil, and Qiskit.

© Santanu Ganguly 2021
S. Ganguly, *Quantum Machine Learning: An Applied Approach*, https://doi.org/10.1007/978-1-4842-7098-1_7

Chapter 6 discussed quantum linear models and the implementation of quantum phase estimation (QPE), specifically for quantum Fourier transfer (QFT)–based algorithms. However, the application of QPE and some other algorithms are not restricted to a limited number of cases; on the contrary, it is common that once data is available to the quantum computing device, it is processed with algorithms such as quantum phase estimation and *matrix inversion* (the HHL algorithm). The *HHL algorithm* has been implemented on various quantum computers to date.

A wide variety of classical data analysis and machine learning protocols operate by performing matrix operations on vectors in a high-dimensional vector space. However, that is *exactly what quantum mechanics is all about*! Quantum mechanics leverages matrix operations on vectors in high-dimensional vector spaces such as Hilbert space. The key ingredient behind these methods is that the quantum state of n quantum bits *or qubits* is a vector in a 2^n-dimensional complex vector space and consequently, performing a quantum logic operation or measurement on qubits multiplies the corresponding state vector by $2^n \times 2^n$ matrices.

Quantum computers perform common linear algebraic operations, such as Fourier transforms [116], finding eigenvectors and eigenvalues [116], and solving linear sets of equations over 2^n-dimensional vector spaces in time that, being polynomial in n, is exponentially faster than their best-known classical counterparts [115]. The latter is known as the Harrow, Hassidim, and Lloyd (HHL) algorithm [115].

HHL Algorithm (Matrix Inversion)

The quantum algorithm for the linear system was first proposed by Harrow, Hassidim, and Lloyd (HHL) [115]. The HHL algorithm underlies many quantum machine learning protocols, but it is a substantially nontrivial algorithm with many conditions. This chapter looks at the implementation of the algorithm to gain a better understanding of how it works and when it works efficiently. The hands-on exercises are derived from references [113] and [114].

The HHL algorithm for inverting linear equations is a fundamental subroutine, underpinning many quantum machine learning algorithms. The algorithm's goal is to solve $Ax = b$ using a quantum computer, where x and b are expressed in "bold" to indicate that they are vectors. The problem of solving for x is attempted by obtaining the *expectation value* of operator A, assumed to be Hermitian, with x, $x^\dagger$ and Ax instead of directly obtaining the value of x. This is considered useful when solving on a quantum computer since quantum systems usually offer probabilities for some measurement. Typically, these operators are Pauli's operators X, Y, Z. The measured probabilities can then be translated to expectation values for these operators.

HHL algorithm quantizes the problem $Ax = b$ by expressing the vector x as a quantum state $|x\rangle$ and vector $b \in C^N$ as a quantum state $|b\rangle$ over $\log_2 N$ qubits. Consequently, the equation $Ax = b$ can be solved by multiplying both sides of the equation by A^{-1} the inverse of A. Then, as per HHL algorithm, we can construct the quantum state proportional to $A^{-1}|b\rangle$. Generally, when A has zero eigenvalues (is not square), then the algorithm can be used to find the state $|x\rangle$ that minimizes [117] $|A|x\rangle - |b\rangle|$.

If $\{|v_i\rangle\}$ and $\{\lambda_i\}$ are eigenvectors and eigenvalues of A, respectively, where $0 < \lambda_i < 1$, then the state $|b\rangle$ can be written as a linear combination of the eigenvectors as follows

$$\{|v_i\rangle\}, |b\rangle = \sum_{i=1}^{N} b_i |E_i\rangle$$

The HHL algorithm aims to obtain $|x\rangle$ in the form $\sum_{i=1}^{N} \alpha_i \frac{1}{\lambda_i}|v_i\rangle$. The HHL procedure finds the eigenvalues of A using the quantum subroutine called *phase estimation*, as explained in Chapter 6.

The algorithm shown in Figure 7-1 uses three sets of qubits: a single ancilla qubit, a register to store eigenvalues of A, and memory qubits (to store$|b\rangle$ and $|x\rangle$).

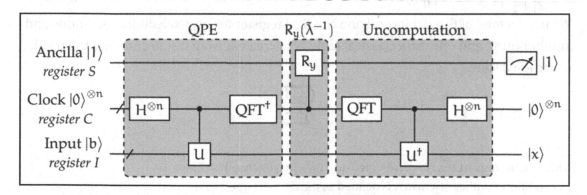

Figure 7-1. *The HHL algorithm is subdivided into QPE (quantum phase estimation) to determine the eigenvalues, controlled rotation R_y to extract the eigenvalue information, and uncomputation to revert the phase estimation*

In Figure 7-1, the following actions are performed.

1. Quantum phase estimation to extract eigenvalues of A

2. Controlled rotations of ancilla qubit

3. Uncompute[1] with inverse quantum phase estimation.

These steps are described as follows.

First, since the eigenvalues λ_i are of binary form $(0. a_1a_2a_3...)$, $|\lambda_i\rangle = |a_1a_2a_3\rangle$. As the spectral theorem says that every Hermitian matrix has an orthonormal basis of eigenvectors, we can write matrix $A \equiv \sum_i \lambda_i |v_i\rangle\langle v_i|$ and $|b\rangle = \sum_n b_n |E_n\rangle$, where $|E_n\rangle$ is an eigenvector of A and eigenvalue $\lambda_n > C$. Here C is a *scaling factor* to prevent the controlled rotation from becoming unphysical, which indicates that from a practical estimation point of view, a choice of *C less than the minimum value* of λ is reasonable; this is formally stated as $C = O\left(\dfrac{1}{\kappa}\right)$, where κ is the *condition number*[2] of matrix A. To visualize this better, let us try our hands at some exercises with the Jupyter notebook `HHL_qiskit.ipynb` from the book's website. The code is adapted from Zhao et al.[3] [113] and is done in Qiskit.

If you are interested in diverse coding options, Google's Cirq tutorials have a comprehensive representation of the HHL algorithm implementation at `https://github.com/quantumlib/Cirq/blob/master/examples/hhl.py` on GitHub [119].

We set up the solution of the equation $A\boldsymbol{x} = \boldsymbol{b}$ by inverting a 2×2 matrix where $A = \dfrac{1}{2}\begin{bmatrix} 3 & 1 \\ 1 & 3 \end{bmatrix}$ anb $b = \begin{bmatrix} 1 \\ 0 \end{bmatrix}$. Matrix A is encoded as a Hamiltonian, and b is in a register. We need a total of five qubits and one classical register for post-selection operations and another qubit and one extra classical register to create a swap test to compare the result to the ideal state.

---

[1]Uncomputation: `https://en.wikipedia.org/wiki/Uncomputation`

[2]A condition number of a matrix is defined as the ratio of largest to smallest singular value and is undefined when the smallest singular value is 0. Generally, a matrix is referred to as "well-conditioned" when its singular values lie between the reciprocal of its condition number $\dfrac{1}{\kappa}$ and 1.

[3]Z. Zhao, A. Pozas-Kerstjens, P. Rebentrost, P. Wittek, "Bayesian Deep Learning on a Quantum Computer," (2018), arXiv:1806.11463 [quant-ph], `https://arxiv.org/abs/1806.11463`, reference [113] of this book.

Listing 7-1a. Libraries and for `HHL_qiskit.ipynb`

```
import numpy as np
from qiskit import QuantumCircuit, ClassicalRegister, QuantumRegister
from qiskit import execute
from qiskit import BasicAer
π = np.pi
q = QuantumRegister(6)
c = ClassicalRegister(2)
hhl = QuantumCircuit(q, c)
```

In the next step, a *phase estimation* is performed to map (or encode) the eigenvalues of the matrix A in an additional register. The quantum phase estimation algorithm performs the following mapping

$$\left(|0\rangle^{\otimes n}\right)^{C} |u\rangle^{I} |0\rangle^{S} \mapsto \left(|\varphi\rangle^{\otimes n}\right)^{C} |u\rangle^{I} |0\rangle^{S} \qquad 7.1$$

where $|u\rangle$ is an eigenvector of some unitary operator U with an unknown eigenvalue $e^{i2\pi\varphi}$ The i indicates the complex plane and not an index.

The superscripts on the kets in equation 7.1 relate to the names of the registers, which store the corresponding states.

In the HHL algorithm, the input register begins with a superposition of eigenvectors (i.e., $|b\rangle = \sum_{n} b_{n} |E_{n}\rangle$. The unitary operator is given by e^{iAt}. To avoid the factor of 2π in the denominator, often, the evolution time t_0 is taken as 2π. In this step, we consider the Hamiltonian simulation gate U which is clock register-controlled. With these assumptions in hand, the initial aim is to create a superposition of A as a Hamiltonian applied for different time durations. Because the eigenvalues are always positioned on the complex unit circle, the eigenstructure is revealed by the differently evolved components in the superposition. Hence, the conditional Hamiltonian evolution can be expressed as $\sum_{\tau=0}^{T-1} |\tau\rangle\langle\tau|^{C} \otimes e^{iA\tau t_{0}/T}$ on the tensor state product $|\psi_0\rangle \otimes |b\rangle$. This operation evolves the state $|b\rangle$ as governed by the Hamiltonian A for the time period τ determined by the state $|\psi_0\rangle$.

Since state $|\psi_0\rangle$ has a superposition of all possible time steps between 0 and T, the result is a superposition of all possible evolutions. We need to make a suitable choice of the number of time steps T and total evolution time t_0 to allow encoding of binary representations of the eigenvalues.

As a final step in the *quantum phase estimation* (QPE) process as part of the HHL algorithm, an inverse Fourier transformation is applied, which writes the phases encoding the eigenvalues of A into new registers.

Application of the phase estimation under A helps compute λ_n. Hence, in this step, we encode the eigenvalues of the matrix A in an additional register. Figure 7-2 shows the phase estimation circuit.

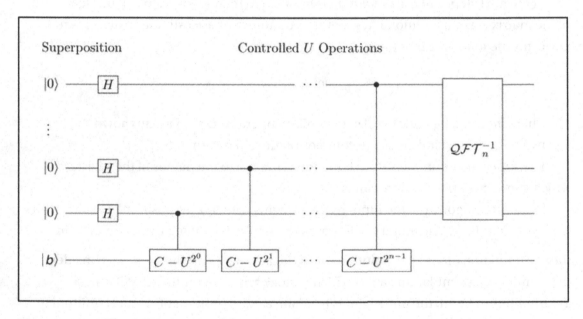

Figure 7-2. *The QPE circuit of the HHL algorithm*

In a 2 × 2 matrix, which has eigenvalues that are powers of 2, we choose $t_0 = 2\pi$ and $T = 4$ to obtain exact results with two controlled evolutions.

Listing 7-1b. Superposition and Quantum Circuit Definition HHL_qiskit.ipynb

```
# Superposition
hhl.h(q[1])
hhl.h(q[2])
# Controlled-U0
```

```
hhl.cu3(-π / 2, -π / 2, π / 2, q[2], q[3])
# hhl.cu1(3 * π / 4, q[2], q[3])
hhl.cp(3 * π / 4, q[2], q[3])
hhl.cx(q[2], q[3])
#hhl.cu1(3 * π / 4, q[2], q[3])
hhl.cp(3 * π / 4, q[2], q[3])
hhl.cx(q[2], q[3])
# Controlled-U1
hhl.cx(q[1], q[3]);
```

Then, the quantum inverse Fourier transformation to write the phase to a register.

Listing 7-1c. Inverse QFT HHL_qiskit.ipynb

```
hhl.swap(q[1], q[2])
hhl.h(q[2])
hhl.cp(-π / 2, q[1], q[2])
hhl.h(q[1]);
```

The state of the system after the QFT decomposition is approximately given by $\sum_i \beta_i |E_i\rangle |\lambda_i\rangle$, where $|b\rangle = \sum_i b_i |E_i\rangle$. Here, $|b\rangle$ is the encoding of the vector b in the eigenbasis of A.

The next step is to perform operations on $|\lambda_i\rangle$ to invert it. In this case, the eigenvalues of A are $\lambda_1 = 2 = 10$ (in binary) and $\lambda_2 = 1 = 01$. The reciprocals of the eigenvalues are $\lambda_1^{-1} = \frac{1}{2} \Rightarrow 2\lambda_1^{-1} = 01$ and $\lambda_2^{-1} = 1 \Longrightarrow 2\lambda_2^{-1} = 10$. In this case, a *SWAP* gate obtains the state $\sum_i b_i |E_i\rangle |2\lambda_i^{-1}\rangle$ which encodes the reciprocals of the eigenvalues.

```
hhl.swap(q[1], q[2]);
```

After applying phase estimation under A to compute λ_n, we rotate the ancilla qubit through an angle of $\sin^{-1} \dfrac{C}{\lambda_n}$ about the Y axis. This controlled rotation of the ancilla qubit is done to get a state close to the following

$$\sum_i b_i |E_i\rangle |2\lambda_i^{-1}\rangle \left[\frac{C}{\lambda_i} |1\rangle + \sqrt{1 - \frac{C^2}{\lambda_i^2}} |0\rangle \right] \qquad 7.2$$

Listing 7-1d. Ancilla Rotation `HHL_qiskit.ipynb`

```
hhl.cu3(0.392699, 0, 0, q[1], q[0])  # Controlled-RY0
hhl.cu3(0.19634955, 0, 0, q[2], q[0]);  # Controlled-RY1
```

When performing quantum computations, you must uncompute all operations except those that store the desired information from the algorithm in the final registers. This step is needed in case the registers are entangled, which would affect the results. In this example, the QPE in step 2 must be uncomputed. After the uncomputation, the final desired state is

$$\sum_i b_i |E_i\rangle |0\rangle \left[\frac{C}{\lambda_i}|1\rangle + \sqrt{1 - \frac{C^2}{\lambda_i^2}}|0\rangle \right]$$

7.3

Listing 7-1e. Uncompute Action on QPE `HHL_qiskit.ipynb`

```
hhl.swap(q[1], q[2])
hhl.h(q[1])
hhl.cp(π / 2, q[1], q[2]) # Inverse(Dagger(Controlled-S))
hhl.h(q[2])
hhl.swap(q[2], q[1])
# Inverse(Controlled-U1)
hhl.cx(q[1], q[3])
# Inverse(Controlled-U0)
hhl.cx(q[2], q[3])
hhl.cp(-3 * π / 4, q[2], q[3])
hhl.cx(q[2], q[3])
hhl.cp(-3 * π / 4, q[2], q[3])
hhl.cu3(-π / 2, π / 2, -π / 2, q[2], q[3])
# End of Inverse(Controlled-U0)
hhl.h(q[2])
hhl.h(q[1]);
```

Next, we perform post-selection by projecting onto the desired state $|1\rangle$. To obtain the expected solution, the correct output state is prepared manually to perform a swap test with the outcome.

Listing 7-1f. Psot-Selection HHL_qiskit.ipynb

```
# Target state preparation
hhl.rz(-π, q[4])
hhl.p(π, q[4])
hhl.h(q[4])
hhl.ry(-0.9311623288419387, q[4])
hhl.rz(π, q[4])
# Swap test
hhl.h(q[5])
hhl.cx(q[4], q[3])
hhl.ccx(q[5], q[3], q[4])
hhl.cx(q[4], q[3])
hhl.h(q[5])
hhl.barrier(q)
hhl.measure(q[0], c[0])
hhl.measure(q[5], c[1]);
```

There are two measurements performed: one of the ancilla register for the post-selection and another one that outputs the result of the swap test. To calculate success probabilities, we define some helper functions.

Listing 7-1g. Success Functions HHL_qiskit.ipynb

```
def get_psuccess(counts):
    try:
        succ_rotation_fail_swap = counts['11']
    except KeyError:
        succ_rotation_fail_swap = 0
    try:
        succ_rotation_succ_swap = counts['01']
    except KeyError:
        succ_rotation_succ_swap = 0
    succ_rotation = succ_rotation_succ_swap + succ_rotation_fail_swap
    try:
        prob_swap_test_success = succ_rotation_succ_swap / succ_rotation
```

```
except ZeroDivisionError:
    prob_swap_test_success = 0
return prob_swap_test_success
```

Finally, the circuit is run on the simulator.

Listing 7-1h. Circuit run `HHL_qiskit.ipynb`

```
backend = BasicAer.get_backend('qasm_simulator')
job = execute(hhl, backend, shots=100)
result = job.result()
counts = result.get_counts(hhl)
print(get_psuccess(counts))
```

This gives an output of 1.

```
1.0
```

A measurement of 1 of the ancilla indicates that the matrix inversion succeeded. This leaves the system in a state proportional to the solution vector $|x\rangle$. In many cases, you are not interested in the single vector elements of $|x\rangle$ but only on certain properties.

If the ancillary qubit is measured and if 1 is observed, then each eigenstate is divided through by λ_n, which affects the inverse. The number of times that the state preparation circuit needs to be applied to succeed, after applying amplitude amplification, is

$O\left(\dfrac{\|A\|}{C}\right)$ which is equivalent to κ, the condition number of the matrix.

The HHL algorithm takes $O[(\log N)^2]$ quantum steps to output $|x\rangle$, compared to the $O(N \log N)$ steps required to find x using the best-known method on a classical computer.

There are several facts to consider when it comes to the HHL algorithm.

- An important factor in the performance of the HHL algorithm is the *condition number κ*. As κ increases, matrix A tends to grow toward a non-invertible matrix, and the solutions become proportionately less stable. Matrices with large condition numbers are said to be ill-conditioned.

- Finding the full answer of x from the quantum state $|x\rangle$ requires $O(N)$ repetitions to reconstruct the N components of x. Generalizations to HHL, such as least-squares fitting, sidestep this problem by allowing the output to have many fewer dimensions than the input.

- The input vector b needs to be prepared, either on a quantum computer or using qRAM (quantum RAM), which may be expensive.

- The matrix must be well-conditioned, and it must be possible to simulate e^{-iA} efficiently.

QUBO

Quadratic unconstrained binary optimization (QUBO), also known as unconstrained binary quadratic programming, unifies various combinatorial optimization problems with a wide range of applications in finance and economics to machine learning that leverage applied optimization models in the quantum computing domain [103]. In recent years QUBO has evolved as a mathematical formulation that can embrace an exceptional variety of important Combinatorial Optimization problems found in industry, science, and government (see Anthony, et al. [104] and Kochenberger, et al. [105]). The power of QUBO solvers can be used to efficiently solve many important problems once they are put into the appropriate QUBO framework.

QUBO model has unchallenged importance as a strong foundation of the *quantum annealing* domain and has become a subject of study in neuromorphic computing. QUBO models lie at the heart of experimentation carried out with quantum computers developed by D-Wave and gate-based quantum computers developed by IBM when this book was written. The consequences of the relationship of QUBO models to quantum computing are being explored in initiatives by organizations such as IBM, Google, Amazon, Microsoft, D-Wave, and Lockheed Martin in the commercial realm and Los Alamos National Laboratory, Oak Ridge National Laboratory, Lawrence Livermore National Laboratory and NASA's Ames Research Center in the public sector, just to name a few. At present, diverse research bodies in the quantum computing communities are engaged in obtaining an in-depth view of the QUBO model and gauge its effectiveness as an alternative to traditional modeling and solution methodologies.

Ising Model

The significance of the QUBO model lies in its capacity to encompass several models in combinatorial optimization. Lucas [110] shows that the QUBO model is equivalent to the Ising model[4] [111]. The *Ising model* is a non-physical, mathematical model of *ferromagnetism* about statistical mechanics, which plays a prominent role in physics. The model uses discrete variables representing the magnetic dipole moments of spin states (which are either +1 or –1). The spin states are organized as a lattice so that each spin can interact with its neighbors (see Figure 7-3).

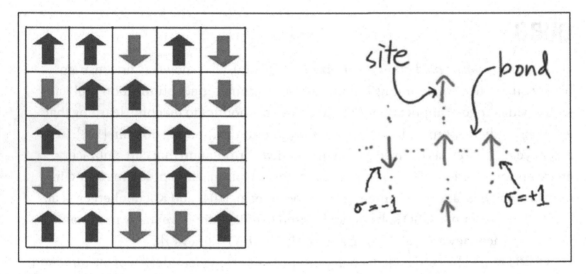

Figure 7-3. *The Ising model*

The Ising model does not correspond to an actual physical system. It's a huge (square) lattice of sites, where each site can be in one of two states. Each site is labeled with an index i, and the two states are referred to as –1 and +1. For example, when we state that spin $\sigma_i = -1$, we mean that the i-th site is in the state –1. Magnetic behavior can be understood by thinking of each atom as a spin, pointing up or down. Spins can want to line up against or with a magnetic field. Neighboring spins affect each other, but not those further away. Since the spin-spin interaction arises between two spins, it is necessary to sum over pairs of sites to find its total interaction to the energy. As a physical analogy, the Ising model can be thought of as a simplified toy magnet.

---

[4]For example, see https://en.wikipedia.org/wiki/Ising_model

The *energy* of the Ising model is defined in terms of its Hamiltonian as

$$H_{ising} = -\sum_i h_i \sigma_i - \sum_{ij} J \sigma_i \sigma_j \qquad 7.4$$

where J represents the spin-spin interaction, h represents the external field, σ is the individual spin on each lattice site. For the Ising model, the spin is given values within the following range: $\sigma_i \in \{-1, +1\}$. From a mathematical point of view, in this scheme, the h's and J's are adjustable constants with *one h* assigned for *each spin* and *one J* assigned for each *interacting pair of spins*.

QUBO from the Ising Model

The adaptation of the Ising model to form QUBOs to facilitate computations in quantum systems was made with the following adjustments.

1. Define QUBO as follows.

$$O\left(c, \alpha_i, \beta_{ij}; q_i\right) = c + \sum_i \alpha_i q_i + \sum_{i<j} \beta_{ij} q_i q_j \qquad 7.5$$

   where the q's are qubit variables 0 and 1 (instead of -1 and $+1$ in the Ising model).

2. Define α's and β's as adjustable constants.

3. Perform conversion between the 0/1 variables (for QUBO) and spins (Ising) via the following.

$$q_i = \frac{1 + \sigma_i}{2} \qquad 7.6$$

Consequently, the broad range of optimization problems solved effectively by state-of-the-art QUBO solution methods are joined by an important domain of problems arising in physics applications. As per Kochenberger and Glover [109], the optimization problems addressed by the QUBO model include the following but *are not restricted* to.

- Maximum independent set problems

- Maximum cut problems

- Graph coloring problems

- Number partitioning problems

- Linear ordering problems

- Clique partitioning problems

- Clustering problems

- Modularity maximization

  - Correlation clustering

  - Other

- Task allocation problems (distributed computer systems)

Regarding how it relates to quantum computing, some papers published by D-Wave talk about optimization and machine learning applications using the Ising model in the context of quantum computing, for example [112].

The QUBO model can be represented by the following expression.

$$maximize\ or\ minimize\ function\ f = x^T Q x \qquad\qquad 7.7$$

where x is a variable that is a binary decision vector. Q is a square matrix whose elements are constants a_{ij}. It is generally assumed that Q is symmetric or in an upper triangular form that can be achieved without any loss of generality.

A *symmetric form* is created when for all i and j, except $i = j$, a_{ij} is replaced by $(a_{ij} + a_{ji})/2$. Coefficients of the linear terms appear on the main diagonal of the Q matrix, and we ignore any constants.

Upper triangular form: for all i and j, with $j > i$, a_{ij} is replaced by $(a_{ij} + a_{ji})$; then all a_{ij} for $j < i$ are replaced by 0. If the matrix is symmetric, then this action doubles the a_{ij} values above the diagonal of the matrix, and all values below the main diagonal are set to 0.

Let's look at an example of a symmetric Q matrix. Consider the following optimization problem.

$$(\text{minimize})\ \ f = -5x - 3y - 8z - 6w + 4xy + 8xz + 2yz + 10zw$$

where variables x, y, z, w are binary; that is, they can only have values 0 and 1.
To solve this optimization problem with QUBO, we observe the following facts.

1. *f* is a quadratic function with binary variables. It has a linear part $(-5x - 3y - 8z - 6w)$ and a quadratic part $(4xy + 8xz + 2yz + 10zw)$.

2. Binary variables satisfy $x = x^2$, $y = y^2$, $z = z^2$ and $w = w^2$, since their values can be either 0 or 1. Hence,

$$-5x - 3y - 8z - 6w = -5x^2 - 3y^2 - 8z^2 - 6w^2$$

3. Now, we are able to write the model in a matrix form as follows.

$$\text{(minimize)} \quad f = (xyzw) \begin{bmatrix} -5 & 2 & 4 & 0 \\ 2 & -3 & 1 & 0 \\ 4 & 1 & -8 & 5 \\ 0 & 0 & 5 & -6 \end{bmatrix} \begin{bmatrix} x \\ y \\ z \\ w \end{bmatrix}$$

4. The resultant matrix can be written in the form of equation 7.7, where x is the column vector of binary variables. Note that the coefficients of the linear terms appear on the main diagonal of the *Q* matrix.

   The *only constraint* is the restriction of either 0 or 1 as values of the decision variables and other than this *QUBO is an unconstrained* model. All the problem data for QUBO is contained in the *Q* matrix.

4. The solution to the model is as follows: $f = -11$; $x = w = 1$; $y = z = 0$.

The programmatic solution to this problem is explored later in this chapter using pyQuil.

Other examples of QUBO optimization problem can be as follows.

QUBO with one variable: $5x + 8$

QUBO with two variables: $10 + 3x - 6y + 9xy$

QUBO with three variables: $2xy + 3xz - yz$

These characteristics make QUBO a particularly attractive modelling framework for combinatorial optimization problems as an alternative to classically constrained formulations. In summary, QUBO models belong to NP-hard class of problems. The practical meaning of this is that exact solvers designed to find "optimal" solutions (e.g., the commercial CPLEX solver (www.ibm.com/uk-en/analytics/cplex-optimizer))

are most likely unsuccessful, except for small problem instances. Using such methods, realistically sized problems can run for days and even weeks without producing high-quality solutions. Fortunately, as we disclose in the sections that follow, impressive successes are being achieved by using methods designed to find high-quality but not necessarily optimal solutions in a modest amount of computer time. These approaches have opened valuable possibilities for joining classical and quantum computing, as you see in this chapter.

Variational Quantum Circuits

In Chapter 6, you saw the inspiration drawn for quantum optimization from the Ising models, which relates to the physics of ferromagnetism. Typically, the Schrödinger equation can be sued to study and simulate the behavior of and the evolution of a single quantum particle. However, real-world phenomena are dependent on the interactions between many different quantum systems. The study of many-body Hamiltonians that model physical systems are the central theme of condensed matter physics, molecular dynamics, and modeling. *Variational methods* are of primary importance in quantum machine learning practice and current research, and hence, we spend a bit of time trying to address the fundamentals of these ideas.

As it stands today, many-body Hamiltonians are inherently hard to model and investigate on classical computers due to the characteristic of the Hilbert space, whose dimension grows *exponentially* with the number of particles in the system. As such, finding the eigenvalues of these operations becomes infeasible for classical computers. This is where a quantum computer becomes very useful. It allows us to study these many-body systems with less overhead as the number of qubits required only grows polynomially. For a perfect quantum computer, the variational tasks should become bread-and-butter and part of daily routine.

However, as attractive as it may seem to use quantum computers for problems related to many-body physics and molecular simulations, current and near-term quantum computers are far from being perfect and suffer from imperfections. This is why it is still a challenge (currently subject of intense research) to run long algorithms that require deep circuits on near-term noisy quantum computers. The preparation of quantum states using low-depth quantum circuits is one of the most promising near-term applications of small quantum computers, especially if the circuit is short enough and the fidelity of gates are high enough to be executed without quantum error correction. Such quantum state preparation can be used in variational approaches, optimizing parameters in the circuit to minimize the energy of the constructed quantum state for a given problem Hamiltonian.

Since 2013, a breed of algorithms (see references [120] and [121]) has been researched and developed, which focus on getting an advantage from imperfect quantum computers. The basic idea is to run a short sequence of gates where some gates are parametrized. The process then reads out the result, adjusts the parameters on a classical computer, and repeats the calculation with the new parameters on the quantum hardware. This way, an iterative loop is created between the quantum and the classical processing units, creating classical-quantum hybrid algorithms as shown in Figure 7-4.

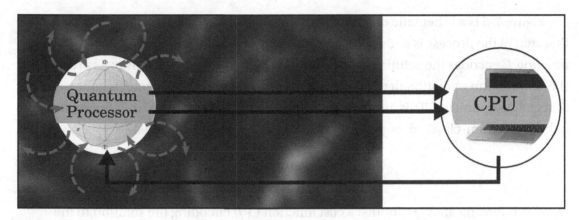

Figure 7-4. *Variational circuits for hybrid quantum-classical systems*

These algorithms are called *variational* to reflect the variational approach to changing the parameters.

Such quantum state preparation can be used in variational approaches, optimizing parameters in the circuit to minimize the energy of the constructed quantum state for a given problem Hamiltonian. For these reasons, variational circuits are also known as *parameterized quantum circuits.*

These variational algorithms, illustrated in Figure 7-5, give us a close approximation of the problems. Though not 100% perfect, they can give us answers which are close to perfection. This is why they are so useful, especially the VQE algorithms.

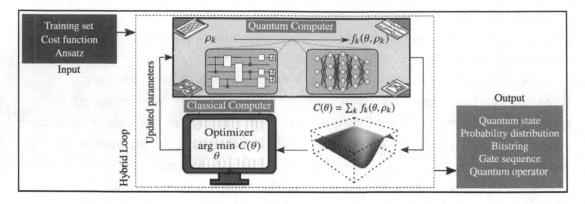

Figure 7-5. *Diagram of a quantum variational algorithm (Source [122])*

Figure 7-5 is a schematic diagram of a *variational quantum algorithm* (VQA). The first step of the process is to develop a VQA to define a cost (or loss) function C. The cost function C encodes the solution to the problem. In the next step, an ansatz is proposed (i.e., a quantum operation depending on a set of continuous or discrete parameters θ that can be optimized. This ansatz is then trained with data from the training set in a hybrid quantum-classical loop to solve the following optimization task

$$\theta^* = \arg \min_{\theta} C(\theta)$$ 7.8

The following are VQA inputs: a cost function $C(\theta)$ encoding the solution to the problem, an ansatz whose parameters are trained to minimize the cost, and optionally (if required), a set of training data used during the optimization. At each iteration of the loop, a quantum computer efficiently estimates the cost or its gradients. The output of this information is fed into a classical computer that leverages the power of optimizers to navigate the cost derivatives and solve the optimization problem in equation 7.8. Once a termination condition is met, the VQA estimates an output of the solution to the problem. The form of the output depends on the precise task. Figure 7-5 indicates some of the most common types of output.

The cost function, introduced in Chapter 4, maps the values of the trainable parameters θ to real numbers. The cost defines a *hypersurface,* usually called the *cost landscape* (see Figure 7-5). The task of the optimizer is to navigate through the cost landscape and obtain the global minima. Without any loss of generality, the cost can be expressed as

$$C(\theta) = \sum_{k} f_k \left[\mathrm{Tr}\left(O_k U(\theta) \rho_k U^{\dagger}(\theta) \right) \right]$$ 7.9

where f_k are functions that encode the task. Discrete and continuous parameters make up θ. $\{O_k\}$ defines a *set* of observables. $U(\theta)$ is a parametrized unitary. $\{\rho_k\}$ consists of input states (from a training set). For a given problem, f_k can differ, which may lead to useful costs.

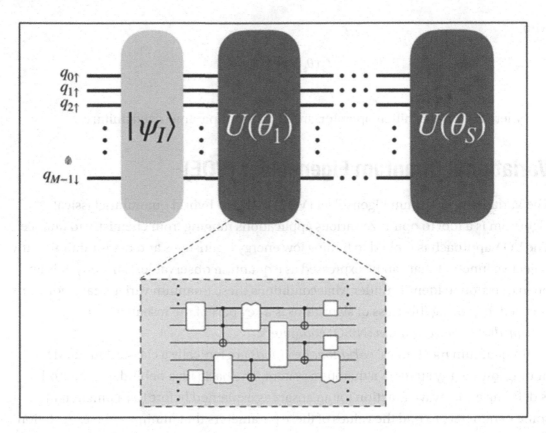

Figure 7-6. *Schematic of an ansatz*

As cost functions are important for VQAs, so is the choice of *ansatz*. The form of the ansatz defines the θ parameters and how they can be trained to minimize the cost. The specific structure of an ansatz generally depends on the task. In some instances, information on the problem can tailor an ansatz, which is a *problem-inspired ansatz*. Some other ansatz architectures are generic and independent of the problem, meaning they can be used even when no relevant information is readily available. For the cost function in equation 7.9, the parameters θ can be encoded in a unitary $U(\theta)$ that is applied to the input states to the quantum circuit. As shown in Figure 7-6, $U(\theta)$ can be generally expressed as the product of S *sequentially* applied unitaries.

$$U(\theta) = U_S(\theta_S)...U_2(\theta_2)U_1(\theta_1)$$

7.10

with,

$$U_S(\theta_S) = \prod_n W_n e^{-i\theta_n \mathbb{H}_n}$$

7.11

where $\mathbb{H}_n$ is a Hermitian operator, and W_n is an unparametrized unitary.

Variational Quantum Eigensolver (VQE)

The Variational Quantum Eigensolver (VQE) [121] is a hybrid quantum-classical algorithm is a tool to optimize various applications ranging from chemistry to finance. The VQA approach is limited to finding low energy eigenstates and helps minimize any objective function that can be expressed as a quantum observable. Currently, it is an open question to identify under what conditions these quantum variational algorithms succeed. Exploring this class of algorithms is a key part of the research for *noisy intermediate-scale quantum* (NISQ) computers.

In quantum mechanics, *variational methods* are typically a classical method for finding low energy states of a quantum system. The basic idea behind this method is defining a trial wave function (or an ansatz as explained before) as a function of some parameters to find the values of these parameters that minimize the expectation value of the energy *with respect to* these parameters. This *minimized ansatz* is then an *approximation to the lowest energy eigenstate*, and the expectation value serves as an upper bound on the energy of the ground state. Hence, realistically, we end up finding the upper bound of the minimum eigenvalue of a specific Hamiltonian and its corresponding eigenvector. VQE helps us determine the ground state energy of a quantum system.

In VQE, the set of parameters that minimize the energy for a given ansatz wave function are the solution to the problem. The algorithm consists of *two* parts: a *quantum circuit* and a *classical optimizer*. The quantum circuit performs two tasks.

- It generates the ansatz wave function of the system with the given variational parameters as input.

- It measures the energy (or expectation value of Hamiltonian) with respect to that of the wave function.

The result of the measurement is then fed into the classical optimizer to generate a new set of variational parameters, which gives a lower energy estimation. Figure 7-7 shows the original VQE process in the schematic from reference [121].

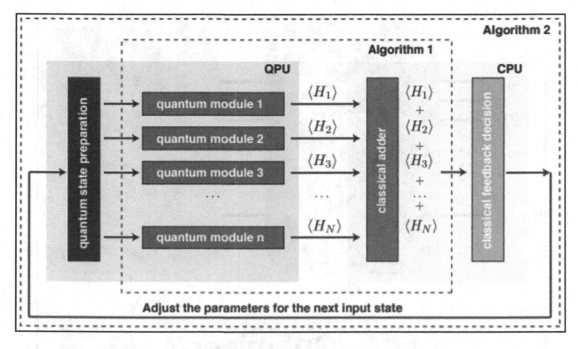

Figure 7-7. *Variational Quantum Eigensolver (Source [121])*

It is known that the eigenvector that minimizes the expectation $\langle H \rangle$ corresponds to the eigenvector of H that has the smallest eigenvalue [121]. Hence, in Figure 7-7, all possible trial wave functions $|\psi\rangle$ can be tried to find the eigenvector that gives the smallest expectation value. How do we create these trial states? In the algorithm, the trial states are created from a parametrized circuit. Different wave functions or ansatz states can be created by changing the parameters [121].

The classical feedback decision is some optimization method that changes the parameters of the quantum state preparation. With different quantum modules, the algorithm calculates expectation values of each Pauli term, and then it sums them by using a classical adder (see Figure 7-7) via a classical computer. The algorithm then returns to classical feedback decision to choose better parameters for the quantum state preparation. VQE repeats this procedure until the optimization method is satisfied with the obtained result.

Figure 7-8 shows the process to simulate the physics of interacting spin-half fermions on a one-dimensional lattice on a quantum computer. The Hamiltonian of the system is described by the Hubbard[5] model [123], where in addition to hopping between lattice sites, fermions with opposite spins interact via an on-site Coulomb potential.

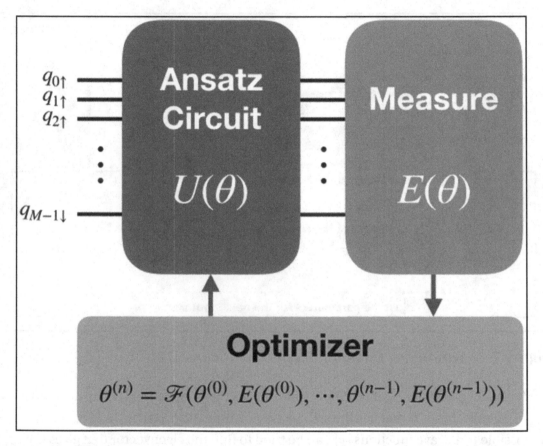

Figure 7-8. *VQE process example of Hubbard model*

As Figure 7-8 shows, the VQE algorithm for the Hubbard model initiates with a set of variational parameters $\theta^{(0)}$. Then, the ansatz circuit represented by the unitary operator $|\psi(\theta^{(0)})\rangle = U(\theta^{(0)}) |\psi_1\rangle$ is created to generate the ansatz wave function $|\psi(\theta^{(0)})\rangle$ on the

---

[5]Hubbard model: https://en.wikipedia.org/wiki/Hubbard_model

quantum computer, where $|\psi_1\rangle$ is the initial wave function, which is usually chosen to be the *ground state* of the non-interacting Hamiltonian H_0. The energy of the system is obtained by measuring the *expectation value* of Hamiltonian with respect to the ansatz, and this is strictly greater than or equal to the *ground state energy* E_0; that is,

$$E\left(\theta^{(0)}\right) = \left\langle \psi\left(\theta^{(0)}\right)\middle| H \middle| \psi\left(\theta^{(0)}\right)\right\rangle \geq E_0$$

In the last step, the classical optimizer generates a new set of parameters $\theta^{(1)}$ dependent on the previous parameters and energies. These new parameters are fed back into the ansatz circuit to generate a *new wave function* with lower energy. Here, the optimizer is modeled as a basic search algorithm where the space of parameters is sampled randomly. The whole process is repeated until the energy $E(\theta^{(n)})$ is sufficiently close to the ground state energy E_0 so that $\theta^{(n)}$ is the solution to the problem, and any information on the system can be extracted by first generating the wave function $|\psi(\theta^{(n)})\rangle$ on the quantum computer and then measuring the desired observables.

Going back to Ising models, physics, and molecular ground state energy, a central task is finding the ground state (lowest energy eigenstate) of a given Hamiltonian H (*not to be confused* with the Hadamard gate!).

$$H|\psi\rangle = \lambda_G|\psi\rangle \qquad\qquad 7.12$$

where λ_G is the ground state eigenvalue and $|\psi\rangle$ is the eigenstate. Studying the ground state gives us information about the low energy properties of the system. Once $|\psi\rangle$ is known, the physical properties can be deduced from the wave function. However, the challenge in equation 7.12 is that generally, the value of λ_G and the eigenstate are unknowns; hence, obtaining an estimate for these is of paramount importance. The energy estimation for the state can be obtained by calculating the expectation value of the Hamiltonian.

$$\langle\psi|H|\psi\rangle = E|\psi\rangle \qquad\qquad 7.13$$

where, E is the energy eigenstate. E_0 is the ground state. The variational principle states that

$$\langle\psi|H|\psi\rangle = E|\psi\rangle \geq E_0 \qquad\qquad 7.14$$

Now, the Hamiltonian H can be described by the properties of any Hermitian matrix as follows.

$$H = \sum_{i=1}^{N} \lambda_i |\psi_i\rangle\langle\psi_i|$$
7.15

So, the expectation value of H can be represented as

$$\langle H \rangle = \langle\psi|H|\psi\rangle = \langle\psi|\left(\sum_{i=1}^{N} \lambda_i |\psi_i\rangle\langle\psi_i|\right)|\psi\rangle$$

$$= \sum_{i=1}^{N} \lambda_i \langle\psi|\psi_i\rangle\langle\psi_i|\psi\rangle$$

$$= \sum_{i=1}^{N} \lambda_i |\langle\psi_i|\psi\rangle|^2$$
7.16

Equation 7.16 indicates that the expectation value of any observable state can be represented by a sum using the eigenvalues associated with H as weights. And, by the postulate of quantum mechanics, it is known that the weights in a linear combination must be greater than or equal to zero. This is what allows the VQE algorithm to work.

For a more in-depth look into Hamiltonians from a computational point of view, a Hamiltonian can be expressed by the sum of tensor products of Pauli operators (Pauli terms).

$$H = 0.2 \cdot X \otimes Y + 0.8 \cdot I \otimes Z + 0.6 \cdot I \otimes X$$

$$= 0.2 \cdot XY + 0.8 \cdot IZ + 0.6 \cdot IX$$
7.17

where, the Pauli operators are expressed in their standard forms.

Given the Hamiltonian of equation 7.17 and a state $|\psi\rangle$, the goal in VQE is to measure the expectation value of H

$$\langle\psi|H|\psi\rangle = \langle H \rangle = 0.2 \cdot \langle\psi|XY|\psi\rangle + 0.8 \cdot \langle\psi|IZ|\psi\rangle + 0.6 \cdot \langle\psi|IX|\psi\rangle$$
7.18

As may be surmised from equation 7.18, the VQE algorithm constructs a quantum circuit for each Pauli term and computes the expectation value of the corresponding Pauli term. Then, the algorithm sums all calculated expectation values of Pauli terms and obtains the expectation value of H. The algorithm runs this routine of estimating the expectation value of H repeatedly for different trial wave functions (ansatz states).

The parameters of the state preparation circuit are controlled by a classical computer. At each step, the classical computer changes the parameters by using some optimization method to create an ansatz state with a smaller expectation value than the ansatz states before. This process allows the classical computer and the quantum computer to work together to archive the algorithm's goal, which is to find the ground state energy. This is why VQE is a hybrid classical-quantum algorithm.

Now that we have looked at the theory and practices of VQE, it is good to explore examples of VQE in different formats, as VQE is one of the most important concepts in quantum machine learning.

VQE with Qiskit

In this example, the task is to determine the lowest eigenvalue of the matrix M using the variational quantum eigensolver method (VQE), where

$$M = \begin{bmatrix} 0 & 0 & 0 & 0 \\ 0 & -1 & 1 & 0 \\ 0 & 1 & -1 & 0 \\ 0 & 0 & 0 & 0 \end{bmatrix}$$

As a recap from Chapter 1, Pauli matrices are given by

$$X \equiv \sigma_1 \equiv \sigma_x \equiv \begin{bmatrix} 0 & 1 \\ 1 & 0 \end{bmatrix}; \; Y \equiv \sigma_2 \equiv \sigma_y \equiv \begin{bmatrix} 0 & -i \\ i & 0 \end{bmatrix}; \; Z \equiv \sigma_3 \equiv \sigma_z \equiv \begin{bmatrix} 1 & 0 \\ 0 & -1 \end{bmatrix}$$

A two-qubit vector can be represented as

$$v \begin{bmatrix} a \\ b \\ c \\ d \end{bmatrix} = a|00\rangle + b|01\rangle + c|10\rangle + d|11\rangle$$

As a first step, the matrix M is decomposed into a sum of two-qubit operators.

$$M = \alpha\left(X_1 \otimes X_2\right) + \beta\left(Y_1 \otimes Y_2\right) + \gamma\left(Z_1 \otimes Z_2\right) + \lambda\left(I_1 \otimes I_2\right) \tag{7.19}$$

where $\otimes$ denotes the tensor product, $\alpha, \beta, \gamma, \lambda$ are unknown coefficients, I is the identity Pauli matrix and the subscripts 1, 2 denote the specific qubit acted upon by the operators. This decomposition is solved using the matrix representation of the tensor product.

To visualize this representation, the Pauli matrices, X, Y, Z, are defined in Listing 7-2a for the VQE_qiskit.ipynb Jupyter notebook file, along with the usual library calls.

Listing 7-2a. Libraries and Pauli Matrices for VQE_qiskit.ipynb

```
from qiskit import *
import numpy as np
# Define the Pauli matrices
X = np.array([[0,1],[1,0]])
Y = np.array([[0,-np.complex(0,1)],[np.complex(0,1),0]])
Z = np.array([[1,0],[0,-1]])
```

In the next step, we create the tensor products of these operators, as shown in Listing 7-2b.

Listing 7-2b. Libraries and Pauli Matrices for VQE_qiskit.ipynb

```
# Compute the tensor product of the Pauli matrices
XX = np.kron(X,X)
YY = np.real(np.kron(Y,Y)) # This will be a real matrix
ZZ = np.kron(Z,Z)
II = np.identity(4)
print("The XX gate is: \n{} \n".format(XX))
print("The YY gate is: \n{} \n".format(YY))
print("The ZZ gate is: \n{} \n".format(ZZ))
```

Listing 7-2b gives us the following output.

```
The XX gate is:
[[0 0 0 1]
 [0 0 1 0]
 [0 1 0 0]
 [1 0 0 0]]

The YY gate is:
[[ 0.  0.  0. -1.]
 [ 0.  0.  1.  0.]
 [ 0.  1.  0.  0.]
 [-1.  0.  0.  0.]]

The ZZ gate is:
[[ 1  0  0  0]
 [ 0 -1  0  0]
 [ 0  0 -1  0]
 [ 0  0  0  1]]
```

Listing 7-2c verifies what we know by *inspection*; that is, M can be decomposed into the following sum of two qubit operators.

$$M = \frac{1}{2}\left(X_1 \otimes X_2 + Y_1 \otimes Y_2 + Z_1 \otimes Z_2 - I_1 \otimes I_2\right)$$ 7.20

Listing 7-2c. Decomposition of M in `VQE_qiskit.ipynb`

```
# This is the decomposition of the "M" matrix
M = (XX+YY+ZZ-II)/2
print('M = \n',M)
print('')
```

In turn, it gives us a confirmation as an output that this is the M matrix.

```
M =
 [[ 0.  0.  0.  0.]
  [ 0. -1.  1.  0.]
  [ 0.  1. -1.  0.]
  [ 0.  0.  0.  0.]]
```

Computing the Expectation Value of the Operators

With the decomposition of M in place, it is time to determine the bound on the *lowest eigenvalue* of M using the *variational theorem*. To achieve this task, we need to know the expectation values of M for different trial wave functions. We know from the basics of quantum mechanics that the expectation value of an operator M for a ground state wave function $|\psi(\theta)\rangle$ is of the following form.

$$F(\theta) = \langle \psi(\theta)|M|\psi(\theta)\rangle$$

where $E(\theta)$ is the expectation value of the matrix M. $|\psi(\theta)\rangle$ is the variational wave function parametrized by the angle θ.

where the simplest computable expectation value is the $Z_1 \otimes Z_2$ operator. When acting on a state given by $|\phi\rangle = (a, b, c, d)$, such that $\langle \phi|\phi\rangle = 1$, the expectation value is given by

$$\langle \phi|Z_1 \otimes Z_2|\phi\rangle = a^2 + d^2 - \left(b^2 + c^2\right)$$

The terms a^2 and d^2 correspond to the probability of the states $|00\rangle$ and $|11\rangle$ respectively and b^2 and c^2 are the probabilities of the states $|01\rangle$ and $|10\rangle$.

In order to formulate simulations, the *probabilities* are calculated as follows.

$$\langle \phi|Z_1 \otimes Z_2|\phi\rangle = \frac{Number\ of\ |00\rangle\ or\ |11\rangle\, states - Number\ of\ |01\rangle\ or\ |10\rangle\, states}{Number\ of\ trials}$$

Taking a leaf out of the decomposition trial to determine the expectation value of other operators for the trial states, we seek to decompose all two-qubit operators into the following form.

$$Q = U^\dagger \left(Z_1 \otimes Z_2\right)U$$

where U is a unitary matrix specific to Q. The following decompositions are relevant to the task.

$$X_1 \otimes X_2 = \left(H_1 \otimes H_2\right)^\dagger \left(Z_1 \otimes Z_2\right)\left(H_1 \otimes H_2\right) \tag{7.21}$$

$$Y_1 \otimes Y_2 = \left(H_1 S_1^\dagger \otimes H_2 S_2^\dagger\right)^\dagger \left(Z_1 \otimes Z_2\right)\left(H_1 S_1^\dagger \otimes H_2 S_2^\dagger\right)$$ 7.22

where H is the Hadamard gate, and

$$S^\dagger = \begin{bmatrix} 1 & 0 \\ 0 & -i \end{bmatrix}$$

Equations 7.20, 7.21, and 7.22 allow us to write the original matrix M as a sum of the operators, as follows.

$$M = \frac{1}{2}\left(H_1 \otimes H_2\right)^\dagger \left(Z_1 \otimes Z_2\right)\left(H_1 \otimes H_2\right)$$

$$+ \frac{1}{2}\left(H_1 S_1^\dagger \otimes H_2 S_2^\dagger\right)^\dagger \left(Z_1 \otimes Z_2\right)\left(H_1 S_1^\dagger \otimes H_2 S_2^\dagger\right) + \frac{1}{2}\left(I_1 \otimes I_2\right) - \frac{1}{2}\left(Z_1 \otimes Z_2\right)$$

From these identities, the expectation value of the matrix M is given as

$$\langle \psi | M | \psi \rangle = \frac{1}{2}\left[\langle \phi |\left(Z_1 \otimes Z_2\right)|\phi\rangle + \langle \varphi |\left(Z_1 \otimes Z_2\right)|\varphi\rangle + \langle \psi |\left(Z_1 \otimes Z_2\right)|\psi\rangle - 1\right]$$

where the new evolved states are

$$|\phi\rangle = \left(H_1 \otimes H_2\right)|\psi\rangle$$

and

$$|\varphi\rangle = \left(H_1 S_1^\dagger \otimes H_2 S_2^\dagger\right)|\psi\rangle$$

Ansatz for the VQE

After the computation of the expectation value of the matrix M, the next step is to define a *trial wave function* as an ansatz.

$$|\Psi(\theta)\rangle = \left(I \otimes X\right)CX\left[R_Z(\theta) \otimes I\right]\left(H \otimes I\right)|00\rangle$$

Listing 7-2d gives the visualization of the ansatz circuit.

Listing 7-2d. Ansatz Circuit Visualization VQE_qiskit.ipynb

```
# Visualize Ansatz-circuit
theta=np.pi
q = QuantumRegister(2)
c = ClassicalRegister(2)
circuit = QuantumCircuit(q,c)
circuit.h(q[0])
circuit.rz(theta,q[0])
circuit.cx(q[0],q[1])
circuit.x(q[1])
circuit.barrier()
circuit.draw()
```

This code snippet gives the following output as our ansatz circuit.

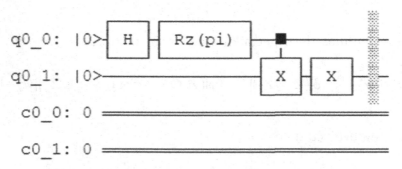

Listing 7-2e defines some functions to facilitate our evaluation of the expectation value of M (as described in the comments).

Listing 7-2e. Function Definition VQE_qiskit.ipynb

```
def anzatz(circuit,q,c,theta):
    '''
    The Anzatz wave function
    |psi > = (I X)(CX)(Rz(theta) I)(H I)|00>
    '''
    circuit.h(q[0])
    circuit.rz(theta,q[0])
    circuit.cx(q[0],q[1])
```

```python
    circuit.x(q[1])
    circuit.barrier()

    return circuit

def measure_ZZ(circuit,N_shots=2**10,simulator_backend='qasm_simulator'):
    '''
    Measures the expectation value of ZZ on the 2-qubit states
    <ZZ> = the number of ( 00 )  and (11) states, minus the number of (01)
    and (10) states
    normalized over the number of trials
    '''
    simulator = Aer.get_backend(simulator_backend)
    result = execute(circuit,backend=simulator,shots=N_shots).result()

    items =result.get_counts().items()

    s = 0
    for key, counts in items:
        s+= (-1)**(int(key[0])+int(key[1]))*counts
    s = s/N_shots

    return s

def hamiltonian(theta,N_shots=2**10):
    '''
    The hamiltonian for the problem that corresponds
    to our matrix M:
    M = (1/2)*(XX+YY+ZZ-II)
    The hamiltonian is computed by separating it into three components
    according to the discussions in the previous cells.
    '''

    q = QuantumRegister(2)
    c = ClassicalRegister(2)

    ## 0.5*XX component
    circuit_xx = QuantumCircuit(q,c)
    circuit_xx = anzatz(circuit_xx,q,c,theta)
    circuit_xx.h(q[0])
    circuit_xx.h(q[1])
```

```
circuit_xx.measure(q,c)
E_XX = 0.5*measure_ZZ(circuit_xx,N_shots=N_shots)

## 0.5*YY component
circuit_yy = QuantumCircuit(q,c)
circuit_yy = anzatz(circuit_yy,q,c,theta)
circuit_yy.sdg(q[0])
circuit_yy.h(q[0])
circuit_yy.sdg(q[1])
circuit_yy.h(q[1])
circuit_yy.measure(q,c)
E_YY= 0.5*measure_ZZ(circuit_yy,N_shots=N_shots)

    # 0.5*ZZ component
circuit_zz = QuantumCircuit(q,c)
circuit_zz = anzatz(circuit_zz,q,c,theta)
circuit_zz.measure(q,c)
E_ZZ = 0.5*measure_ZZ(circuit_zz,N_shots=N_shots)

# The - 1/2 comes from the fact that <psi|II|psi> = 1
# it is always a constant

return (E_XX+E_YY+E_ZZ-0.5)
```

The quantum circuit simulations are run in the next step, as shown in Listing 7-2f.

Listing 7-2f. Minimum Expectation Value of M VQE_qiskit.ipynb

```
# Generate several thetas, and find the best set of parameters
import matplotlib.pyplot as plt
# The number of trials to run
N_shots = 2**10

# Generate theta grid
theta = np.linspace(0.0,2*np.pi,200)
E = np.zeros(len(theta))

# Calculate the expectation value of the Hamiltonian for different theta
for k in range(len(theta)):
    E[k] =  hamiltonian(theta=theta[k],N_shots=N_shots)
```

```
# Plot the results
plt.title('Expectation value vs Angle',fontsize=15)
plt.ylabel('Expectation value of H',fontsize=15)
plt.xlabel('Theta (radians)',fontsize=15)
plt.plot(theta,E,label='Number of shots: {}'.format(N_shots))
plt.legend()
plt.show()

# Determine the lowest bound from varying theta
print('='*100)
print('The minimum bound on the lowest eigenvalue of M is E0={},\n'.
format(min(E)))
print('The parameter that corresponds to this energy is theta={:0.2f}
Rads'.format(theta[np.argmin(E)]))
print('='*100)
```

This code snippet gives the following output as our expectation value.

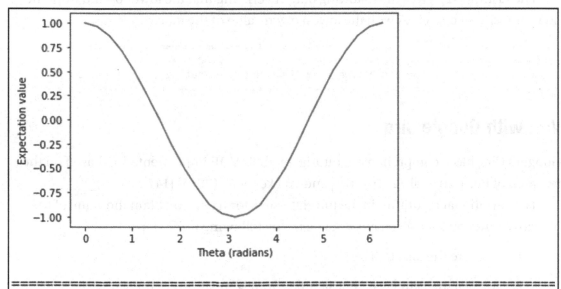

```
================================================================
The minimum bound on the lowest eigenvalue of M is E0=-1.0,

The parameter that corresponds to this energy is theta=3.14 Rads
================================================================
```

The minimum value of the minimum energy of the Hamiltonian in the output occurs at approximately $\theta = \pi$ in the wave function ansatz with −2 as a lower bound of expectation value. The energy is shown in arbitrary units.

Now that the lower bound for the smallest eigenvalue of M is found to be $E0 = -2.0$, the minimum eigenvalue can be calculated directly to verify our solution.

Listing 7-2g. Ansatz Circuit Visualization VQE_qiskit.ipynb

```
w, v = np.linalg.eig(M)

print('='*100)
print('Minimum eigenvalue using classical computer: Lambda={:0.2f}'.
format(min(w)))
print('The minimum bound of the Eigenvalue from the quantum computing
simulations is E={:0.2f}'.format(min(E)))
print('='*100)
```

The Listing 7-2g gives the following output verifying that the lower bound from the quantum circuit coincides with the lowest eigen value of the matrix.

```
=========================================================================================
Minimum eigenvalue using classical computer: Lambda=-2.00
The minimum bound of the Eigenvalue from the quantum computing simulations is E=-2.00
=========================================================================================
```

VQE with Google Cirq

Google's Cirq has a comprehensive tutorial on their VQE implementation based on the research of Peruzzo et al. (2014) [121] and Wecker et al. (2015) [147].

The variational algorithm in the tutorial works for trying to obtain the value of the objective function for a given ansatz state by the following three steps.

1. Prepare the ansatz state.

2. Make a measurement that samples from some terms in the Hamiltonian.

3. Go to 1.

They also point out the important aspect that you cannot always *directly measure* the Hamiltonian without *quantum phase estimation* (see Chapter 6). Hence, one often relies on the linearity of expectation values to measure parts of the Hamiltonian in step 2. One always needs to repeat the measurements to obtain an estimate of the expectation value.

The tutorial shows that you can use a quantum computer to obtain estimates of the objective function for the ansatz. This can then be used in an outer loop to obtain parameters for the lowest value of the objective function. For these values, you can then use that best ansatz to produce samples of solutions to the problem, which obtain a hopefully good approximation for the lowest possible value of the objective function. The tutorial is complete with the code in full at `https://quantumai.google/cirq/ tutorials/variational_algorithm` [148]. You are highly encouraged to visit this page and try out the VQE implementation example in Cirq.

For ease of accessing the material of this tutorial, the associated opensource Jupyter notebook code for [148] has been included in this chapter as `variational_algorithm_ Cirq.ipynb` with permission from Google.

VQE with Rigetti Forest

VQE programming in Qiskit was discussed earlier in the chapter. Given the importance of VQE in quantum machine learning and molecular modeling, it is helpful to see how other platforms implement it. We look first at a Rigetti Forest implementation of VQE similar to the Qiskit exercise and then a molecular modeling exercise leveraging VQE. We leverage VQE for various QML applications and libraries related to various platforms as we go forward in this book.

To start, we open three terminals in our Forest virtual environment, just as we did in Chapter 5. We start QVM in server mode, Quilc in server mode, and Jupyter notebook with the following commands, each in its separate terminal.

- QVM: `qvm -S`

- Quilc: `quilc -S`

- Jupyter notebook: `jupyter notebook`. Select Python 3 to create a new notebook instance.

Let's start by looking at a basic VQE implementation using Forest. To run some optimization tasks, we need to import Rigetti's Grove library.

As usual, we first import the required libraries, define a matrix with the Pauli Z-gate and define an ansatz, as shown in Listing 7-3a for VQE_pyquil.ipynb.

Listing 7-3a. Ansatz Circuit Visualization VQE_pyquil.ipynb

```
# Variational-Quantum-Eigensolver
import numpy as np
from scipy.optimize import minimize
from pyquil.api import QVMConnection
import pyquil.api as api
from pyquil.gates import *
from pyquil import Program, get_qc
from grove.pyvqe.vqe import VQE

# Create connection with QVM
qvm = QVMConnection()
qc = get_qc('2q-qvm')

# Define a matrix
from pyquil.paulis import sZ
H = sZ(0)
# Define an ansatz
from pyquil.gates import RY
def ansatz(params):
    return Program(RY(params[0], 0))
```

In the next step, we define functions for calculating expectation values the same as we did in Qiskit with 10,000 measurements for our trial ansatz. This is shown in Listing 7-3b.

Listing 7-3b. Functions VQE_pyquil.ipynb

```
# Function calculating expectation value
def expectation(params):
    # Define number of measurements
    samples = 10000
```

```
    # Define program
    prog = ansatz(params)
    # Measure
    ro = prog.declare('ro', 'BIT', 1) # Classical registry storing the
    results
    prog.measure(0, ro[0])
    # Compile and execute
    prog.wrap_in_numshots_loop(samples)
    prog_exec = qc.compile(prog)
    ret = qc.run(prog_exec)
    # Calculate expectation
    freq_is_0 = [trial[0] for trial in ret].count(0) / samples
    freq_is_1 = [trial[0] for trial in ret].count(1) / samples
    return freq_is_0 - freq_is_1

# Test of expectation value function
test = expectation([0.0])
print(test)
```

Listing 7-3b generates a 1.0 as output confirming our estimation.

Listing 7-3c, plots the VQE graph of expectation vs. parameter values (i.e., values of θ).

Listing 7-3c. Plot the VQE curve VQE_pyquil.ipynb

```
# Draw expectation alue against parameter value
params_range = np.linspace(0.0, 2 * np.pi, 25)
data = [expectation([params]) for params in params_range]
import matplotlib.pyplot as plt
plt.xlabel('Parameter value (Thetas)')
plt.ylabel('Expectation value')
plt.plot(params_range, data)
plt.show()
```

We get the following output which is very similar to the one obtained for the Qiskit case.

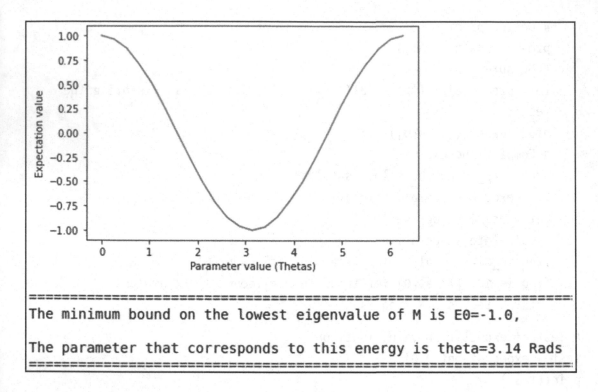

```
=================================================================
The minimum bound on the lowest eigenvalue of M is E0=-1.0,

The parameter that corresponds to this energy is theta=3.14 Rads
=================================================================
```

The minimum value of the minimum energy of the Hamiltonian in the output occurs at $\theta = \pi$ in the wave function ansatz with -1 as a lower bound of expectation value. The energy is shown in arbitrary units.

As seen in both examples of VQE (in Qiskit and Forest), there is a practical limit to the number of expectation values that can be calculated against parameters. Calculation of expectation values of much larger Hamiltonians, which require more parameter values, is a challenge. This is addressed by using optimization algorithms. For example, the Nelder-Mead [149] optimization algorithm can be applied via the SciPy optimize package, as shown in Listing 7-3d.

Listing 7-3d. Nelder-Mead Optimization VQE_pyquil.ipynb

```
# Eigenvalue Optimization
initial_params = [0.0]
minimum = minimize(expectation, initial_params, method='Nelder-Mead',
                   options={'initial_simplex': np.array([[0.0], [0.05]]),
                   'xatol': 1.0e-2})
print(minimum)
```

The Nelder-Mead[6] method [149] operates with a *simplex*. A simplex is a triangle in two dimensions and a tetrahedron in 3-D. The Wikipedia page illustrates how a simplex moves toward a minimum; it keeps changing its size and shape to become smaller near the minimum. The search is successful if two conditions are satisfied.

- The *size* of the simplex is at most `xatol`, which is part of the SciPy library. It defines the *accuracy* of the result in terms of *absolute error*. For example, if xatol is 0.01 and the method returns the location of the minimum as [1.33, 2.56], then there is hope (but not certainty) that the *actual minimum* has coordinates somewhere within 1.32–1.34 and 2.55–2.57.

- The difference of function values at the vertices of the simplex is at most `fatol`.

The meaning of this boils down to the simplex becoming smaller and the values of the objective function at its vertices remaining almost the same.

Listing 7-3d gives us the following output where `fun` is the minimum eigenvalue.

```
final_simplex: (array([[3.15    ],
       [3.14375]]), array([-1., -1.]))
          fun: -1.0
      message: 'Optimization terminated successfully.'
         nfev: 30
          nit: 14
       status: 0
      success: True
            x: array([3.15])
```

So far, we have dealt with noise-free QVM formulations. Rigetti Forest also offers options to add noise to the simulations. Because a VQE algorithm is meant to be used on near-term quantum computing platforms, it is helpful to consider a QVM run with a particular noise model. Listing 7-3e is a code snippet to set up noisy QVM in pyQuil.

---

[6]Nelder-Mead: https://en.wikipedia.org/wiki/Nelder%E2%80%93Mead_method

Listing 7-3e. Noisy-QVM VQE_pyquil.ipynb

```
pauli_channel = [0.1, 0.1, 0.1] #10% chance of each gate at each timestep
noisy_qvm = api.QVMConnection(gate_noise=pauli_channel)

#Verify Simulator is indeed noisy
p = Program(X(0), MEASURE(0, 1))
noisy_qvm.run(p, [0, 1], 10)
```

Listing 7-3e introduces a 10% chance of each gate at each timestamp. An example output of the code measuring the |0⟩ state shows the bit 0 being measured, which indicates that the QVM simulator is indeed noisy.

```
[[1, 1],
 [1, 1],
 [0, 0],
 [1, 1],
 [1, 1],
 [0, 0],
 [0, 0],
 [1, 1],
 [1, 1],
 [1, 1]]
```

A plot for the noisy VQE can be generated by running a for loop over the range of angle θ and the expectation values. To achieve this, Forest's grove libraries must be installed in the same directory as the QVM and Quilc with the following command.

```
$ pip install quantum-grove
```

The installation command and the initial output are shown in the following screenshot where the grove library is being installed in a virtual environment called qml.

```
(qml) sgangoly@ubuntu:~$ pip install quantum-grove
Collecting quantum-grove
  Downloading quantum-grove-1.7.0.tar.gz (59 kB)
     |████████████████████████████████| 59 kB 1.4 MB/s
```

The actual generation of the VQE plot with a noisy QVM is left as an exercise for you. Refer to the Grove documentation at `https://grove-docs.readthedocs.io/en/latest/vqe.html`.

VQE for Molecular Systems

As a quick example of demonstrating the real-life value of a VQE-type algorithm, the following is an exercise to garner an accurate approximation for the lowest energy state of a molecular system. The ability to increase efficiency in running such algorithms has a direct positive impact on improvements in drug and medicine development and methods of materials engineering.

This code is adapted from the Grove documentation at `https://grove-docs.readthedocs.io/en/latest/vqe.html`. Grove allows us to define our own Hamiltonian and obtain results using the Pauli matrices for sigma Z and sigma Y. Listing 7-3e demonstrates how a choice of Hamiltonian may result in different measurements. In theory, if the decomposition of Pauli matrices for the Hydrogen molecule Hamiltonian could be determined then, we would be able to substitute the Hamiltonians used with a combination of those that apply to the molecule. Hence, either a graph of the variation of the parameter could be generated and the result found by eye, or we could try a different minimization method using the data received from the VQE function.

Listing 7-3f gives you a taste of VQE, helping resolve a real-world molecular simulation exercise where we loop the expectation value 10,000 times.

Listing 7-3f. Molecular System VQE_pyquil.ipynb

```
# Molecular Systems Solutions
from pyquil.paulis import sY
initial_angle = [0.0]
#Hamiltonian is sigma Y
hamiltonian = sY(0)
vqe_inst = VQE(minimizer=minimize,
               minimizer_kwargs={'method': 'nelder-mead'})
#ansatz is rotation by X
def small_ansatz(params):
    return Program(RX(params[0], 0))
```

```
#looping from 0 to 2 pi
angle_range = np.linspace(0.0, 2 * np.pi, 20)
data = [vqe_inst.expectation(small_ansatz([angle]), hamiltonian, 10000, qvm)
        for angle in angle_range]

plt.xlabel('Angle [radians]')
plt.ylabel('Expectation value')
plt.plot(angle_range, data)
plt.show()
```

Listing 7-3f produces the following plot as output for the expectation values of the molecular Hamiltonian against variation of parameter θ.

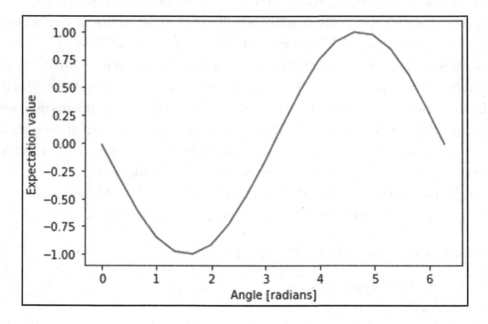

In summary, the VQE is a hybrid quantum-classical algorithm [121] that finds the smallest eigenvalue and corresponding eigenvector of a given Hamiltonian. One of the main applications of the algorithm is finding the ground state energy of molecules. Currently, in the NISQ era, our quantum computing abilities are limited to working with very noisy qubits because they are not efficiently isolated from their environment. This restriction gives a huge advantage to algorithms such as VQE that utilize small depth circuits.

Applications such as simulating large quantum systems or solving large-scale linear algebra problems are immensely challenging for classical computers due to their extremely high computational cost. Quantum computers promise to unlock these applications, although fault-tolerant quantum computers will not likely be available for several years. Currently, quantum devices have serious constraints, including limited qubit numbers and noise processes that limit circuit depth.

VQE in molecular simulation realistically utilizes quantum computers' features because it is a "near term" quantum computing algorithm. This means that it does not depend on thousands of qubits, reliable and scalable error correction, long coherence times, or large circuit depths. These are the algorithm types that have high potential to improve the efficiency of applications running on quantum platforms.

QAOA

The Quantum Alternating Operator Ansatz (QAOA)—also known as Quantum Approximate Optimization Algorithm—was originally introduced by Farhi et al. [69] in 2014 to obtain approximate solutions for combinatorial optimization problems. It was later generalized as a standalone ansatz [124] shown to be computationally universal [125]. QAOA is one of the most important examples of the VQE family of algorithms.

The QAOA is a shallow-circuit variational algorithm for quantum computers that was inspired by quantum annealing. The evolution of a quantum system is governed by the Schrödinger equation $i\hbar \dfrac{d|\psi(t)\rangle}{dt} = H|\psi(t)\rangle$; $|\psi(t)\rangle = e^{-iHt/\hbar}|0\rangle$ (where $\hbar$ is the reduced Planck constant) describes the state a system is in after Hamiltonian H has been applied to it for a certain time t. Given these conditions, a *Hamiltonian simulation* outputs a sequence of computational gates that implement a unitary $U = e^{-iHt}$, where we have taken $\hbar = 1$ in proper units for ease of calculation.

Total Hamiltonians are Hermitian operators that are usually a sum of a large number of individual Hamiltonians. As an example, a Hamiltonian H can be equal to $(H_1 + H_2)$, which, in turn, can be expressed using the *Lie product formula*[7] as follows.

$$e^{-i(H_1+H_2)/t} = \lim_{n\to\infty}\left(e^{-iH_1 t/n}e^{-iH_2 t/n}\right)^n$$

7.23

---

[7]Reference: https://en.wikipedia.org/wiki/Lie_product_formula

Since the limit of equation 7.23 is infinite, the series needs to be truncated when implementing this formula on a quantum computer. The truncation procedure introduces error in the simulation that can be bound by a maximum simulation error ϵ, defined by the condition

$$\epsilon > \left\| e^{-iHt} - U \right\|$$

This process of truncation, shown in Figure 7-9, is known as *Trotterization*. It's widely used to simulate non-commuting Hamiltonians on quantum computers. The general Trotterization formula is given by

$$e^{-iHt} = \left(e^{-iH_o t/n} e^{-iH_1 t/n} \ldots e^{-iH_c t/n} \right)^n + O\left(some\ polynomials \right) \qquad 7.24$$

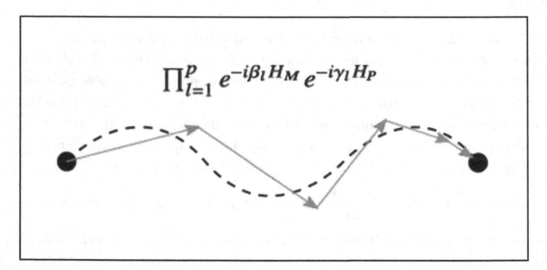

Figure 7-9. *Trotterization Source [122]*

The importance of Trotterization and equation 7.20 can be illustrated in a simple example. Suppose we want to simulate a Hamiltonian $H = X_0 + Z_0$, where X and Z are Pauli matrices and the subscripts are the labels for the qubits that the Hamiltonians are applied to. In this case, it is not possible to simulate each Hamiltonian *separately* because Pauli gates X and Z do not commute. This is a case where Trotterization is used where the whole Hamiltonian evolves by repeatedly switching between evolving between X_0 and Z_0 for small periods of time.

In the QAOA framework, the adiabatic pathway is discretized in some p steps, where p is a measure of precision and each discrete time step l has two parameters: β_l and γ_l. As in the generic VQE, the classical variational algorithms optimize these parameters based on the observed energy at the end of a run on the quantum hardware. Hence, the QAOA can be interpreted as a *Trotterized* adiabatic transformation, where the order p of the Trotterization determines the precision of the solution.

The goal of the QAOA algorithm is to map an input state $|\psi_0\rangle$ to the ground-state of a given problem, Hamiltonian H_p by sequentially applying a problem unitary $U_p = e^{-iH_p\beta_l}$ and a mixer unitary $U_m = e^{-iH_m\gamma_i}$ [127]. This gives the following ansatz, which is *in the form* of equations 7.10 and 7.11.

$$U(\beta,\gamma) = \prod_{l=1}^{p} e^{-iH_p\beta_l} e^{-iH_m\gamma_i} \qquad 7.25$$

where, $\theta = (\beta,\gamma)$

Fundamentally, the desired action is to discretize (i.e., separate and distinguish individually) the *time-dependent* Hamiltonian given by $H(t) = (1 - t)H_0 + tH_1$ under adiabatic conditions. This is achieved by *Trotterizing* the unitary. For instance, for time step t_0, the unitary can be split as $U(t_0) = U(H_0,\beta_0)U(H_1\gamma_0)$. We can continue doing this for subsequent time steps, eventually splitting up the evolution to p such chunks.

$$U = U(H_0,\beta_0)U(H_1\gamma_0)...U(H_0,\beta_p)U(H_1\gamma_p) \qquad 7.26$$

The Hamiltonian H_0 is referred to as the *driving or mixing* Hamiltonian, and H_1 as the *cost* Hamiltonian. The simplest of the mixing Hamiltonians is given by $H_0 = -\sum_i \sigma_i^x$. By alternating between the two Hamiltonians, the mixing Hamiltonian drives the state toward an equal superposition, whereas the cost Hamiltonian tries to find its own ground state. At the end of optimizing the parameters, the discretized evolution generates an approximation for the adiabatic pathway.

Note that decomposing the unitaries into native gates may result in a circuit of longer depth due to the many-body terms in the Hamiltonian and limited device connectivity. However, one of the strengths of QAOA is that the feasible subspace for certain problems is smaller than the full Hilbert space, and this restriction may result in a better-performing algorithm.

Hands-on QAOA with Rigetti Forest

Now that we have had a look at a theoretical approach to QAOA, it is good to have a practical hands-on approach to it. We do this with Rigetti Forest and QVM. As before, we start QVM and Quilc in server mode and fire up a Jupyter notebook with the following commands, each in its separate terminal.

As usual, we first import the required libraries, define a matrix with the Pauli Z-gate and define the number of qubits, as shown in Listing 7-4a for QAOA_pyquil.ipynb.

Listing 7-4a. Libraries and Qubits QAOA_pyquil.ipynb

```
#Libraries
import numpy as np
from functools import partial
from pyquil import Program, api
from pyquil.paulis import PauliSum, PauliTerm, exponential_map, sZ
from pyquil.gates import *
from pyquil import Program, get_qc
from pyquil.api import ForestConnection
from pyquil.api import WavefunctionSimulator
from pyquil.api import QVMConnection
from scipy.optimize import minimize
np.set_printoptions(precision=3, suppress=True)
qvm = QVMConnection()

# Number of qubits
n_qubits = 2
```

Next, we define the mixing Hamiltonian on some qubits. To achieve this, we use the Pauli-X operator, the coefficient of which indicates the strength of the transverse field at the given qubit. This operator acts trivially on all qubits, except for the given one. The following is the definition of the mixing Hamiltonian over two qubits.

Listing 7-4b. Libraries and Qubits QAOA_pyquil.ipynb

```
Hm = [PauliTerm("X", i, 1.0) for i in range(n_qubits)]
```

As an example, the Ising problem can be minimized. The Ising problem is defined by the cost Hamiltonian $H_C = -\sigma_1^Z \otimes \sigma_2^Z$, whose minimum is reached whenever $\sigma_1^Z = \sigma_2^Z$.

Listing 7-4c. Define Ising Problem QAOA_pyquil.ipynb

```
J = np.array([[0,1],[0,0]]) # weight matrix of the Ising model. Only the
coefficient (0,1) is non-zero.

Hc = []
for i in range(n_qubits):
    for j in range(n_qubits):
        Hc.append(PauliTerm("Z", i, -J[i, j]) * PauliTerm("Z", j, 1.0))
```

In the next step, we compute $e^{-iH_m\gamma}$ and $e^{-iH_C\beta}$ for the iterations. To achieve this, we can use the pyQuil exponential_map function to build two functions that take γ and β and return $e^{-iH_m\gamma}$ and $e^{-iH_C\beta}$.

Listing 7-4d. Define Ising Problem QAOA_pyquil.ipynb

```
exp_Hm = []
exp_Hc = []
for term in Hm:
    exp_Hm.append(exponential_map(term))
for term in Hc:
    exp_Hc.append(exponential_map(term))
```

Then, we set the value for p and initialize the γ_i and β_i parameters.

Listing 7-4e. Initializing Values QAOA_pyquil.ipynb

```
n_iter = 10 # number of iterations of the optimization procedure
p = 1
β = np.random.uniform(0, np.pi*2, p)
γ = np.random.uniform(0, np.pi*2, p)
```

The initial state is a uniform superposition of all the states $|q_1q_2...q_n\rangle$ and can be created using Hadamard gates on all the initial qubits at state $|0\rangle$.

Listing 7-4f. Initializing Values QAOA_pyquil.ipynb

```
initial_state = Program()
for i in range(n_qubits):
    initial_state += H(i)
```

Next, we create the circuit by composing the different unitary matrices given by the function evolve.

Listing 7-4g. Circuit Definition QAOA_pyquil.ipynb

```
def create_circuit(β, γ):
    circuit = Program()
    circuit += initial_state
    for i in range(p):
        for term_exp_Hc in exp_Hc:
            circuit += term_exp_Hc(-β[i])
        for term_exp_Hm in exp_Hm:
            circuit += term_exp_Hm(-γ[i])

    return circuit
```

Next, an evaluate_circuit function is created that takes a single vector beta_gamma which is the concatenation of γ and β and returns $\langle\psi|H_C|\psi\rangle = H_C|\psi\rangle$, where ψ is defined by the circuit created with this function.

Listing 7-4h. Circuit Evaluation QAOA_pyquil.ipynb

```
def evaluate_circuit(beta_gamma):
    β = beta_gamma[:p]
    γ = beta_gamma[p:]
    circuit = create_circuit(β, γ)
    return qvm.pauli_expectation(circuit, sum(Hc))
```

Finally, we optimize the angles.

Listing 7-4i. Optimize Angles QAOA_pyquil.ipynb

```
result = minimize(evaluate_circuit, np.concatenate([β, γ]),
method='L-BFGS-B')
result
```

This gives us the following output.

```
      fun: (-0.9999999999999868+0j)
 hess_inv: <2x2 LbfgsInvHessProduct with dtype=float64>
      jac: array([-0., -0.])
  message: 'CONVERGENCE: NORM_OF_PROJECTED_GRADIENT_<=_PGTOL'
     nfev: 30
      nit: 8
     njev: 10
   status: 0
  success: True
        x: array([0.785, 4.32 ])
```

The following is an explanation of the results.

A circuit is created using the optimal parameters found.

```
circuit = create_circuit(result['x'][:p], result['x'][p:])
```

The statevector_simulator backend in order to display the state created by the circuit.

```
wf_sim = api.WavefunctionSimulator(connection=fc)
state = wf_sim.wavefunction(circuit)
print(state)
```

This gives an output.

```
(-0.5+0.5j)|00> + (-2.89e-08+4.55e-08j)|01> + (-2.89e-08+4.55e-08j)|10> + (-0.5+0.5j)|11>
```

This output shows that the state is approximately
$(0.5 - 0.5i)(|00\rangle + |11\rangle) = e^{i\varphi} \frac{1}{\sqrt{2}}(|00\rangle + |11\rangle)$. Here, φ is a phase factor and it does not affect
the probabilities. This result corresponds to a uniform superposition of the two solutions

of the classical problem: $(\sigma_1 = 1, \sigma_2 = 1)$ and $(\sigma_1 = -1, \sigma_2 = -1)$. Finally, we evaluate the operators σ_1^Z and σ_2^Z.

```
print(qvm.pauli_expectation(circuit, PauliSum([sZ(0)])))
print(qvm.pauli_expectation(circuit, PauliSum([sZ(1)])))
```

This gives us the following output.

$$0j$$
$$0j$$

The output shows that both are approximatively equal to zero, which is an expected outcome given that both spins states are –1 and 1 half of the time. The output corresponds to a typical quantum behavior, where

$$\text{Expectation of } \left(\sigma_1^Z \sigma_2^Z\right) \neq \left[\text{Expectation of}\left(\sigma_1^Z\right)\right]\left[\text{Expectation of}\left(\sigma_2^Z\right)\right]$$

QAOA Solution for a QUBO

At this point, it is good to look at an example of solving a QUBO problem with QAOA. QUBOs have received significant interest in recent years to solve discrete combinatorial optimization tasks, which are widespread in the logistics industry, amongst others. QUBOs have also become popular for being implementation-friendly on near-term quantum hardware.

We looked at calculating a QUBO example earlier in this chapter. This exercise focuses on solving that same QUBO problem using Rigetti QVM and SciPy libraries. As a recap, the following was our example of the optimization problem.

$$(\text{minimize}) \quad f = -5x - 3y - 8z - 6w + 4xy + 8xz + 2yz + 10zw$$

where variables x, y, z, w are binary; that is, they can only have values 0 and 1. To solve this optimization problem with QUBO, we created the model in a matrix form as follows.

$$(\text{minimize}) \quad f = (xyzw) \begin{bmatrix} -5 & 2 & 4 & 0 \\ 2 & -3 & 1 & 0 \\ 4 & 1 & -8 & 5 \\ 0 & 0 & 5 & -6 \end{bmatrix} \begin{bmatrix} x \\ y \\ z \\ w \end{bmatrix}$$

> **Note**   that the coefficients of the linear terms appear on the main diagonal of the *Q* matrix. The *only constraint* is the restriction of either 0 or 1 as values of the decision variables; otherwise, QUBO is an *unconstrained* model. All the problem data for QUBO is contained in the *Q* matrix.

To solve this, we start with importing some libraries and defining the communication with the QVM as in Listing 7-5a of QUBO_pyquil.ipynb available in Jupyter notebook format.

Listing 7-5a. Libraries QUBO_pyquil.ipynb

```
# Import Libraries and communicate with QVM
import numpy as np
import matplotlib.pyplot as plt
from pyquil.paulis import PauliSum
from pyquil.api import WavefunctionSimulator
from pyquil.api import ForestConnection
from pyquil.api import QVMConnection
from scipy.optimize import minimize
from pyquil.unitary_tools import lifted_pauli
from scipy.optimize import minimize
np.set_printoptions(precision=3, suppress=True)
qvm = QVMConnection()
```

Next, we define the matrix form using a func_y function and the binary value mapping of the coefficients using another function, num_2_bin. Then we plot the results.

Listing 7-5b. QUBO Solution QUBO_pyquil.ipynb

```
def func_y(config):
    matr = np.array([[-5, 2, 4, 0], [2, -3, 1, 0], [4, 1, -8, 5], [0, 0, 5, -6]])
    return config @ matr @ config

def num_2_bin(num):
    bin_str = np.binary_repr(num, width=4)
    bin_arr = np.array([int(x) for x in bin_str])
    return bin_arr
```

Listing 7-5b produces the following output plot for a minimized f for the corresponding values of variables.

Figure 7-10 shows the solution to the model is as follows: $f = -11$; $x = w = 1$; $y = z = 0$. $x = 1, y = 0, z = 0, w = 1$ is the bit-string 1001; that is, the binary representation of the number 9, whose corresponding function value is $f = -11$.

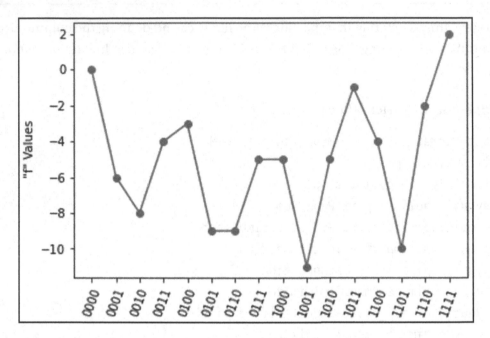

Figure 7-10. *QUBO solution with QAOA*

Supervised Learning: Quantum Support-Vector Machines

Data classification is one of the most important machine learning tools as it can identify, group, and study new data pertaining to use cases as varied as computer vision problems, medical imaging, drug discovery, handwriting recognition, and geostatistics and many other fields. In machine learning, one of the most common data classification methods is support-vector machines (SVM). SVM is a supervised learning method and particularly useful because it allows classification into one of two categories of an input set of training data by using a hyperplane to separate the two classes. The support vectors used along with the hyperplane are essentially data points used to maximize the distance between classes.

Chapters 5 and 6 discussed mapping classical data into feature maps. *Quantum support-vector machines* (qSVMs) use feature maps to map data points to quantum circuits. qSVMs are the subject of several studies both theoretically and experimentally [129]. qSVMs use qubits instead of classical bits to solve problems. Many qSVMs [130], [131], [132] and *quantum-inspired* SVM [133] algorithms leverage the *kernel trick* in the quantum domain. A great research paper on qSVM and supervised learning is a work by Havlicek et al. [135].

Generating a qSVM-based exercise requires the following.

1. Scaling, normalization, and principal component analysis

2. Generating a kernel matrix

3. Estimating the kernel for a new set of data points (test data) for QSVM classification

In the QSVM classification phase, *classical SVM* generates the *separating hyperplane* rather than a quantum circuit, and here the quantum computer is used twice. In the first instance, the kernel is estimated for all pairs of training data. The second time, the kernel is estimated for a new date or test data.

In a continuation from classical SVM covered in Chapter 2, we inspect the requirements to translate that knowledge into the quantum domain.

- To achieve this task, the first consideration is about how to translate the classical data point x into a quantum data point $|\psi(x)\rangle$. This can be achieved by creating a circuit $C(\psi(x))$ encompassing $\psi()$ where ψ is any classical function applied to the classical data.

- Next, a parametrized quantum circuit $V(\vartheta)$ is built with parameters ϑ to process the data.

- A measurement is performed that returns a classical value -1 or $+1$ for each classical input x that identifies the label of the classical data.

- Finally, we leverage *quantum variational circuit* (QVC) concepts and define an ansatz for the problem in the form $V(\vartheta)C(\psi(x))\,|0\rangle$.

Note   When this book was written, there was no definitive experimental proof that QSVM offers a distinct quantum advantage. However, the authors of reference [135] argue that there is an advantage if we use feature maps because otherwise, we would not need a quantum computer to construct the kernel. For this reason and because most libraries today offer a distinct built-in function for quantum SVM, hands-on code for this is not explored here. If you are interested in trying out hands-on code for qSVM, refer to the Qiskit documentation at `https://qiskit.org/documentation/tutorials/machine_learning/01_qsvm_classification.html` [136].

When this book was written, establishing use cases to state the advantage of qSVM over classical SVM is very much an area of active research. For example, the researchers at the CERN Large Hadron Collider (LHC) in Switzerland are actively working to see if the enormous computing challenge associated with the demands of their experiments can be met with the help of quantum computers. To this end, they have successfully employed qSVM for a ttH (H to two photons), Higgs coupling to top quarks analysis at LHC [138].

Many data analysis and machine learning techniques involve *optimization*. To this end, using D-Wave processors to solve combinatorial optimization problems through quantum annealing has been of deep interest for many years.

Quantum Computing with D-Wave

The D-Wave Systems, founded in 1999, produces quantum computing devices that leverage quantum annealing that has been used to perform deep quantum learning protocols for over a decade.

Quantum computing, as it exists today, is defined in different paradigms. The common format is quantum computing leveraging the gate model. That is an extension of the classic gate model where computing gates are applied on qubit registers to perform arbitrary transformations of quantum states made up of qubits. You are familiar with the fundamental approach of gate-model computation from classical computing. The gate model in the quantum computing domain is a natural extension.

Another common paradigm is quantum annealing, often referred to as adiabatic quantum computing, although there are subtle differences. Quantum annealing solves specific problems by making it relatively easier for an engineering challenge to scale it up. Their latest system, Advantage, has 5000 superconducting qubits in 2020, compared to the less than 100 qubits on gate-model quantum computers. D-Wave has been building superconducting quantum annealers for over a decade, and this company currently holds the record for the number of qubits.

Quantum annealing, the computational model on which D-Wave quantum computers are based, works by gradually evolving a many-body quantum system from one that is easy to characterize to one that is hard to characterize. Efficient open-system sampling in this model requires fast evolution of the quantum system.

Figure 7-11 illustrates the annealing process.

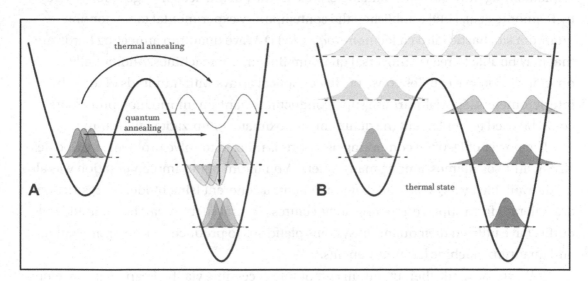

Figure 7-11. *Quantum annealing (Adapted from [137])*

A quantum state can be tunneled when approaching a resonance point before decoherence induces thermalization. The shades of blue illustrate the occupation of energy levels. In Figure 7-11, diagram A shows the criteria that a quantum state must traverse a local minimum in thermal annealing, whereas a coherent quantum state can tunnel when brought close to resonance. Diagram B in Figure 7-11 illustrates that coherent effects decay through interaction with an environment, causing the probability distribution of the occupancy of a system's energy levels to follow a Gibbs distribution.

An essential feature of deep quantum learning is that it does not require a large, general-purpose quantum computer. Quantum annealers are not *universal quantum computers* (UQC) but are much easier to construct and scale up than any UQC (see diagram A in Figure 7-11). Quantum annealers are well suited for implementing deep quantum learners and are commercially available via D-Wave since as far back as 2011. The D-Wave quantum annealer is a tunable transverse Ising model that can be programmed to yield the thermal states of classical systems and certain quantum spin systems.

From a hardware perspective, D-Wave implements flux qubits that use superconducting loops of superconducting metal. This metal loop allows current to flow clockwise or anticlockwise, which is how information is encoded on the qubits. Quantum Boltzmann machines [139], with more general tunable couplings capable of implementing universal quantum logic, are currently at the design stage [140]. A study by Biamonte et al. [140] established the contemporary experimental target for non-quantum stochastic (also called *non-stoquastic*) D-Wave quantum annealing hardware that may be able to realize universal quantum Boltzmann machines. Additionally, on-chip silicon waveguides construct linear optical arrays with hundreds of tunable interferometers. Special-purpose superconducting quantum information processors could be used to implement the quantum approximate optimization algorithm.

The power of D-Wave computing platforms is utilized to solve problems by Google, NASA, and Los Alamos, among many others. A quantum optics implementation was also made available by a QNN cloud that implements a coherent Ising model. Its restrictions are different from superconducting architectures. Several such research and industrial outfits have utilized their quantum systems platforms to produce proven optimization and quantum machine learning benefits.

D-Wave has made their quantum computing accessible via the Leap quantum cloud service for users in 38 countries (as it stands when this book was written). Users can browse to `https://cloud.dwavesys.com/leap/` and request access to free but limited usage time and resources on their real-life quantum hardware. Figure 7-12 shows the D-Wave Leap interface after successful login.

Figure 7-12. *D-Wave Leap website at* `https://cloud.dwavesys.com/leap/`

More information about their systems and documentation are available at `https://cloud.dwavesys.com/leap/resources` [142]. Libraries and example code are available in their Ocean SDK on GitHub at `https://github.com/dwavesystems` [141]. When this book was written, there were more than 250 user-developed quantum applications on D-Wave (`www.dwavesys.com`), including airline scheduling, election modeling, quantum chemistry simulation, automotive design, preventative healthcare, logistics, and much more.

Fundamentally, quantum annealing works by first encoding a problem into a search space of candidate solutions, of which the most optimal solution is the one that requires the *lowest energy*. D-Wave's processor finds this optimal solution by going through all the possible solutions simultaneously with superposition to find the one with the lowest energy.

373

Programming the D-Wave Quantum Annealing System

When programming the D-Wave quantum annealing system, a user maps a problem into a search for the "lowest energy valley," corresponding to the best possible outcome. The quantum processing unit considers all the possibilities simultaneously to determine the lowest energy required to form those relationships. The solutions are values that correspond to the optimal configurations of qubits found or the lowest points in the energy landscape. These values are returned to the user program over the network (reference D-Wave website at [143]). Because a quantum computer is probabilistic rather than deterministic, the computer returns many very good answers in a short amount of time— literally thousands of samples in one second. This provides not only the best solution found but also other efficient alternatives from which to choose. D-Wave's quantum computers are particularly useful for solving optimization and NP-hard problems such as the *traveling salesperson problem* (TSP) and Max-Cut.

Application development is facilitated by D-Wave's open source Ocean SDK (`https://github.com/dwavesystems/dwave-ocean-sdk`), available on GitHub and in Leap (`www.dwavesys.com/take-leap`), which has built-in templates for algorithms, and the ability to develop new code with the familiar programming language Python. Figure 7-13 illustrates the overall architecture of a D-Wave quantum computing platform along with their Ocean software stack.

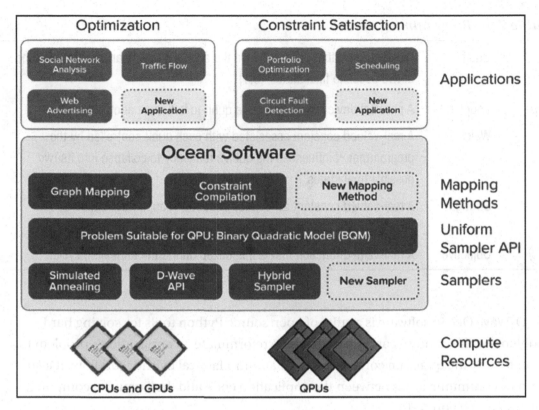

Figure 7-13. *D-Wave platform and Ocean software stack (Source [143])*

The program is modeled after a mathematical expression with a generic structure as follows (reference D-Wave documentation [141], [142].

$$Obj\left(a_i, b_{ij}; q_i\right) = \sum_i a_i q_i + \sum_{ij} b_{ij} q_i q_j$$

7.27

The components of equation 7.27 used to define the programming model ([141], [142]) are listed in Table 7-1.

Table 7-1. *Programming Model*

q_i	Qubit	A quantum bit that participates in the annealing cycle and settles into one of two possible final states: {0,1}
$q_i q_j$	Coupler	A physical device that allows one qubit to influence another qubit.
a_i	Weight	A real-valued constant associated with each qubit controlled by the programmer. It influences the qubit's tendency to collapse into its two possible final states.
b_{ij}	Strength	A real-valued constant associated with each coupler controlled by the programmer. It controls the influence exerted by one qubit on another.
Obj	Objective	A real-valued function that is minimized during the annealing cycle.

D-Wave Ocean software is a suite of open source Python tools for solving hard problems with quantum computers that help reformulate an application's problem for a solution by the quantum computer or a quantum-classical hybrid workflow. It also handles communications between the application code and the quantum computer, as illustrated in Figure 7-14.

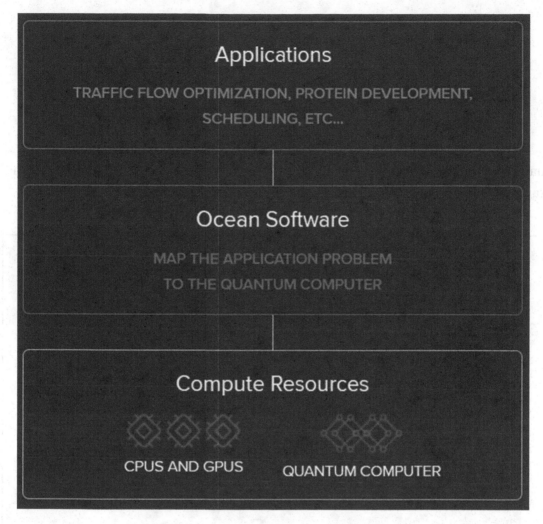

Figure 7-14. Programmability of Ocean software stack (Adapted from [142])

The following is a structural guideline on how a problem runs on a D-Wave quantum computer.

- The problems are formulated as an Ising model or QUBO model.

- The constants a's and b's are turned into voltages, currents, and magnetic fields as the problem are programmed onto the D-Wave QPU (quantum processing unit).

- The qubit spins begin in their superposed states.

- The qubit spins evolve, exploring problem space.

- By the end of the annealing cycle, the system is in the ground state, or a low excited state, of the submitted problem.

- The states of the spins are read, optional postprocessing is applied, and delivered back to the user.

- This can be done hundreds or thousands of times per second

Figure 7-15 describes the full-stack workflow of end-to-end programming in a quantum annealing system [38] using the NP-hard traveling salesperson problem as an example. TSP is described with hands-on coding later in this chapter.

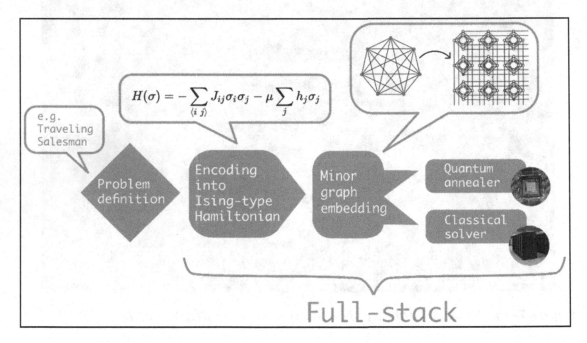

Figure 7-15. *Full-stack quantum annealing workflow (Adapted from [38])*

D-Wave's system utilizes QUBO. As a quick recap, the following are the QUBO properties leveraged by D-Wave's programming process.

- The highest power of variable is 2.

- No constraint is applied to the variables.

- Variables values are within {0,1}.

- Minimize or maximize an objective function.

Earlier in the chapter, the theory behind QUBO included an example using the *symmetric form*. D-Wave programming usually leverages the *upper triangular form* of defining a QUBO. In Figure 7-16, diagram A shows a matrix representation of a QUBO. Diagram B shows a graphical representation of the same expression: $2x + 3y - 4z + xy - 2.5xz + 7yz$.

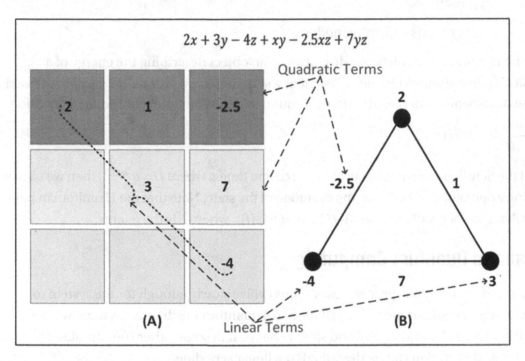

Figure 7-16. QUBO on D-Wave (Adapted from [142])

The following are the rules in the *matrix* representation in diagram A in Figure 7-16.

- Quadratic terms appear on off-diagonals.

- For D-Wave's Ocean, the upper diagonal is sufficient.

- Constant term and variable names are generally omitted in this representation.

- Linear terms appear on the diagonal.

When expressing a QUBO graphically, as in diagram B in Figure 7-16, we abide by the following rules.

- Quadratic terms appear on edges.

- Constant term and variable names are generally omitted in this representation.

- Linear terms appear on nodes.

The Hamiltonian, which was described as an object describing the energy of a classical or quantum system, also describes a system that evolves with time as expressed by the time-dependent (i.e., the state evolution depends on time) Schrödinger equation:

$$i\hbar \frac{d|\psi(t)\rangle}{dt} = H|\psi(t)\rangle .$$

If the Schrödinger equation is solved for some time t, where $U = e^{-iHt/\hbar}$, then we obtain a unitary operator that assists in the evolution of the state. Note that the Hamiltonian here describes the energy of the system: $H|\psi(t)\rangle = E|\psi(t)\rangle$, where E is the energy.

Adiabatic Quantum Computing

Typically, in an adiabatic process, conditions evolve slowly enough for the system to adapt to a new configuration. For instance, in a quantum mechanical system, we can start from some Hamiltonian H_i and slowly change it to some other Hamiltonian H_{i+1}. The simplest evolution can be described by a linear schedule.

$$H(t) = (1 - t)H_i + tH_{i+1}, 0 \le t \le 1$$

The time dependence of the Hamiltonian makes the derivation of the solution of the Schrödinger equation quite complicated. The adiabatic theorem states that a slow evolution (change) in the time-dependent Hamiltonian should help keep the resulting dynamics simple. Hence, if we start close to an eigenstate, the system remains close to an eigenstate. This implies that if the system would be started in the ground state E_0 under certain *conditions*, then the system should stay at E_0 or at a state very close to E_0.

The difference in energy level between two states (for example, the ground state and the first excited state) is called the *gap* or *energy gap*. If $H(t)$ has a nonnegative gap for each t during the transition and the change occurs appropriately slowly, then the system should stay in the ground state. If the time-dependent gap is given by Δt, then the approximation of the speed limit scales as $\frac{1}{1\Delta t^2}$.

The adiabatic theorem allows something quite unusual: the ground state of an easy-to-solve quantum many-body system can be attained, and the Hamiltonian *changed* to a system that we are interested in. As an example, it is possible to start with the simple Hamiltonian $H_0 = -\sum_i \sigma_i^x$ as its ground state is the equal superposition. As a hands-on example utilizing D-Wave's dimod API,[8] we try this on two sites, as shown in the Jupyter notebook Annealing_QC_DWave.ipynb.

Listing 7-6a. Libraries and Arrays Annealing_QC_DWave.ipynb

```
import numpy as np
np.set_printoptions(precision=3, suppress=True)

X = np.array([[0, 1], [1, 0]])
IX = np.kron(np.eye(2), X)
XI = np.kron(X, np.eye(2))
H_0 = - (IX + XI)
λ, v = np.linalg.eigh(H_0)
print("Eigenvalues:", λ)
print("Eigenstate for lowest eigenvalue", v[:, 0])
```

The following are descriptions of some of the functions used in the code block from NumPy documentation [155].

- numpy.kron() or np.kron(): Kronecker product of two arrays in NumPy. Computes the Kronecker product, a composite array made of blocks of the second array scaled by the first.

- numpy.linalg.eigh() or numpy.linalg.eigh(): Returns the eigenvalues and eigenvectors of a complex Hermitian (conjugate symmetric) or a real symmetric matrix.

---

[8]A dimod is a shared API for samplers (see https://github.com/dwavesystems/dimod). It provides a binary quadratic model (BQM) class that contains Ising and quadratic unconstrained binary optimization (QUBO) models used by samplers such as the D-Wave system; a discrete quadratic model (DQM) class and higher-order (non-quadratic) models; and reference examples of samplers and composed samplers.

Listing 7-6a gives the following output as eigenvalue and eigenstate calculation.

```
Eigenvalues: [-2. -0.  0.  2.]
Eigenstate for lowest eigenvalue [-0.5 -0.5 -0.5 -0.5]
```

Once the eigenvalues and eigenstates are known, the Hamiltonian can be slowly turned into a classical Ising model to obtain the global solution. Adiabatic quantum computing process exploits this phenomenon and can perform calculations with the final Hamiltonian being in the same *general* form in equation 7.4.

$$H = -\sum_i h_i \sigma_i - \sum_{ij} J \sigma_i \sigma_j$$

A quantum computing system that implements the Hamiltonian by respecting the speed limit and guaranteeing the finite gap can be argued to be approximately equivalent to the gate model.

Quantum Annealing

A major limitation to adiabatic quantum computing is the calculation of the speed limit which is non-trivial. Calculating the speed limit is often harder than solving the main problem of finding the ground state of the Hamiltonian of interest. In addition, there are engineering constraints, such as decoherence of the qubits and temperature control. *Quantum annealing* addresses these challenges by circumnavigating the speed limit challenge. It repeats the *annealing* (or the transition) repeatedly until several samples are collected, and the spin configuration with the lowest energy is selected as the preferred solution. However, there is usually no guarantee that this is the ground state.

Quantum annealing has a slightly different software stack than gate-model quantum computers, as shown in Figure 7-15. Instead of a quantum circuit, the level of abstraction is in the form of the classical Ising model, and *the problem of interest* is formulated in the same form. A classical solver for the Ising model is the *simulated annealer*.

Listing 7-6b. Energy Annealing_QC_DWave.ipynb

```
import dimod
J = {(0, 1): 1.0, (1, 2): -1.0}
h = {0:0, 1:0, 2:0}
model = dimod.BinaryQuadraticModel(h, J, 0.0, dimod.SPIN)
sampler = dimod.SimulatedAnnealingSampler()
response = sampler.sample(model, num_reads=10)
print("Energy of samples:")
print([solution.energy for solution in response.data()])
```

Here, we use the `dimod` API, which contains a binary quadratic model (BQM) class that contains Ising and quadratic unconstrained binary optimization (QUBO) models used by D-Wave samplers to get the following output as energy of the samples.

```
Energy of samples:
[-2.0, -2.0, -2.0, -2.0, -2.0, -2.0, -2.0, -2.0, -2.0, -2.0]
```

To obtain the required solution in an annealing-based quantum system, we need to match the problem's connectivity to the hardware during programming. To achieve this, it is necessary to find a graph minor embedding that combines several physical qubits into a logical qubit.

The part of the minor embedding problem is NP-hard (explained in the next section). As such, probabilistic methods find an embedding. For example, for some generations of the quantum annealer that D-Wave produces, qubits are "oriented" on the QPU (quantum processing unit) vertically or horizontally and have unit cells containing a $K_{4,4}$ *bipartite fully connected graph,* with two remote connections from each qubit going to qubits in neighboring unit cells. A unit cell with its local and remote connections indicated is shown in Figure 7-17.

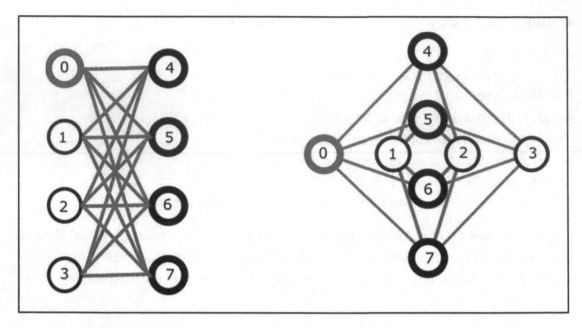

Figure 7-17. *Unit cell in Chimera graph (Source D-Wave documentation [155])*

Figure 7-17 is called the Chimera graph. In each of these renderings [155] there are two sets of four qubits. Each qubit connects to all qubits in the other set but to none on its own, forming a $K_{4,4}$ graph; for example, the green qubit labeled 0 connects to qubits 4 to 7 in bold.

The Chimera graph is available as a NetworkX graph in the dwave_networkx package. Our example draws a smaller version of Figure 7-17, utilizing the chimera_graph() function and consisting of 2 × 2 unit cells.

Listing 7-6c. Cells in Chimera graph Annealing_QC_DWave.ipynb

```
import matplotlib.pyplot as plt
import dwave_networkx as dnx
%matplotlib inline
connectivity_structure = dnx.chimera_graph(2, 2)
dnx.draw_chimera(connectivity_structure)
plt.show()
```

This code snippet produces the following output in Figure 7-18.

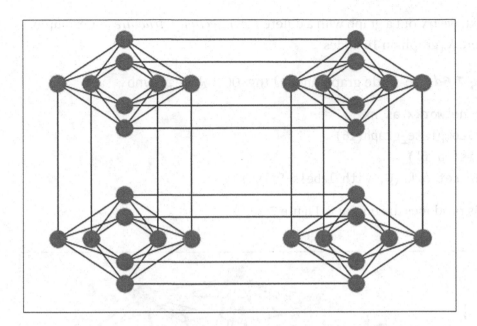

Figure 7-18. *2 × 2 Unit cells in Chimera graph*

We could have produced a different graph—for example, 4 × 4 unit cells, as shown in Figure 7-19 with `connectivity_structure = dnx.chimera_graph(4, 4)`. However, the 2 × 2 unit cells in Figure 7-18 are used for visual clarity.

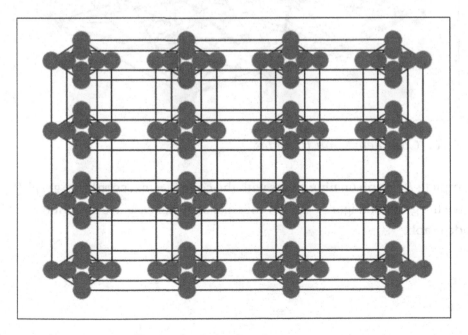

Figure 7-19. *4 × 4 Unit cells in Chimera graph*

Next, we try out a graph with a different *connectivity structure*, for example, the complete K_n graph on 16 nodes.

Listing 7-6d. 16-node graph Annealing_QC_DWave.ipynb

```
import networkx as nx
G = nx.complete_graph(16)
plt.axis('off')
nx.draw_networkx(G, with_labels=False)
```

This produces the output in Figure 7-20.

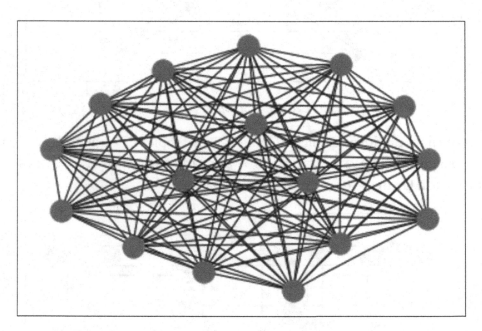

Figure 7-20. *Complete K_n on 16 nodes*

A complete K_n graph on nine nodes with the code G = nx.complete_graph(9) would give the graph in Figure 7-21. For the sake of visual ease, we continue with this nine-node graph.

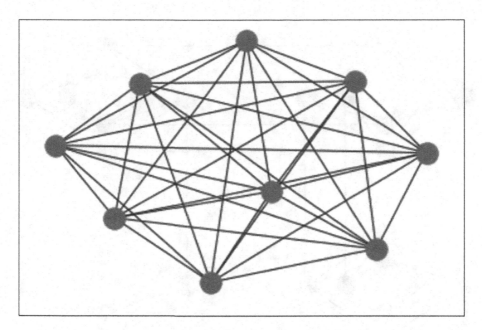

Figure 7-21. *Complete K_n on 9 nodes*

In our next step, we try out embedding the graph for Figure 7-21 and plot it.

Listing 7-6e. Embed 9 Node Graph Annealing_QC_DWave.ipynb

```
import minorminer
embedded_graph = minorminer.find_embedding(G.edges(), connectivity_
structure.edges())
dnx.draw_chimera_embedding(connectivity_structure, embedded_graph)
plt.show()
```

Listing 7-6e gives us the following output as the $K_n = K_9$ graph embeddings of the nine nodes. If this snippet is run more than once, then we may see a change in the structure; that is, a different embedding may be obtained. None of these results are wrong. They are all possible correct graphs.

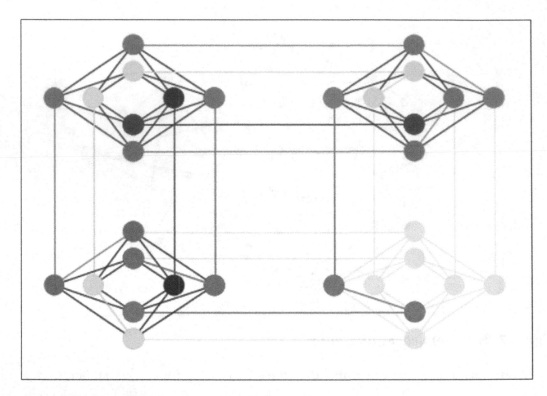

Figure 7-22. *Embedding of the 9 nodes*

In Figure 7-22, qubits that have the same color correspond to a logical node in the original problem defined by the K_9 graph of Figure 7-21. The qubits combined to form a chain. Although the problem has nine variables (nodes), almost all 32 available on the toy Chimera graph have been used. Listing 7-6f gives the maximum chain length.

Listing 7-6f. Maximum Chain Length `Annealing_QC_DWave.ipynb`

```
max_chain_length = 0
for _, chain in embedded_graph.items():
    if len(chain) > max_chain_length:
        max_chain_length = len(chain)
print(max_chain_length)
```

The output is the value 4.

The `chain` on the hardware is implemented via strong couplings between the elements; this chain is twice as strong as what the user can set. However, long chains can still break, resulting in occasional inconsistent results. In general, shorter chains are preferred to minimize the wastage of qubits and obtain more reliable results.

If you are interested in exploring D-Wave's power, go to www.youtube.com/playlist? list=PLPvKnT7dgEsujrrP7i_mgbkivBN3J6A8u for a free, three-part quantum computing tutorial course by Dr. Joel Gottlieb (formerly of D-Wave) on the fundamentals of practical quantum computing.

Solving NP-Hard Problems

For a problem to be computationally *feasible*, it must be computable in what is known as *polynomial time*. Polynomial time refers to the number of computational steps required to find a solution, which must be a polynomial function n^c for a problem of size n and some constant c. Any problem that takes more than a polynomial number of steps to compute; for example, an exponential c^n number of steps is considered infeasible, as the computation time grows too quickly with the problem size, requiring more time than the age of the universe to compute a problem of sizes as small as $n \simeq 100$.

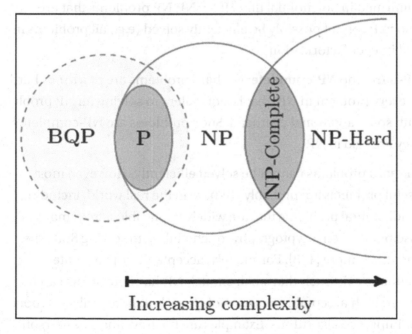

Figure 7-23. Complexity of problems

Problems are grouped into complexity classes based on how feasible they are. The most relevant complexity classes, along with their relationships, are shown in Figure 7-23.

- **BQP (bounded-error quantum polynomial time)**: Problems can be solved efficiently in polynomial time using a quantum computer. The BQP class includes problems from P and at least some from NP-intermediate. However, the BQP class *does not include* NP-hard or NP-complete problems. Examples include integer factorization (NP-intermediate) and simulation of quantum systems.

- **P (polynomial time)**: Problems in P are defined as those that can be *solved efficiently* in polynomial time. These problems are considered feasible with classical computers. Examples are integer multiplication, sorting, and searching through a list.

- **NP (non-deterministic polynomial time)**: Problems in NP are those that can be *verified efficiently*; that is, a valid solution can be confirmed in polynomial time. If $P \neq NP$, NP problems that are *not* also in P cannot possibly be efficiently solved (e.g., all problems in P and integer factorization).

- **NP-hard** and **NP-complete**: NP-hard problems are *at least* as hard as every problem in NP; that is, equivalent to solving an NP problem with some additional overhead. Such problems are NP-complete if they are also in NP.

NP-hard problems cannot be solved efficiently. However, most useful problems are probably NP-hard in the real world, including fundamental problems that are widely applicable across many disciplines, from cryptography to scheduling to solving Sudokus, and much more [153]. Fortunately, acceptable approximate solutions exist such that small enough NP-hard problems can be solved with acceptable amounts of effort, but there is always room for improved algorithms. Examples are the traveling salesperson problem, halting problem, map coloring, and Max-Cut. Of these, we try out the Max-Cut problem in hands-on exercises. We also look at coding a more complex traveling salesperson problem with D-Wave compilers in Chapter 8.

Unsupervised Learning and Optimization

Earlier in this chapter, we looked at aspects of supervised learning and qSVMs. *Unsupervised learning* implies a lack of labels (refer to Chapters 2 and 3) where we seek structure in the data without intuition about the characteristic of that structure. A common example of unsupervised learning is clustering, where the goal is to identify instances that group together in some high-dimensional space. Unsupervised learning, in general, is harder as a problem as compared to supervised learning. Deep learning injected a huge boost to galvanize supervised learning. It has induced significant advances in unsupervised learning. However, there remains plenty of room for improvement. In the following code examples, we explore ways of mapping an unsupervised learning problem to graph optimization, which can be solved on a quantum computer.

The first problem we consider is mapping of clustering to optimization. Suppose there are $\{x_i\}_{i=1}^N$ points that are members of a higher-dimensional space $\mathcal{H}^D$. In order to form clusters, we need to be able to define, distinguish, and group points that are near each other and points that are far away from each other. To get a better understanding, let's formulate a dataset with two classes: the first ten instances belong to the first class, and the second 10 instances belong to the second class.

Listing 7-7a. Data Points 3-D unsupervisedLearning.ipynb

```
import numpy as np
import matplotlib.pyplot as plt
from mpl_toolkits.mplot3d import Axes3D
%matplotlib inline

n_instances = 20
class_1 = np.random.rand(n_instances//2, 3)/10
class_2 = (0.6, 0.1, 0.05) + np.random.rand(n_instances//2, 3)/10
data = np.concatenate((class_1, class_2))
```

```
colors = ["red"] * (n_instances//2) + ["green"] * (n_instances//2)
fig = plt.figure()
ax = fig.add_subplot(111, projection='3d', xticks=[], yticks=[], zticks=[])
ax.scatter(data[:, 0], data[:, 1], data[:, 2], c=colors);
```

This gives us a visualization of our data points in three dimensions.

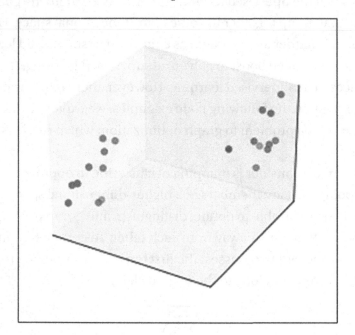

As an additional example, if we wanted our data points to be in a two-dimensional space, the following code snippet would generate that for five instances per class.

Listing 7-7b. 2-D Data Points

```
import matplotlib.pyplot as plt
import numpy as np
%matplotlib inline
np.set_printoptions(precision=3, suppress=True)
np.random.seed(0)

# Generating the data
c1 = np.random.rand(5, 2)/5
c2 = (-0.6, 0.5) + np.random.rand(5, 2)/5
data = np.concatenate((c1, c2))
```

```
plt.subplot(111, xticks=[], yticks=[])
plt.scatter(data[:, 0], data[:, 1], color='navy')
```

Listing 7-7b gives the following output.

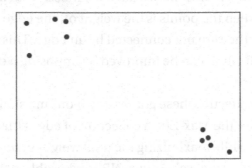

The Euclidean distance measurement is the simplest case as measured in the higher-dimensional space. The pairwise distances can be calculated between the data points with Listing 7-7c.

Listing 7-7c. Gram Matrix unsupervisedLearning.ipynb

```
import itertools
d = np.zeros((n_instances, n_instances))
for i, j in itertools.product(*[range(n_instances)]*2):
    d[i, j] = np.linalg.norm(data[i]-data[j])
print("The distances are ", d)
```

This prints the distances between the points, as shown in the following truncated output.

```
The distances are  [[0.         0.08371355 0.0696934  0.09737453 0.04235822 0.03049551
  0.08484548 0.06464411 0.04068431 0.09187931 0.60419337 0.64422134
  0.65372866 0.61127771 0.64244994 0.65455482 0.68082458 0.59827498
  0.68953772 0.65727225]
 [0.08371355 0.         0.01545735 0.03799086 0.07991677 0.060498
  0.09624356 0.05109954 0.09892387 0.03933224 0.52857902 0.57310589
  0.57973387 0.54003001 0.5718616  0.58141924 0.60923563 0.52661042
  0.61552924 0.58448598]
 [0.0696934  0.01545735 0.         0.0377093  0.07153253 0.04588019
  0.09090702 0.04000061 0.08899346 0.03675314 0.54159887 0.58499063
  0.59212424 0.55137773 0.5831622  0.59371546 0.62081129 0.53784985
  0.62767871 0.59658564]
 [0.09737453 0.03799086 0.0377093  0.         0.10797898 0.07193932
  0.12004233 0.05371193 0.12361236 0.02627776 0.52270439 0.56480477
```

The matrix generated by Listing 7-7c is called the *Gram* matrix or the *kernel matrix*. The Gram matrix contains information about the topology of the points in the higher-dimensional space. The Gram matrix can be interpreted as the weighted adjacency matrix of a graph: two nodes represent two data instances. As defined in the Gram matrix, the distance between the points is the weight on the edge that connects them. If the distance is zero, then they are not connected by an edge. This is a dense graph with many edges where visualization can be improved by imposing a distance function that gets exponentially smaller.

We may wonder how effective these graphs are in our mission to find clusters. This is where we could consider the Max-Cut, a collection of edges that would split the graph in exactly two if removed while maximizing the total weight of these edges, as shown by Otterbach et al., 2017 [159]. It is a well-known NP-hard problem that maps to an Ising model naturally.

Max-Cut with Annealing (D-Wave)

Max-Cut is an NP-hard graph theory problem, as shown by Richard Karp [156] in his seminal paper alongside 20 other NP-hard problems published in 1972. Max-Cut has been extensively researched by computer scientists, data scientists, and mathematicians. The problem's NP-hard quality cannot be solved in polynomial time, but it can be verified in polynomial time. As there are no polynomial-time algorithms for NP-hard problems, there has been a great amount of research into approximation algorithms and heuristics. The Maximum Cut (Max-Cut) graph problem simply stated is partitioning the nodes of a graph into two subsets such that the number of edges between the subsets is maximized.

The Max-Cut problem has a wide variety of real-world applications, such as computer networking. Suppose there are various servers, each with different types of interfaces (i.e., some have 10 Gb ethernet ports, some have USB ports, some Bluetooth). Our goal is to group these computers into two groups for two different projects to guarantee that two groups can connect. However, in real life, the connections often fail due to a mismatch in physical and logical systems. The aim here is to calculate the best possible way to stay connected.

The maximum cut theory allows us one way to approach this. To create a working model of our server network, we can visualize a graph formed by the servers as nodes (or vertices), and edges are drawn between the servers so that they can connect, as shown in Figure 7-24.

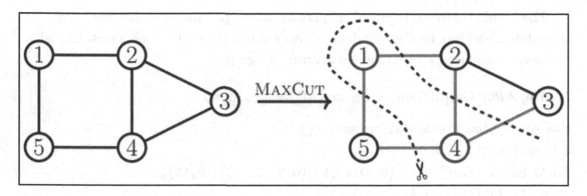

Figure 7-24. *Max-Cut problem*

On the left side of Figure 7-24 is a graph and on the right side is a possible Max-Cut solution for the same graph. The two partitions are $P_1 = \{1, 4\}$ and $P_2 = \{2, 3, 5\}$. The Max-Cut is of the size $f(x) = 5$ with $x = (1, -1, -1, 1, -1)$. It is important to note that this example does have multiple solutions (partitions with the same sized cut), and this is an example of one of them.

If we look for a maximum cut in the graph, we are looking to split the nodes into two groups so that there are as many edges as possible between the groups. Then, in our server network, we have two groups with as many connections as possible between the two groups. Now, if one connection goes down, we have several more to use as a path to redundancy. This solution helps create a more resilient network by providing redundant connections between groups if one connection fails.

Quantum annealing is a powerful way to visualize an NP-hard Max-Cut problem. To see how this is done, we look at a 12-node Max-Cut problem using D-Wave's dimod and NetworkX (https://docs.ocean.dwavesys.com/projects/dwave-networkx/en/latest/), among others.

Listing 7-8a. Max-Cut Libraries and Preliminaries maxcut_DWave.ipynb

```
import numpy as np
import matplotlib.pyplot as plt
import dwave_networkx as dnx
import networkx as nx
from dimod import ExactSolver, SimulatedAnnealingSampler
```

The SimulatedAnnealingSampler() is a dimod sampler that uses the simulated annealing algorithm. Listing 7-8b defines the 12 nodes and their connections, and calls the dwave_networkx() [154] function to create the graph.

Listing 7-8b. Graph Generation maxcut_DWave.ipynb

```
sampler = SimulatedAnnealingSampler()
G = nx.Graph()
G.add_edges_from([(0,6),(0,7),(0,8),(0,9),(0,10),(0,11),
(1,6),(1,7),(1,8),(1,9),(1,10),(1,11),
(2,6),(2,7),(2,8),(2,9),(2,10),(2,11),
(3,6),(3,7),(3,8),(3,9),(3,10),(3,11),
(4,6),(4,7),(4,8),(4,9),(4,10),(4,11),
(5,6),(5,7),(5,8),(5,9),(5,10),(5,11)])
pos = nx.spring_layout(G)
nx.draw(G,pos,with_labels=True)
nx.draw_networkx_nodes(G,pos)
plt.show()
candidate = dnx.maximum_cut(G, sampler)
print (candidate, " is the maxcut")
S = dnx.maximum_cut(G, ExactSolver())
print (S, " is the maxcut")
```

The code snippet gives us the following output as our Max-Cut solution. The solution may vary from run-to-run due to the probabilistic nature of the computation. This does not mean that any solution is wrong. They are all correct!

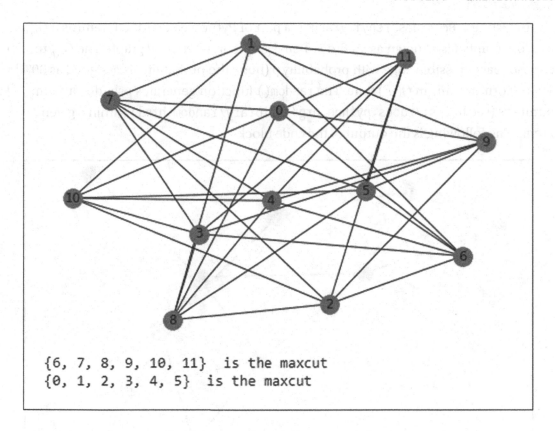

```
{6, 7, 8, 9, 10, 11}  is the maxcut
{0, 1, 2, 3, 4, 5}  is the maxcut
```

We generated the graph with well-defined edges and nodes. As an exercise, we can randomize the selection of the nodes and observe the quality of the solution.

Listing 7-8c. Graph Generation `maxcut_DWave.ipynb`

```
seed = random.randrange(1, 1000)
G1 = nx.erdos_renyi_graph(n=12, p=0.3, seed=seed)
pos = nx.spring_layout(G1)
nx.draw(G1,pos,with_labels=True)
plt.show()
sampler = SimulatedAnnealingSampler()
candidate = dnx.maximum_cut(G1, sampler)
print (candidate, " is the maxcut")
S = dnx.maximum_cut(G1, ExactSolver())
print (S, " is the maxcut")
```

In Listing 7-8c, `erdos_renyi_graph`[9] is a part of Python NetworkX. It returns a $G_{n,p}$ random graph, also known as an Erdős-Rényi graph or a *binomial* graph. The $G_{n,p}$ model chooses each possible edge with probability p (here, the probability is assigned as 30%). This algorithm runs in $O(n^2)$ time. The `random()` function generates pseudo-random numbers (see `https://docs.python.org/3/library/random.html`) within a given range. The following is the output of the code block.

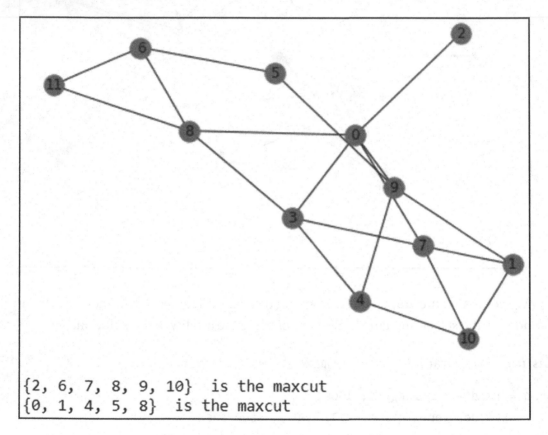

```
{2, 6, 7, 8, 9, 10}  is the maxcut
{0, 1, 4, 5, 8}  is the maxcut
```

The solution appear more arbitrary than the previous one due to the randomness in node selection. The solution may vary from run-to-run due to the probabilistic nature of the computation.

---

[9]Reference: `https://networkx.org/documentation/networkx-1.10/reference/generated/networkx.generators.random_graphs.erdos_renyi_graph.html`

Max-Cut with QAOA (pyQuil)

Now that you have an insight into Max-Cut and have seen it at work on D-Wave quantum annealing, the same Max-Cut problem can also be solved on Rigetti Forest or other gate-model quantum computers. To achieve this, translate the couplings and the on-site fields to match the programming interface, as shown in Listing 7-9 (Jupyter notebook `maxcut_QAOA_pyquil.ipynb`). As before, open three separate terminals to start the QVM server (`qvm -S`), the Quilc server (`quilc -S`), and the Jupyter notebook (`jupyter notebook`). Import the necessary libraries and define the network edges.

Listing 7-9a. Assign the Edges `maxcut_QAOA_pyquil.ipynb`

```
# Import Libraries and define QVM  Connection
import numpy as np
from grove.pyqaoa.maxcut_qaoa import maxcut_qaoa
import pyquil.api as api
import matplotlib.pyplot as plt
import networkx as nx
qvm_connection = api.QVMConnection()

# assign network edges
nw_edges=[(0,1),(1,2),(2,3),(3,4),(4,5),(5,6),(7,0),(7,8),(8,9),(9,2),
(9,10),(10,11),(11,4),(11,12),(12,6),(0,10),(3,12)]
```

Then we run the optimization on the QVM.

Listing 7-9b. Run Optimization on QVM `maxcut_QAOA_pyquil.ipynb`

```
steps = 2
inst = maxcut_qaoa(graph=nw_edges, steps=steps)
betas, gammas = inst.get_angles()
steps = 2
inst = maxcut_qaoa(graph=nw_edges, steps=steps)
betas, gammas = inst.get_angles()
t = np.hstack((betas, gammas))
param_prog = inst.get_parameterized_program()
```

```
prog = param_prog(t)
wf = qvm_connection.wavefunction(prog)
wf = wf.amplitudes
wlist = []
for state_index in range(2**len(inst.qubits)):
    ww = np.conj(wf[state_index])*wf[state_index]
    wlist.append([inst.states[state_index], ww.real])
wlist.sort(key=lambda x: float(x[1]),reverse=True)
```

The following is the truncated output.

```
Parameters: [1.82113153 1.33738605 4.04846415 3.36494378]
E => -6.862911860306613
Parameters: [1.92357018 1.22872344 4.07256215 3.55422186]
E => -7.6438792379061296
Parameters: [1.74999358 1.28096508 4.23522365 3.69618043]
E => -8.626865106985527
Parameters: [1.74999358 1.28096508 4.23522365 3.69618043]
E => -8.48358187726647
Parameters: [1.81046084 1.18849737 4.37529328 3.60811354]
E => 0.01277210654007C
```

We can obtain a list of partitioning solutions with the weights via the wlist command (output not shown for brevity but is included in the notebook).

Next, look for the Max-Cut partitions.

Listing 7-9c. Max-Cut Partitions maxcut_QAOA_pyquil.ipynb

```
mxcutlist0 = []
mxcutlist1 = []
for i in range(len(wlist[0][0])):
    if wlist[0][0][i] == '0':
        mxcutlist0.append(i)
    else:
        mxcutlist1.append(i)
print("maxcut0 is ",mxcutlist0)
print("maxcut1 is ",mxcutlist1)
```

The output for the partitions is as follows.

```
maxcut0 is  [0, 4, 7, 11]
maxcut1 is  [1, 2, 3, 5, 6, 8, 9, 10, 12]
```

We can generate the Max-Cut plot and partitioning from the following code snippet.

Listing 7-9d. Max-Cut Graph maxcut_QAOA_pyquil.ipynb

```
G1 = nx.Graph()
G1.add_edges_from(nw_edges)
pos = nx.spring_layout(G1)
#gd.draw_custom(G2,pos)
nx.draw(G1,pos,with_labels=True)
plt.show()
print("The first set from maxcut partioning is ",mxcutlist0)
print("The second set from maxcut partioning is ",mxcutlist1)
```

The following output shows the graph for our Max-Cut solution in pyQuil and the partitions.

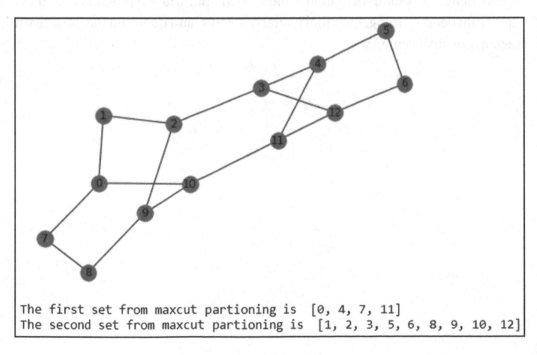

```
The first set from maxcut partioning is  [0, 4, 7, 11]
The second set from maxcut partioning is  [1, 2, 3, 5, 6, 8, 9, 10, 12]
```

The output graph may vary from run-to-run due to the probabilistic nature of the computation.

A common real-life example of the Max-Cut problem is the application of very-large-scale integrated (VLSI) circuit design in pin pre-assignments and layer preferences. Another common example is the application of finding ground states of spin glasses with exterior magnetic fields. Both cases were explored by Barahona et al. (2016) [157].

Solving problems such as Max-Cut is an example of the power of annealing-based quantum computers such as D-Wave's. It is surmised that a QAOA based Max-Cut solution requires hundreds of qubits on a gate-based quantum computer for any potential quantum speed-up benefit (Guerreschi et al. 2019) [158].

Summary

This chapter looked at some very important and commonly used algorithms and techniques in quantum computing and machine learning, such as VQE, QUBO, QAOA, and HHL. It introduced D-Wave's annealing system, the world's first commercially produced quantum computer. The methodologies to solve similar problems using various platforms and techniques of optimization are helpful in the next chapter, where we explore hands-on examples of more complicated optimization problems, such as TSP, quantum deep learning, quantum neural networks, and quantum machine learning leveraging Xanadu's PennyLane.

CHAPTER 8

Deep Quantum Learning

Where the mind is without fear

and the head is held high

Where knowledge is free.

—Rabindranath Tagore

It is human nature to stretch the limits of our knowledge through learning. Cultivating and developing the art of optimizing each little corner of our lives through continuous learning has been the backbone of human progress through the history of civilization. We benefit from efforts to advance various parts of our daily lives. Chapter 7 dealt with quantum annealing, that while optimization theories originate from mathematical concepts, they can also be transformed and built into physics. For example, physics has the *principle of least action*, the *principle of minimum entropy generation*, and the *variational principle*. Physics offers *physical annealing*, which preceded computationally simulated annealing. Physics also offers the *adiabatic principle*, which, in its quantum form, is *quantum annealing*. Hence, physical machines can solve the mathematical problem of optimization, including constraints—a property that can be extended to using quantum systems to treat classical and quantum data.

Note Chapter 7 solved the Max-Cut NP-hard problem. This chapter looks at solving a more complex NP-hard problem with classical input data utilizing D-Wave's qbsolve: the traveling salesperson problem. We also glean from the insights gained from classical neural networks in Chapters 2 and 3 and grow into options of quantum deep learning with the help of libraries offered by Xanadu's PennyLane and Google's TensorFlow Quantum.

© Santanu Ganguly 2021
S. Ganguly, *Quantum Machine Learning: An Applied Approach*, https://doi.org/10.1007/978-1-4842-7098-1_8

To solve today's real-life problems, quantum computers need to read, interpret, and analyze input datasets. Input data to a quantum system can be any data source in a natural or artificial quantum system.

The three generic stages of execution in a classical data-driven quantum computing process were shown in Figure 5-6 and are shown again here for reference.

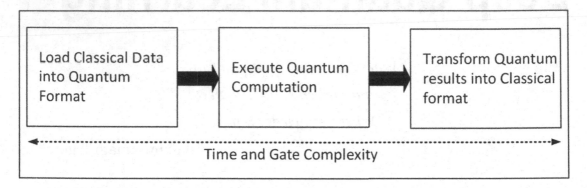

Reproduced Figure 5-6. The three stages of execution in a quantum computer

Formulations of quantum circuits for loading classical data into quantum states for processing by a quantum computer are an active focus of current research. Since different quantum algorithms have varying constraints on how the classical input data is loaded and formatted into the corresponding quantum states, several data loading circuits are in use in the field today (e.g., basis encoding, amplitude encoding, etc.). As we progress through this chapter, you see how classical data is encoded and treated through quantum systems.

As a continuation of Chapter 7, we look at using D-Wave's qbsolve to solve a well-known NP-hard problem: the traveling salesperson.

Optimized Learning by D-Wave

We explored some functionalities and problems that can be solved by quantum annealing based systems from D-Wave in Chapter 7. The classical optimization problem, given by equation 7.27 (reproduced here for ease of reading), gave us the following mathematical expression, which can be used to build a programming model on D-Wave platforms.

$$Obj\left(a_i, b_{ij}; q_i\right) = \sum_i a_i q_i + \sum_{ij} b_{ij} q_i q_j$$

The equation 7.27 is quantized via an Ising model Hamiltonian and used to achieve quantum-enhanced optimization by mapping on annealing driven platforms. For this purpose, equations 7.4, 7.5, and 7.27 formulate the following Hamiltonian (D-Wave documentation [144]).

$$H(s) = A(s) \sum_i \sigma_i^x + B(s) \left[\sum_i a_i \sigma_i^z + \sum_{ij} b_{ij} \sigma_i^z \sigma_j^z \right]$$ 8.1

where $A(s)$ and $B(s)$ are Lagrange multipliers.[1] In programming terms, $A(s)$ is also called the gamma parameter (explained later).

In this context, it is helpful to be familiar with D-Wave's qbsolve (https://github.com/dwavesystems/qbsolv) [161]. qbsolve is a decomposing, hybrid quantum/classical solver. It finds a minimum value of a large quadratic unconstrained binary optimization (QUBO) problem by splitting it into pieces, solved either via a D-Wave system or a classical tabu solver. It is an open source software tool designed for problems too large and/or too dense to run on D-Wave quantum computer that divides problems into chunks and iterates into sub-QUBOs. The qbsolve process can be run on a classical computer.

qbsolve takes a QUBO file format or Q matrix as input for simulated annealing, or QPU, and gives bit strings as output (see Figure 8-1).

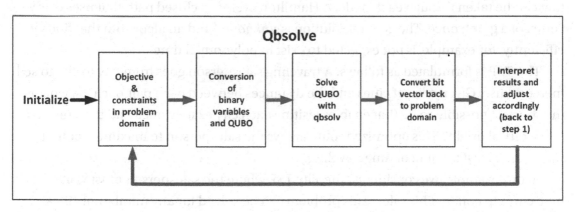

Figure 8-1. *qbsolve hybrid algorithm workflow*

---

[1]Lagrange Multiplier: https://en.wikipedia.org/wiki/Lagrange_multiplier#:~:text=In%20mathematical%20optimization%2C%20the%20method,chosen%20values%20of%20the%20variables)

Figure 8-1 shows the workflow for the qbsolve algorithm with the following overall steps.

- Reads QUBO instance

- Partitions into smaller subQUBOs

- Solves subQUBOs (this can be done on CPU or QPU)

- Combines the results

A great example of qbsolve in action is the traveling salesperson problem.

Traveling Salesperson Problem (qbsolve)

The traveling salesperson problem (TSP)[2] [162] is a well-known and notorious NP-hard problem that has excited mathematicians for centuries and computer scientists for decades. From the context of applications, a TSP type of problem is important in the financial and marketing industries. TSP is considered important due to its *hardness* and similarity to some other relevant combinatorial optimization problems that arise in practice.

The mathematical formulation based on early analysis was proposed by Irish mathematician W. R. Hamilton and British mathematician Thomas Kirkman in the early 19th century. Mathematically, the problem can be abstracted in terms of graphs (e.g., Max-Cut). The TSP on the nodes of a graph asks for the shortest Hamiltonian cycle that can be taken through each node. A Hamilton cycle is a closed path that uses every vertex of a graph once. The general solution is *unknown,* and an algorithm that finds it efficiently, for example, is not expected to exist in polynomial time.

The TSP is formulated as follows: A traveling salesperson goes from city to city to sell merchandise. Given a list of cities and the distances between each pair of cities, what is the shortest possible route that enables visiting each city exactly once and then returns to the city of origin? This optimized route allows the salesperson to maximize potential sales in the least amount of time traveling.

In our example, we consider a nine-city TSP where the salesperson must visit nine cities to conduct his sales. This problem can be solved for any number of cities. We choose the following nine US capital cities and define a data file (`data_9cities.txt`) with their location parameters (latitudes and longitudes) and index them for programming ease (this file can be found on the book's website).

---

[2]The traveling salesperson problem: https://en.wikipedia.org/wiki/
Travelling_salesman_problem#cite_note-6

```
0,Colorado,Denver,39.7391667,-104.984167
1,Connecticut,Hartford,41.767,-72.677
2,Delaware,Dover,39.161921,-75.526755
3,Illinois,Springfield,39.783250,-89.650373
4,Indiana,Indianapolis,39.790942,-86.147685
5,Massachusetts,Boston,42.2352,-71.0275
6,Michigan,Lansing,42.7335,-84.5467
7,New Hampshire,Concord,43.220093,-71.549127
8,New Jersey,Trenton,40.221741,-74.756138
```

We define another data file (`data_distance9.txt`) with the distances between each pair of those capital cities giving a total of $9 \times 8 = 72$ distances (this file can be found on the book's website). The following is an example of a problem path.

$$(0->1)+(1->2)+(2->3)+(3->4)+(4->5)+(5->6)+(6->1)$$

or

$$(0->2)+(2->1)+(3->2)+(3->4)+(4->5)+(5->6)+(6->1)$$

Any other combination can be used if the path and time traveled are optimized for the salesperson to maximize their sales. The following two constraints are considered in our problem when formulating the QUBU.

- The salesperson finishes at the same place he started.

- He visits each city exactly once.

Initially, we need to think about ways to model the problem mathematically. To do this, we define our *binary variables* first.

$$a*b = \begin{cases} 1, & \text{if the trip } \textit{includes} \text{ segment } \mathbf{0->1} \\ 0, & \text{if the trip } \textit{does not include} \text{ segment } \mathbf{0->1} \end{cases}$$

$$\vdots \quad \vdots \quad \vdots \qquad \vdots \quad \vdots \quad \vdots \qquad \vdots \quad \vdots \quad \vdots \qquad \vdots \quad \vdots \quad \vdots \qquad \vdots \quad \vdots \quad \vdots$$

$$h*i = \begin{cases} 1, & \text{if the trip } \textit{includes} \text{ segment } \mathbf{7->8} \\ 0, & \text{if the trip } \textit{does not include} \text{ segment } \mathbf{8->7} \end{cases}$$

The constraints allow us a mathematical formulation to minimize our distance operator $\mathcal{D}$.

$$\mathcal{D}(ab*ab)+\mathcal{D}(ac*ac)+\cdots+\mathcal{D}(hi*hi)$$

Keeping in mind our constraints, each city appears in the QUBO exactly twice—once arriving and once departing. For example,

for City 0 (Denver), we have

$$ab+ac+ad+ae+af+ag+ah+ai-2$$

for City 1 (Hartford), we have

$$ab+bc+bd+be+bf+bg+bh+bi-2$$

and so and so forth until City 8, which is Trenton.

Now that we have defined the preliminaries, we must formulate the constraints and *objective* (*Obj*) (equation 7.27).

$$Obj \equiv min\left(\mathcal{D}(ab*ab)+\mathcal{D}(ac*ac)+\cdots+\mathcal{D}(hi*hi)\right)$$

where, *min* stands for *minimize*. The action of *minimization* seeks the shortest distance between the sides, which occurs at equality. The square is imposed to eliminate negative distances. The *constraints* are given by

$$\left(ab+ac+ad+ae+af+ag+ah+ai-2\right)^2$$

$$\left(ab+bc+bd+be+bf+bg+bh+bi-2\right)^2$$

$$\vdots \quad \vdots \quad \vdots \quad \vdots \quad \vdots \quad \vdots \quad \vdots \quad \vdots \quad \vdots$$

$$\left(ai+bi+ci+di+ei+fi+gi+hi-2\right)^2$$

Next, we combine the objective and constraints to formulate our QUBO.

$$\text{QUBO} = objectives + \gamma\left(constraints\right)$$

where, γ is a Lagrange parameter that can be tuned for optimum results.

Equipped with this, we start our exercise with the Jupyter notebook file TSP_9City_ Dwave.ipynb downloadable from the book's website. The first code snippet is libraries and visualization tools.

Listing 8-1a. Libraries and Pauli Matrices for TSP_9City_Dwave.ipynb

```
## Libraries & visualization tools for us to use.
import networkx as nx
import pylab
import matplotlib.pyplot as plt
from matplotlib.pyplot import pause
import sys
from bokeh.sampledata import us_states
from bokeh.plotting import *
from bokeh.models import Plot, Range1d, MultiLine, Circle, HoverTool,
TapTool, BoxSelectTool
from bokeh.io import output_notebook
from bokeh.palettes import Spectral4
from bokeh.models.graphs import from_networkx, NodesAndLinkedEdges,
EdgesAndLinkedNodes
output_notebook()
%matplotlib inline
import argparse
import re
```

Here *bokeh* refers to a Python library for creating interactive visualizations for modern web browsers. It is used in our exercise to create a map of the United States and the cities traveled.

The next task is to set the number of cities in the problem and an indexing function. The binary variables are $x_{i,j} = 1$ if and only if city i is visited at stop j. The QUBO dictionary/matrix is $N^2 \times N^2$, hence, these variables must be assigned row and column indices. The function $x(a, b)$ assigns a QUBO matrix index to the $x_{a,b}$ variable.

Listing 8-1b. Matrix for `TSP_9City_Dwave.ipynb`

```
# Number of cities in our problem
N = 9
# Function to compute index in Q for variable x_(a,b)
def x(a, b):
    return (a)*N+(b)
```

In this example, there are two *Lagrange parameters*.

- *A* or `gamma` is the standard Lagrange parameter that indicates how important our constraints are in the quality of our solution. This can be fine-tuned to obtain optimum results. This controls whether satisfying our constraints or finding the shortest distance is more important. For this problem, it is crucial that we visit every city on our route, so we should set this parameter to be larger than the greatest distance between two cities.

- *B* indicates how important the objective function is in our QUBO. Generally, we define *B* = 1 and modify the value in *A*.

It is also possible to adjust the following parameters.

- **chain strength** (`chainstrength`): This tells the embedding function how strongly to tie together chains of physical qubits to make one logical qubit. This should be larger than any other values in the QUBO.

- **number of runs** (`numruns`): This tells the system how many times to run our problem. Due to the probabilistic nature of the D-Wave QPU, we should run the problem many times and look at the best solutions found.

This code will not run on the physical QPU but use `qbsolv` offline or classically. `numruns` run `qbsolv` often, but *chain strength* does not affect the samples unless we modify the code to use the QPU.

Listing 8-1c. Tunable Parameters TSP_9City_Dwave.ipynb

```
# Gamma = A
A = 6500
B = 1
chainstrength = 3000
# Number of runs
numruns = 100
```

Next, utilizing city locations data_9cities.txt, the first data file, we generate a map of the United States. The cities that the salesperson traveled to are pinpointed in red.

Listing 8-1d. Map of the cities TSP_9City_Dwave.ipynb

```
us_states = us_states.data.copy()
# Delete states Hawaii and Alaska from mainland map
del us_states["HI"]
del us_states["AK"]
# separate latitude and longitude points for the borders
#   of the states.
state_xs = [us_states[code]["lons"] for code in us_states]
state_ys = [us_states[code]["lats"] for code in us_states]
with open('data_9cities.txt', "r") as myfile:
    city_text = myfile.readlines()
    myfile.close()
cities = [',']*N
states = [',']*N
lats=[]
longs=[]
for i in city_text:
    index, state, city,lat,lon = i.split(',')
    cities[int(index)] = city.rstrip()
    states[int(index)] = state
    lats.append(float(lat))
    longs.append(float(lon))
```

```
# init figure
p = figure(title="Find shortest route that visits each city",
           toolbar_location="left", plot_width=550, plot_height=350)
# Draw state lines
p.patches(state_xs, state_ys, fill_alpha=0.0,
    line_color='blue', line_width=1.5)
# The scatter markers
p.circle(longs, lats, size=10, color='red', alpha=1)
show(p)
```

This code snippet produces the US map shown in Figure 8-2, highlighting the nine cities courtesy of bokeh.

Figure 8-2. *Map of the nine cities visited*

Now that we have the base map, we upload the inter-city distances via the file data_distance9.txt.

Listing 8-1e. Inter-City Distances TSP_9City_Dwave.ipynb

```
# Input file containing inter-city distances
fn = "data_distance9.txt"

# check that the user has provided input file
try:
  with open(fn, "r") as myfile:
    distance_text = myfile.readlines()
    myfile.close()
except IOError:
  print("Input distance file missing")
  exit(1)
```

The distances are stored in an $N \times N$ matrix. The matrix has zeros on the diagonal. The distance from one city (e.g., 0) to the next city (e.g., 1) are in entry $\mathcal{D}(a,b)$ since distances are the same when traveling in either direction between two cities. The matrix is symmetric.

Listing 8-1f. Create the Matrix TSP_9City_Dwave.ipynb

```
# Initialize matrix of correct size with all 0's
D = [[0 for z in range(N)] for y in range(N)]
# Read in distance values and enter in matrix
for i in distance_text:
  if re.search("^between", i):
    m = re.search("^between_(\d+)_(\d+) = (\d+)", i)
    citya = int(m.group(1))
    cityb = int(m.group(2))
    D[citya][cityb] = D[cityb][citya] = int(m.group(3))
```

In the next step, we start developing the QUBO for this problem by creating an empty Q matrix with all values initialized at 0.

Listing 8-1g. Q-matrix TSP_9City_Dwave.ipynb

```
Q = {}
for i in range(N*N):
    for j in range(N*N):
        Q.update({(i,j): 0})
```

The next code block defines the constraint that each row has exactly *one* 1 in the permutation matrix. For row 1, this constraint is

$$A\left(\sum_{j=1}^{9}x_{1,j}-1\right)^2 \equiv \sum_{j=1}^{9}-Ax_{1,j}+\sum_{j=1}^{9}\sum_{k=j+1}^{9}2Ax_{1,j}x_{1,k}$$

Listing 8-1h. Permutation Matrix TSP_9City_Dwave.ipynb

```
for v in range(N):
    for j in range(N):
        Q[(x(v,j), x(v,j))] += -1*A
        for k in range(j+1, N):
            Q[(x(v,j), x(v,k))] += 2*A
```

The next step is to develop the QUBO. The main objective is to minimize the distance traveled.

Consider the case where the traveling salesperson visits city u at stop 3 and city v at stop 4. Then, the distance he travels in this segment is $\mathcal{D}(u,v)x_{u,3}x_{v,4}$. This adds $\mathcal{D}(u,v)$ to the total distance if he visits city u at stop 3 and city v at stop 4 and adds 0 to our total distance otherwise. Hence, for every pair of cities, u and v, we add $\sum_{j=1}^{9}\mathcal{D}(u,v)x_{u,j}x_{v,j+1}$.

This adds the distance traveled directly from cities u and v to the overall route. It needs to be added for every u and v option in *both directions*.

Listing 8-1i. Add Distance objective TSP_9City_Dwave.ipynb

```
for u in range(N):
    for v in range(N):
        if u!=v:
            for j in range(N):
                Q[(x(u,j), x(v,(j+1)%N))] += B*D[u][v]
```

Running the Problem

Now we run qbsolv offline. qbsolv[3] is a problem decomposition tool that can either run classically offline (on our laptops, for example) or in a hybrid manner in a classical-QPU combination.

Listing 8-1j. Run qbsolve TSP_9City_Dwave.ipynb

```
from dwave_qbsolv import QBSolv
resp = QBSolv().sample_qubo(Q)
```

In the next step, we do the *post processing*. Once qbsolv is run, we collect the results and report back the best answer found. Here we list the lowest energy solution found, the order the cities are visited in the salesperson's route, and the total mileage required for this route. We include some validity checks in this output, indicating if a stop is being assigned to more than one city and indicating if a stop has no city assigned.

Listing 8-1k. Add Distance Objective TSP_9City_Dwave.ipynb

```
# First solution is the lowest energy solution found
sample = next(iter(resp))
# Display energy for best solution found
print('Energy: ', next(iter(resp.data())).energy)
# Print route for solution found
route = [-1]*N
for node in sample:
    if sample[node]>0:
        j = node%N
        v = (node-j)/N
        if route[j]!=-1:
            print('Stop '+str(i)+' used more than once.\n')
        route[j] = int(v)
# Compute and display total mileage
mileage = 0
```

---

[3]qbsolv is based on the paper "A Multilevel Algorithm for Large Unconstrained Binary Quadratic Optimization," Wang, Lu, Glover, and Hao (2012). [163]

```
for i in range(N):
    mileage+=D[route[i]][route[(i+1)%N]]
print('Mileage: ', mileage)
print('\nRoute:\n')
for i in range(N):
    if route[i]!=-1:
        print(str(i) + ':   ' +cities[route[i]]+ ',' + states[route[i]] + '\n')
    else:
        print(str(i) + ':  No city assigned.\n')
```

This code snippet, run on qbsolve, gives us the output we want with the lowest energy, total mileage, and route.

```
Energy:  -114297.0
Mileage:  2703

Route:

0:  Springfield,Illinois

1:  Denver,Colorado

2:  Trenton,New Jersey

3:  Dover,Delaware

4:  Hartford,Connecticut

5:  Boston,Massachusetts

6:  Concord,New Hampshire

7:  Lansing,Michigan

8:  Indianapolis,Indiana
```

Note that the sequence of the cities visited may change from run to run due to the probabilistic nature of the calculation. You are encouraged to change the Lagrangian gamma value (here gamma=A) from lower to higher and observe the change in the lowest energy calculated. Generally, the value of parameter A is set to be larger than the greatest distance between two cities.

Let's run a few checks to verify that our result is valid.

- If every city appears exactly once in our list, then our route list consists of the numbers 0, 1, 2...$(N-1)$ in some order, which adds up to $N(N-1)/2$. If this sum is not correct, our route is invalid.

- Verify if every city has an assigned stop and vice versa. If not, we print a message to the user to make them aware.

Listing 8-1l. Verify If Route Is Valid TSP_9City_Dwave.ipynb

```
alert = 0
if sum(route)!=N*(N-1)/2:
    print('Route invalid.\n')

for i in range(N):
    if route[i]==-1:
        print('Stop '+str(i)+' has no city assigned.')

if alert==0:
    print("Route valid.")
```

In our case, the code block verifies that the route is valid.

```
                    Route valid.
```

The final code block visualizes the route found.

Listing 8-1m. Visualization of the Route TSP_9City_Dwave.ipynb

```
Path = nx.Graph()
coord={}
coord[route[0]]=(longs[route[0]],lats[route[0]])
Path.add_node(cities[route[0]],pos=coord[route[0]],label=cities[route[0]])

for i in range(N-1):
    e=(cities[route[i]],cities[route[i+1]])
    Path.add_edge(*e)
    coord[route[i+1]]=(longs[route[i+1]],lats[route[i+1]])
    Path.add_node(cities[route[i+1]],pos=coord[route[i+1]],label=cities[rou
te[i+1]])
```

```
e=(cities[route[N-1]],cities[route[0]])
Path.add_edge(*e)

fig, ax = plt.subplots(figsize=(120,60))
margin=0.15
fig.subplots_adjust(margin, margin, 1.-margin, 1.-margin)
ax.axis('equal')
nx.draw(Path, nx.get_node_attributes(Path, 'pos'), with_labels=True,
width=10, edge_color='b', node_size=200,font_size=72,font_weight='bold',
ax=ax)
plt.show()
```

The last code block finally gives the path visualization on the map as output.

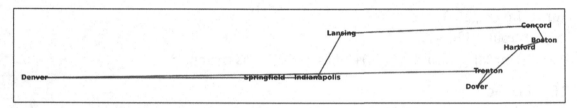

You are encouraged to modify the data files (data_9cities.txt and data_distance9.txt) to several more cities with their coordinates, additional city-to-city distances and try out the visualization several times. As the number of cities increase, the visualization may start to vary from run to run.

If you are interested in experiencing real qubits compute code, D-Wave offers a limited free user quota on Leap (https://cloud.dwavesys.com/leap/)—cloud-based access to their hardware platforms. As explained in Chapters 1 and 7, users from 38 countries (at present) can register for free accounts at Leap and utilize a plethora of training materials, documentation, developer resources, and much more. Once you log in to the Leap platform, they need to scroll down on the first landing screen. There is a box with API information on the bottom-left corner that can be copied and used to make an API call to utilize the D-Wave's actual QPU.

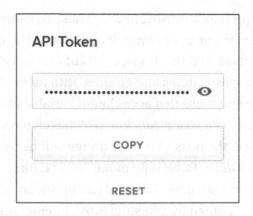

To make the call to the QPU using qbsolve for our TSP problem in this section, one would use the following code block.

```
from dwave.system.composites import EmbeddingComposite
from dwave.system.samplers import DWaveSampler

sampler = DWaveSampler(solver={'qpu': True})  # Some accounts need to
replace this line with the next:
sampler = EmbeddingComposite(DWaveSampler(token = 'my_token',
solver=dict(name='solver_name')))
print("Connected to QPU.")
```

The two code lines in blue are used on an either/or basis. For most users, the second line requiring an API token is the one that works. The API token is copied from the Leap interface and entered in <my_token> to secure a connection to the actual QPU. When using qbsolve, the value of the name is given by name='qbsolve'.

Let's move forward to deep learning methodologies in the quantum domain. *Quantum deep learning* has several implementations today; besides D-Wave, there is Google's TensorFlow Quantum and Xanadu's PennyLane.

Quantum Deep Neural Networks

Classical deep neural networks are efficient tools for machine learning and serve as the fundamental base for developing deep quantum learning methods. The simplest deep neural network to quantize is the Boltzmann machine [139]. Classical Boltzmann machines (Ackley et al. 1985 [168]; Du and Swamy 2019 [169]) offer a powerful framework for modeling probability distributions. These types of neural networks use an undirected graph structure to encode relevant information. The respective information is stored in bias coefficients and connection weights of network nodes, which are typically related to binary spin-systems and grouped into those that determine the output: the visible nodes; and those that act as latent variables: the hidden nodes. The network structure is linked to an *energy function* which facilitates the *definition of a probability distribution* over the possible node configurations by using a concept from statistical mechanics (i.e., *Gibbs states* (Boltzmann 1877 [170]; Gibbs 1902 [171])).

The classical Boltzmann machine[4] consists of components with tunable interactions. The Boltzmann machine is trained by adjusting those interactions so that the thermal statistics of the bits described by a *Boltzmann–Gibbs distribution* (see Figure 7-11b) reproduces the data's statistics. To quantize the Boltzmann machine, the neural network is expressed as a set of interacting quantum spins corresponding to a tunable Ising model. The neurons used as input to the Boltzmann machines are initialized into a fixed state; following this, the system can thermalize, and the output qubits are read as results. An attractive feature of deep quantum learning is that it does not require a large, general-purpose quantum computer.

Previously we discussed special-purpose quantum information processors, such as quantum annealers, which are well suited for constructing deep quantum learning networks [164, 165, 166]. This chapter also looks at quantum systems based on programmable photonic circuits as they also are useful for creating efficient quantum deep learning networks.

The classical neural networks are related to the processing of neurons, the transformation performed by the neuron, the interconnection between neurons, the neural network dynamics, and the learning rules. The neural network learning rules govern the change of neural network connection strengths. The neural networks are differentiated by training in a supervised or unsupervised manner. Neural networks can use single systems to save different class data and to classify the stimulus in a distributed way. Hence neural networks can be very useful in creating classification systems.

The coherence in quantum states, which is called *superposition*, is deemed to resemble neurons in classical neural networks. The measurement function in quantum computing is analogous to interconnections in classical neural networks. The gain function in a neural network resembles the property of entanglement in quantum computing.

---

[4]Classical Boltzmann Machine: https://en.wikipedia.org/wiki/Boltzmann_machine [167]

A quantum neural network (QNN) is similar to brain function and help create new information systems. QNNs are used to solve classically challenging problems that require exponential capacity and memory. In a nutshell, QNNs are considered the natural next step in the evolution of neurocomputing systems.

The links between quantum computing and machine learning have been developing for the last several years. However, many well-defined "textbook" QML algorithms were designed with fault-tolerant quantum computers in mind. A different set of algorithms, tools, and strategies are required in the current era of noisy intermediate-scale quantum (NISQ) devices. It is worth presenting the key ideas for machine learning in the NISQ era and surveying the current state of play.

Quantum Learning with Xanadu

Xanadu is a full-stack quantum startup based in Toronto, Canada, that is advancing a quantum photonic processor with an open source full-stack quantum software platform called Strawberry Fields. The company has also developed a cross-platform Python library for differentiable programming of quantum computers [146] called PennyLane (`https://pennylane.readthedocs.io/en/stable/introduction/pennylane.html`). PennyLane and Strawberry Fields are implemented in Python. Xanadu has also started publishing results using their hardware and software in diverse fields such as quantum chemistry, graph theory, quantum machine learning, and others.

PennyLane[5] [145] is an open source software framework built around quantum differentiable programming to achieve machine learning tasks with quantum computers. It seamlessly integrates classical machine learning libraries with quantum simulators and hardware, giving users the power to train quantum circuits. PennyLane, while being QML capable, also has several versatile capabilities.

- Supports hybrid quantum and classical models allowing users to connect quantum hardware with PyTorch, TensorFlow, and NumPy, thereby boosting quantum machine learning capabilities

- Allows built-in automatic differentiation of quantum circuits

- Provides optimization and machine learning tools

---

[5]PennyLaneAI/pennylane is licensed under the Apache License 2.0.

- Hardware agnostic—the same quantum circuit model can be run on different backends—and allows plugins for access to diverse devices, including Strawberry Fields, Amazon Braket, IBM Q, Google Cirq, Rigetti Forest, Microsoft QDK, and ProjectQ

Installing PennyLane is very simple. It requires Python 3.6 or later versions. The recommended Python installation is Anaconda Python 3 (www.anaconda.com/download/). Since this book has professed Anaconda right from the start, we should be fine on that front. The rest is easy: install the latest release of PennyLane via the following command.

```
pip install pennylane --upgrade
```

This command should successfully and painlessly install PennyLane on top of your Python 3.6+ environment. PennyLane has some dependencies, such as NumPy, SciPy, network, which it checks for while installing (and if required, does install them automatically), as shown in Figure 8-3.

```
Collecting pennylane
  Downloading PennyLane-0.14.1.tar.gz (404 kB)
  |                                     | 404 kB 1.1 MB/s
Requirement already satisfied, skipping upgrade: numpy in c:\users\santagan\anaconda3\lib\site-packages (from pennylane)
  (1.18.1)
Requirement already satisfied, skipping upgrade: scipy in c:\users\santagan\anaconda3\lib\site-packages (from pennylane)
  (1.4.1)
Requirement already satisfied, skipping upgrade: networkx in c:\users\santagan\anaconda3\lib\site-packages (from pennyla
ne) (2.4)
Collecting autograd
  Downloading autograd-1.3.tar.gz (38 kB)
Collecting toml
  Downloading toml-0.10.2-py2.py3-none-any.whl (16 kB)
Collecting appdirs
  Downloading appdirs-1.4.4-py2.py3-none-any.whl (9.6 kB)
Collecting semantic_version==2.6
  Downloading semantic_version-2.6.0-py3-none-any.whl (14 kB)
Requirement already satisfied, skipping upgrade: decorator>=4.3.0 in c:\users\santagan\anaconda3\lib\site-packages (from
  networkx->pennylane) (4.4.1)
Requirement already satisfied, skipping upgrade: future>=0.15.2 in c:\users\santagan\anaconda3\lib\site-packages (from a
utograd->pennylane) (0.18.2)
Building wheels for collected packages: pennylane, autograd
  Building wheel for pennylane (setup.py) ... done
  Created wheel for pennylane: filename=PennyLane-0.14.1-py3-none-any.whl size=481998 sha256=6755ad9a1085decfc2fe87e1968
d6442e884c00735c4c939e0fe875bada99d8a
```

Figure 8-3. *PennyLane installation process*

Xanadu is developing a photonic quantum computer: a device that processes information stored in quantum states of light. Photonic quantum computers use continuous degrees of freedom—such as the amplitude and phase of light—to encode information. This continuous or analog structure makes photonic devices an attractive platform for quantum versions of neural networks.

QNNs are quantum circuits or algorithms that closely resemble the structure of classical neurons and neural networks (see Chapter 3) while extending and generalizing them with powerful quantum properties. There is published research on QNNs that takes advantage of photonics [84], and there are several other proposals in the literature.

The proposal of resembling QNNs with photonics is exciting for some of the research community. Killoran et al., in 2018, released a paper [84] where they propose a photonic circuit that consists of a sequence of repeating building blocks or *layers*. These layers can be built with the output of one layer serving as the input to the next. These photonic layers are similar to the layers that appear in classical neural networks (see Chapter 3). Classical nets take an input x, multiply it by a weight matrix W, add a bias b, and pass the result through a nonlinear activation function such as sigmoid, tanh, and ReLU.

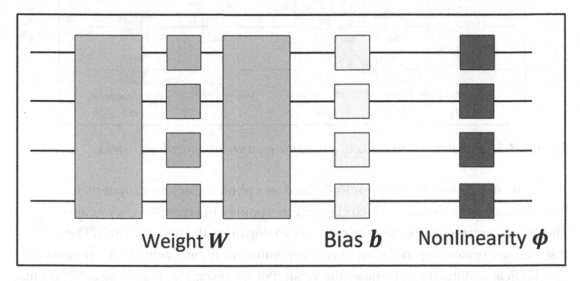

Figure 8-4. *Fundamental units of classical neural networks*

Figure 8-4 shows the fundamental structure of a classical neural network (NN), which performs the following transformation.

$$z = f(x) \rightarrow \phi(\mathbf{w}.\mathbf{x} + b)$$ 8.2

Typically, in equation 8.2, weight **w** is expressed by two orthogonal matrices and a diagonal matrix.

The quantum NN layer resembles the functionality of the classical NN represented in Figure 8-4 by using photonic quantum gates, which include interferometers made from phase shifters and beamsplitters, squeezing and displacement gates, and a fixed nonlinear transformation. These are the same gates that build a photonic quantum computer; hence, the QNN architecture is interpreted to have the computational ability of quantum computers.

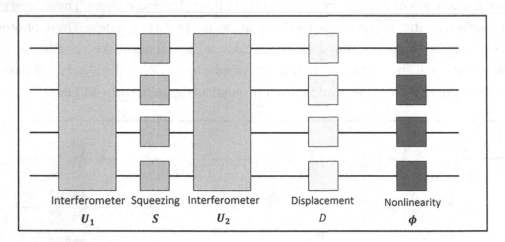

Figure 8-5. *Fundamental units of photonic quantum neural networks*

Figure 8-5 shows basic QNN units defined by a photonic quantum system. It has colored gates to indicate which classical component in Figure 8-4 they relate to. The interferometers and squeezing gates are analogous to the weight matrix. The displacement gates map to the bias and the quantum nonlinear activation functions to the classical nonlinearity. The quantum version of NN resembles the classical NN quite closely and can be used to run the classical version. By controlling the quantum net in a way that discards any quantum "weirdness," such as superposition and entanglement. PennyLane can combine existing classical nodes to create and explore hybrid models, as shown in Figure 8-6 (from PennyLane documentation [145]).

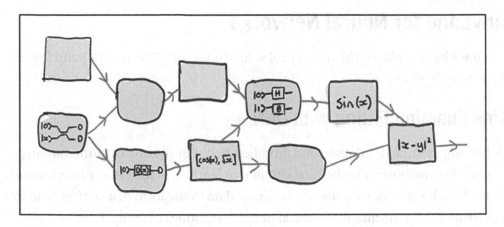

Figure 8-6. *Example structure of hybrid classical-quantum models with PennyLane*

While we discuss quantum deep learning, we need to keep in mind that we are still not in an era where abundant quantum data can be trained and tested on quantum NN models. Much of the real-life data today is classical in nature. Hence, we need to ask ourselves how do we encode this classical data and input it into a quantum machine?

There are many ways that classical data can be embedded into a quantum circuit: Different vendors of quantum computing platforms have made available templates to achieve this. For example, PennyLane has `qml.templates` module. PennyLane supports various quantum frameworks and quantum hardware via an array of plugins. For example, photonic layers require the `pennylane-sf` plugin for Strawberry Fields.

To construct the layers of the quantum nodes (or QNodes as they are called in PennyLane terminology) the templates of preconstructed qubit layers can be used besides manually placing individual gates. The following are some of the preconstructed layers.

- `RandomLayers`

- `StronglyEntanglingLayers`

- `BasicEntanglerLayers`

- `SimplifiedTwoDesign`

- `CVNeuralNetLayers`

Repeated layers like classical NNs offer the following advantages.

- It provides control over the depth of circuit.

- Layers can be designed to be hardware-friendly.

- It is a model for universal quantum computation.

425

PennyLane for Neural Networks

Now that we have explored the theoretical side of quantum NNs (QNNs) and PennyLane, it is time to get coding.

Cosine Function Fitting with QNN

The following example is adapted and modified from PennyLane documentation [172]. This example constitutes a variational circuit that learns a *fit* for a one-dimensional function. To achieve this, we generate synthetic data in the form of a `cosine` function with noise and train the quantum neural network (QNode) to apply them.

The following explains how to do this.

1. Create a `device` with `default.qubit`.

2. Create a QNode that takes an input x and trainable weights and outputs a y.

3. Define a range for x to train over ($-\pi$ to π) and calculate $y = \cos x$.

4. Choose a PennyLane optimizer.

5. Run the optimization loop.

The variational circuit used in the following example is the *continuous-variable quantum neural network* model described in Killoran et al. (2018) [84].

In the first step, we `import` PennyLane (after installing it), the wrapped version of NumPy, provided by PennyLane, and an optimizer called `AdamOptimizer` (https://pennylane.readthedocs.io/en/stable/code/api/pennylane.AdamOptimizer.html). The `device` used for this is the Strawberry Fields simulator, which can be installed via the Strawberry Fields plugin for PennyLane as follows.

```
$ pip install pennylane-sf
```

To start with this exercise, we open the Jupyter notebook called `qnn_cosine_pl.ipynb` available from the download link of the website for this book.

Listing 8-2a. Libraries and Optimizer `qnn_cosine_pl.ipynb`

```
# Import PennyLane, the wrapped version of NumPy provided by PennyLane, and
  an optimizer.
import pennylane as qml
from pennylane import numpy as np
from pennylane.optimize import AdamOptimizer
import matplotlib.pyplot as plt
import pylab
import pickle
```

`AdamOptimizer` is a method for stochastic optimization. It is based on the principles of the gradient descent optimizer with adaptive learning rate [84], offering Adaptive Moment Estimation using a step-dependent learning rate. Chapter 3 covered gradient descent for classical NNs, one of the most common training algorithms for feedforward networks. Gradient descent optimizers use the slope (or the gradient) to guide the parameter space. Mathematically, gradient descent, given by equation 3.5, is reproduced here.

$$w_{ij}^{n+1} = w_{ij}^n - \eta \nabla E = w_{ij}^n - \eta \frac{\partial C(w_1, w_2)}{\partial w_{ij}}$$

8.3

where, the current weight w_{ij}^{n+1} is the updated weight, w_{ij}^n is the weight of the previous step down toward the direction of the steepest gradient, η is the user-defined hyperparameter corresponding to the learning rate or step-size, E is the cost function and C is the mathematical equivalent of E in a differential form with a derivative with respect to w_{ij} to obtain the gradient.

The `pickle` module implements binary protocols for serializing and deserializing a Python object structure. *Pickling* is the process whereby a Python object hierarchy is converted into a byte stream. *Unpickling* is the inverse operation, whereby a byte stream (from a binary file or bytes-like object) is converted back into an object hierarchy. `Pickle` can be used to load the data into another Python script later.

In the next step, we create a noisy cosine function for sampling.

Listing 8-2b. Noisy Cosine Function Definition qnn_cosine_pl.ipynb

```
Xlim = 5
noise = np.random.normal(0,0.1,100) # generate noise to add to the function
values (Y-values)
# define functions
X = np.arange(-Xlim, Xlim, 0.1)
Y = np.cos(X)+noise
```

This code block generates two lists, X and Y, with the X-axis and Y-axis data. Let's now write the Y-axis data into a file for future use.

Listing 8-2c. Write Noisy Cosine Data to a File qnn_cosine_pl.ipynb

```
# write the data out to a file
cosdata = open('cosdata.md', 'wb')
pickle.dump(Y, cosdata)
cosdata.close()
plt.plot(X[0:200], Y[0:200])
```

This snippet writes the data of *y* into a file named cosdata.md. Equipped with this, we plot some of the data with the plt.plot() function to generate the following plot of the noisy cosine function.

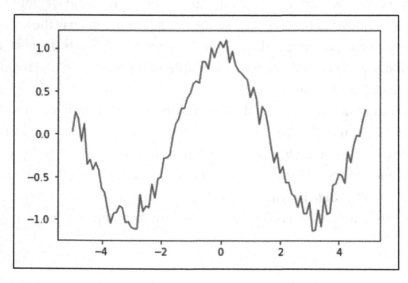

Now we have some noisy cosine data to train our QNN on. The device used is the Strawberry Fields simulator with only one quantum mode (or wire). For this exercise to work, we need to have the Strawberry Fields plugin for PennyLane installed as specified before.

Note   If the Strawberry Fields plugin for PennyLane is *not* installed, you get the following error: DeviceError: Device does not exist. Make sure the required plugin is installed.

Listing 8-2d. Strawberry Fields plugin `qnn_cosine_pl.ipynb`

```
dev = qml.device("strawberryfields.fock", wires=1, cutoff_dim=10)
```

PennyLane's Strawberry Field plugin accesses a Fock[6] state simulator backend. This simulator represents quantum states in the Fock basis $|0\rangle, |1\rangle, |2\rangle, ..., |D-1\rangle$, where D is the user-given value for `cutoff_dim` that *limits* the dimension of the Hilbert space. The advantage of this representation is that any continuous-variable operation can be represented. Note the following.

- The simulations are approximations whose accuracy increases with the cutoff dimension `cutoff_dim`.

- For M modes or wires and a cutoff dimension of D, the Fock state [288] simulator needs to keep track of at least M^D values. Hence, the simulation time and required memory grow much faster with the number of modes than in qubit-based simulators.

- It is often useful to keep track of the normalization of a quantum state during optimization to make sure the circuit does not "learn" to push its parameters into a regime where the simulation is vastly inaccurate.

In the next step, we define the QNode leveraging our explorations of variational circuits in Chapter 7. For a single quantum mode, each layer of the *variational circuit* is defined with an *input layer*, a *bias*, and a *nonlinear transformation*.

---

[6]Fock State: https://en.wikipedia.org/wiki/Fock_state#:~:text=In%20quantum%20
mechanics%2C%20a%20Fock,the%20Soviet%20physicist%20Vladimir%20Fock.

Listing 8-2e. Define the QNN Layer `qnn_cosine_pl.ipynb`

```
def layer(v):
    # Matrix multiplication of input layer
    qml.Rotation(v[0], wires=0)
    qml.Squeezing(v[1], 0.0, wires=0)
    qml.Rotation(v[2], wires=0)

    # Bias
    qml.Displacement(v[3], 0.0, wires=0)

    # Element-wise nonlinear transformation
    qml.Kerr(v[4], wires=0)
```

The variational circuit in the quantum node first encodes the input into the displacement of the mode and then executes the layers. The output is the expectation of the x-quadrature [288].

Listing 8-2f. Encode Input `qnn_cosine_pl.ipynb`

```
@qml.qnode(dev)
def quantum_neural_net(var, x=None):
    # Encode input x into quantum state
    qml.Displacement(x, 0.0, wires=0)
    # "layer" subcircuits
    for v in var:
        layer(v)
    return qml.expval(qml.X(0))
```

As an objective, we take the square of the loss between target labels and model predictions.

Listing 8-2g. Prediction Modeling `qnn_cosine_pl.ipynb`

```
def square_loss(labels, predictions):
    loss = 0
    for l, p in zip(labels, predictions):
        loss = loss + (l - p) ** 2
    loss = loss / len(labels)
    return loss
```

Next, we define the cost function. The outputs from the variational circuit are computed in the cost function. Function fitting is a regression problem, and we interpret the *expectations* from the quantum node as *predictions* without applying postprocessing such as thresholding.

Listing 8-2h. Cost Function `qnn_cosine_pl.ipynb`

```
def cost(var, features, labels):
    preds = [quantum_neural_net(var, x=x) for x in features]
    return square_loss(labels, preds)
```

In the next step, we load the noisy data into the cost function. Before training a model, let's examine the data (see Listing 8-2i).

Listing 8-2i. Cost Function `qnn_cosine_pl.ipynb`

```
plt.figure()
plt.scatter(X, Y)
plt.xlabel("x", fontsize=18)
plt.ylabel("f(x)", fontsize=18)
plt.tick_params(axis="both", which="major", labelsize=16)
plt.tick_params(axis="both", which="minor", labelsize=16)
plt.show()
```

Listing 8-2i gives the following plot as output.

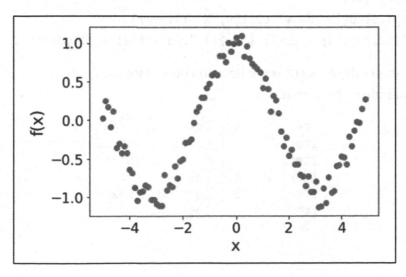

The network's weights (called var here) are initialized with values sampled from a normal distribution. We use four layers. Performance has been found to plateau at around six layers.

Listing 8-2j. Define the Number of Layers qnn_cosine_pl.ipynb

```
np.random.seed(0)
num_layers = 4
var_init = 0.05 * np.random.randn(num_layers, 5)
print(var_init)
```

This code snippet gives the following as output for the layers.

```
[[ 0.08820262  0.02000786  0.0489369   0.11204466  0.0933779 ]
 [-0.04886389  0.04750442 -0.00756786 -0.00516094  0.02052993]
 [ 0.00720218  0.07271368  0.03805189  0.00608375  0.02219316]
 [ 0.01668372  0.07470395 -0.01025791  0.01565339 -0.04270479]]
```

Finally, we engage the AdamOptimizer. We run the optimizer and update the weights for *only* 50 steps for this exercise. This runs for some time.

Listing 8-2k. Run the Optimizer qnn_cosine_pl.ipynb

```
opt = AdamOptimizer(0.01, beta1=0.9, beta2=0.999)
var = var_init
for it in range(50):
    var = opt.step(lambda v: cost(v, X, Y), var)
    print("Iter: {:5d} | Cost: {:0.7f} ".format(it + 1, cost(var, X, Y)))
```

Running this code block starts the iteration output for the 50 steps. The following is a truncated example of the output.

```
Iter:     1 | Cost: 1.1368855
Iter:     2 | Cost: 0.9848804
Iter:     3 | Cost: 0.9151763
Iter:     4 | Cost: 0.8904604
Iter:     5 | Cost: 0.8909280
Iter:     6 | Cost: 0.9044083
```

More steps in the code block lead to a much better fit but runs for a longer time. You are encouraged to run this; for example, 500 or 700 steps by tuning the step value in `for it in range(<number_of_steps>)`.

Next, we collect the *predictions* of the trained model for 50 values in the range [0, 30] and plot the function that the model has "learned" from the noisy data (maroon dots).

Listing 8-2l. Collect Predictions and Plot Function `qnn_cosine_pl.ipynb`

```
x_pred = np.linspace(-3, 3, 50)
predictions = [quantum_neural_net(var, x=x_) for x_ in x_pred]
# plot the function
plt.figure()
plt.scatter(X, Y)
plt.scatter(x_pred, predictions, color="maroon")
plt.xlabel("x")
plt.ylabel("f(x)")
plt.tick_params(axis="both", which="major")
plt.tick_params(axis="both", which="minor")
plt.show()
```

This produces the following output.

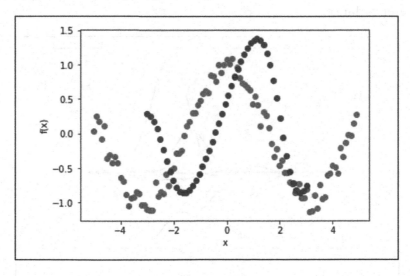

The model has learned to smooth the noisy data. We can use PennyLane to look at typical functions that the model produces without being trained at all. Higher number steps in Listing 8-2k leads to a much better fit but runs for a longer time.

In the last step, we plot the functions with the variance hyperparameter.

Listing 8-2m. Functions Against Hyperparameter `qnn_cosine_pl.ipynb`

```
variance = 1.0
plt.figure()
x_pred = np.linspace(-5, 5, 50)
for i in range(7):
    rnd_var = variance * np.random.randn(num_layers, 7)
    predictions = [quantum_neural_net(rnd_var, x=x_) for x_ in x_pred]
    plt.plot(x_pred, predictions, color="black")
plt.xlabel("x")
plt.ylabel("f(x)")
plt.tick_params(axis="both", which="major")
plt.tick_params(axis="both", which="minor")
plt.show()
```

This produces the last output of this exercise, plotting the function behavior for hyperparameter variance.

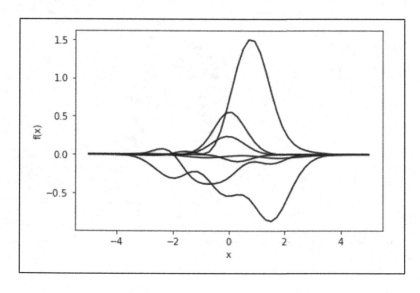

The shape of these functions varies significantly with the variance hyperparameter for the weight initialization. Setting this hyperparameter to a small value produces almost linear functions since all quantum gates in the variational circuit perform the identity transformation. Larger values produce smoothly oscillating functions with a period that depends on the number of layers used—generally, the more layers, the smaller the period.

Binary Classifier with PennyLane

Chapter 2 looked at a classifier (for classical data analysis) using the Iris dataset. We ran a support vector classifier inside `svmIris.ipynb`, where after 10000 iterations, we obtained a confusion matrix and ~98% accuracy. Chapter 3 looked at the same Iris dataset for a classical deep NN Classifier in `DNNIris.ipynb` where we ran 100 epochs to obtain an accuracy of 100%.

It will be interesting to see how the quantum NN algorithms may differ, if at all, from the classical experience with the same dataset. The data used here was used in Chapters 2 and 3: a publicly available dataset called *Iris* available from Python's `sklearn`. The Jupyter notebook titled `qml_iris_pl.ipynb` is available from the book's website. To start, we import all the necessary libraries from Python and PennyLane along with PennyLane's `GradientDescentOptimizer` to perform the optimization. We also use the `StronglyEntanglingLayers` template to build our NN layer.

Listing 8-3a. Functions Against Hyperparameter `qml_iris_pl.ipynb`

```
from itertools import chain
from sklearn import datasets
from sklearn.utils import shuffle
from sklearn.preprocessing import minmax_scale
from sklearn.model_selection import train_test_split
import sklearn.metrics as metrics

import pennylane as qml
from pennylane import numpy as np
from pennylane.templates.embeddings import AngleEmbedding
```

```
from pennylane.templates.layers import StronglyEntanglingLayers
from pennylane.init import strong_ent_layers_uniform
from pennylane.optimize import GradientDescentOptimizer
from pennylane.devices.default_qubit import DefaultQubit
```

The GradientDescentOptimizer is a basic optimizer that is a base class for other gradient-descent-based optimizers. A step of the gradient descent optimizer computes the new values via the rules defined in equation 8.3.

StronglyEntanglingLayers is a template for layers consisting of single-qubit rotations and entanglers, inspired by the circuit-centric classifier design addressed in the paper by Schuld et al., 2020 [59].

In the next step, we import data, perform preprocessing on it (shuffle, select classes, and normalize it), and divide the data into an 80%–20% split for the training and test datasets, respectively.

Listing 8-3b. Load Data, Preprocess and Split qml_iris_pl.ipynb

```
# Import data, pre-processing, data splitting
# load the dataset
iris = datasets.load_iris ()

# shuffle the data
X, y = shuffle (iris.data , iris.target , random_state = 0)

# select only 2 first classes from the data
X = X[y <=1]
y = y[y <=1]

# normalize data
X = minmax_scale (X, feature_range =(0 , np.pi ))

# split data into train + validation and test
X_train_val , X_test , y_train_val , y_test = train_test_split
(X, y, test_size =0.2)
```

The next step is to build the quantum classifier. To achieve this, we define the number of qubits to match the number of features, define a device with the DefaultQubit() function, set up a function for the variational classifier and the cost function.

Listing 8-3c. Building the Quantum Classifier `qml_iris_pl.ipynb`

```
# Building the Quantum Classifier
# number of qubits is equal to the number of features
n_qubits = X.shape [1]

# quantum device handle
dev = DefaultQubit(n_qubits)

# quantum circuit
@qml.qnode (dev )
def circuit (weights , x= None ):
    AngleEmbedding (x, wires = range (n_qubits ))
    StronglyEntanglingLayers (weights , wires = range ( n_qubits ))
    return qml.expval ( qml.PauliZ (0))

# variational quantum classifier
def variational_classifier (theta, x= None ):
    weights = theta [0]
    bias = theta [1]
    return circuit ( weights , x=x) + bias

def cost (theta , X, expectations ):
    e_predicted = \
        np.array ([ variational_classifier (theta , x=x) for x in X])
    loss = np.mean (( e_predicted - expectations )**2)
    return loss
```

`StronglyEntanglingLayers()` has the argument `weights`, which contain the weights for each layer.[7] The number of layers is derived from the first dimension of `weights`.

---

[7]Documentation for StronglyEntanglingLayers: https://pennylane.readthedocs.io/en/ stable/code/api/pennylane.templates.layers.StronglyEntanglingLayers.html

Listing 8-3d. Prepare to Training the Data `qml_iris_pl.ipynb`

```
# number of quantum layers
n_layers = 3

# split into train and validation
X_train , X_validation , y_train , y_validation = \
train_test_split ( X_train_val , y_train_val , test_size =0.20)

# convert classes to expectations : 0 to -1, 1 to +1
e_train = np. empty_like ( y_train )
e_train [ y_train == 0] = -1
e_train [ y_train == 1] = +1

# select learning batch size
batch_size = 5

# calculate numbe of batches
batches = len ( X_train ) // batch_size

# select number of epochs
n_epochs = 10
```

This code snippet runs for ten epochs. I recommend that you first run it for five epochs by setting n_epochs = 5, and then run it for ten epochs. The difference in epoch runs is reflected in the final accuracy calculation.

In the next step, we impose random weights on the data, train the variational classifier, engage the optimizer to start the learning process, and split the training data into batches as we did in Chapters 2 and 3 for classical systems.

Listing 8-3e. Train the Data! `qml_iris_pl.ipynb`

```
# draw random quantum node weights
theta_weights = strong_ent_layers_uniform ( n_layers , n_qubits , seed =42)
theta_bias = 0.0
theta_init = ( theta_weights , theta_bias ) # initial weights

# train the variational classifier
theta = theta_init
# start of main learning loop
```

```
# build the optimizer object
pennylane_opt = GradientDescentOptimizer ()

# split training data into batches
X_batches = np.array_split (np.arange (len ( X_train )), batches )
for it , batch_index in enumerate ( chain (*( n_epochs * [ X_batches ]))):
    # Update the weights by one optimizer step
    batch_cost = \
        lambda theta : cost (theta , X_train [ batch_index ], e_train [
                            batch_index ])
    theta = pennylane_opt.step ( batch_cost , theta )
    # use X_validation and y_validation to decide whether to stop
# end of learning loop
```

Finally, now that we have got all our tools defined and ready, we can explore the inference.

Listing 8-3f. Inference qml_iris_pl.ipynb

```
# Inference
# convert expectations to classes
expectations = np. array ([ variational_classifier (theta , x=x) for x in
X_test ])
prob_class_one = ( expectations + 1.0) / 2.0
y_pred = ( prob_class_one >= 0.5)

print ( metrics . accuracy_score ( y_test , y_pred ))
print ("Accuracy: {:.2f} %".format(metrics.accuracy_score ( y_test , y_pred
)*100))
print ( metrics.confusion_matrix ( y_test , y_pred ))
```

This code block outputs the following for an n_epochs value of 10.

```
1.0
Accuracy: 100.00 %
[[12  0]
 [ 0  8]]
```

However, if we reduce the epochs to a value of 5 by setting n_epochs = 5, we get the following output.

```
0.75
Accuracy: 75.00 %
[[8 4]
 [1 7]]
```

As expected in classical cases, we get a more accurate result for a higher number of epochs.

QNN with TensorFlow Quantum

A QNN is interpreted as any quantum circuit with trainable continuous parameters. QNN is a machine learning model that allows quantum computers to classify various datasets and, among them, image data. The image data used is classical data, but classical data cannot reach a superposition state. So, to carry out this protocol on quantum systems, the data must be made readable into a quantum device that provides superposition. The field of QNN is relatively new as research and work in this field continue to grow. We have previously seen some examples of PennyLane code to resolve specific QNN problems. This section looks at QNN for *image classification*.

QNN uses a supervised learning method to predict image data, as shown by Benedetti, Lloyd et al. (2019) [186]. Farhi and Neven in 2019 proposed [185] a QNN that could represent labeled data, classical or quantum, and be trained by supervised learning. The quantum circuit consists of a sequence of parameter-dependent unitary transformations that acts on an input quantum state. For binary classification, a single Pauli operator is measured on a designated readout qubit. The measured output is the QNN's predictor of the binary label of the input state. The authors of the paper [185] wrote, "We continue to use the word *neural* to describe our network since the term has been adopted by the machine learning community recognizing that the connection to neuroscience is now only historical." For MNIST like image classification, the authors randomly assigned two-qubit unitaries in the circuit. The Reed-Muller representation [187] used for subset parity is another way to think about the two-qubit unitaries.

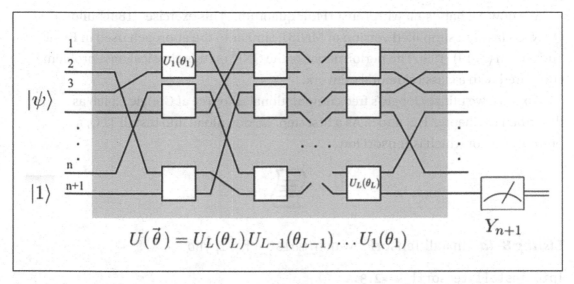

$$U(\vec{\theta}) = U_L(\theta_L)\,U_{L-1}(\theta_{L-1})\ldots U_1(\theta_1)$$

Figure 8-7. *Schematic of a QNN on a quantum processor (Source Farhi and Neven, 2019 [185])*

Figure 8-7 illustrates a schematic of the QNN proposed by Farhi and Neven [185] on a quantum processor. $|\psi, \ldots, 1\rangle$ are the input states which are prepared and then fed into the QNN. These input states go through a sequence of *few* qubit unitaries given by $U_i(\theta_i)$ dependent on parameter θ_i which, in turn, get adjusted during the learning process such that the measurement of Y_{n+1} on the measurement, readout tends to produce the desired label for $|\psi\rangle$.

TensorFlow Quantum (TFQ), proposed by Broughton et al. in 2020 [191], is a Python framework for hybrid quantum-classical machine learning primarily focused on modeling quantum data. TFQ is an application framework developed to allow quantum algorithms researchers and machine learning applications researchers to explore computing workflows that leverage Google's quantum computing offerings, all from within TensorFlow. Google AI released TFQ in March 2020 as an open source library for rapid prototyping of quantum machine learning models. TFQ depends on Cirq, an open source platform for implementing quantum circuits on near-term computers. Cirq includes fundamental structures, such as qubits, gates, circuits, and calculation operators, which are needed to define quantum computations. The concept behind the Cirq is to provide a simple programming model that abstracts the fundamental building blocks of quantum applications.

We now get hands-on with TensorFlow quantum.[8] This exercise[9] [188] builds a QNN to classify a simplified version of MNIST, similar to the approach used in Farhi and Neven (2019) [185]. The performance of the QNN on this classical data problem is compared with a classical neural network.

To start, we utilize Google's free computational resource at Google Colab as described in Chapter 1, 2, and 3. As a first step, we download and install TFQ, a prerequisite for which is TensorFlow.

Listing 8-4a. Install TensorFlow `qnn_mnist_tfq.ipynb`

```
!pip install tensorflow==2.3.1
```

Output (truncated).

```
Collecting tensorflow==2.3.1
  Downloading https://files.pythonhosted.org/packages/eb/18/374;
  |████████████████████████████| 320.4MB 46kB/s
Collecting numpy<1.19.0,>=1.16.0
  Downloading https://files.pythonhosted.org/packages/d6/c6/58e;
  |████████████████████████████| 20.1MB 1.5MB/s
Collecting tensorflow-estimator<2.4.0,>=2.3.0
  Downloading https://files.pythonhosted.org/packages/e9/ed/585;
  |████████████████████████████| 460kB 44.5MB/s
```

This step may exhibit the following message about restarting the runtime for TensorFlow.

```
WARNING: The following packages were previously imported in this runtime:
  [numpy]
You must restart the runtime in order to use newly installed versions.

RESTART RUNTIME
```

In this case, click the `Restart Runtime` button.
Next, install TFQ.

---

[8]For this exercise the code is adapted from TensorFlow Quantum documentation: `https://www.tensorflow.org/quantum/tutorials/mnist`

[9]Licensed under the Apache License, Version 2.0 (the "License")

Listing 8-4b. Install TensorFlow Quantum `qnn_mnist_tfq.ipynb`

```
!pip install tensorflow-quantum
```

Output (truncated).

```
Collecting tensorflow-quantum
  Downloading https://files.pythonhosted.org/packages/
  |████████████████████████████████| 5.9MB 6.4MB/s
Collecting sympy==1.5
  Downloading https://files.pythonhosted.org/packages/
  |████████████████████████████████| 5.6MB 21.1MB/s
Collecting cirq==0.9.1
  Downloading https://files.pythonhosted.org/packages/
  |████████████████████████████████| 1.6MB 45.1MB/s
Requirement already satisfied: mpmath>=0.19 in /usr/l(
Collecting freezegun~=0.3.15
  Downloading https://files.pythonhosted.org/packages/
```

```
      Successfully uninstalled sympy-1.7.1
Successfully installed cirq-0.9.1 freezegun-0.3.15 sympy-1.5 tensorflow-quantum-0.4.0
```

Next, we import TensorFlow and the module dependencies.

Listing 8-4c. Import Libraries `qnn_mnist_tfq.ipynb`

```
import tensorflow as tf
import tensorflow_quantum as tfq
import cirq
import sympy
import numpy as np
import seaborn as sns
import collections

# visualization tools
%matplotlib inline
import matplotlib.pyplot as plt
from cirq.contrib.svg import SVGCircuit
```

Now, we are ready to load the data. In this exercise, we build a binary classifier to distinguish between the digits 1 and 7, following Farhi and Neven [185]. This section covers the data handling that.

1. Loads the raw data from Keras.

2. Filters the dataset to only 1s and 7s.

3. Downscales the images so they can fit in a quantum computer.

4. Removes any contradictory examples.

5. Converts the binary images to Cirq circuits.

6. Converts the Cirq circuits to TFQ circuits.

In the next step, we load the raw MNIST dataset distributed with Keras.

Listing 8-4d. Load the Data qnn_mnist_tfq.ipynb

```
(x_train, y_train), (x_test, y_test) = tf.keras.datasets.mnist.load_data()
```

```
# Rescale the images from [0,255] to the [0.0,1.0] range.
x_train, x_test = x_train[..., np.newaxis]/255.0, x_test[...,
np.newaxis]/255.0
```

```
print("Number of original training examples:", len(x_train))
print("Number of original test examples:", len(x_test))
```

Output.

```
Downloading data from https://storage.googleapis.com/tensorflow/tf-keras-datasets/mnist.npz
11493376/11490434 [==============================] - 0s 0us/step
Number of original training examples: 60000
Number of original test examples: 10000
```

Next, we filter the dataset to keep only the 1s and 7s, remove the other classes. At the same time, convert the label, *y*, to boolean: True for 1 and False for 7.

Listing 8-4e. Filter the Data qnn_mnist_tfq.ipynb

```
def filter_17(x, y):
    keep = (y == 1) | (y == 7)
    x, y = x[keep], y[keep]
    y = y == 1
    return x,y
```

```
x_train, y_train = filter_17(x_train, y_train)
x_test, y_test = filter_17(x_test, y_test)

print("Number of filtered training examples:", len(x_train))
print("Number of filtered test examples:", len(x_test))
```

Output of the filtering process reads out the numbers for training and test split.

```
Number of filtered training examples: 13007
Number of filtered test examples: 2163
```

And we try out an example to see what the data looks like.

Listing 8-4f. Example `qnn_mnist_tfq.ipynb`

```
print(y_train[0])
plt.imshow(x_train[0, :, :, 0])
plt.colorbar()
```

This code block gives us the following output.

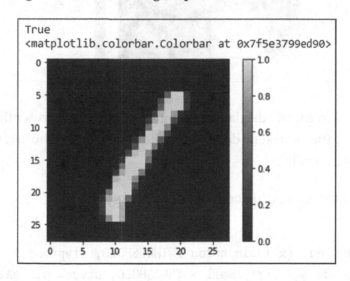

Next, we downscale the image to 4 × 4 as an image size of 28 × 28 is much too large for current quantum computers and display it after resizing it.

Listing 8-4g. Example qnn_mnist_tfq.ipynb

```
x_train_small = tf.image.resize(x_train, (4,4)).numpy()
x_test_small = tf.image.resize(x_test, (4,4)).numpy()

print(y_train[0])
plt.imshow(x_train_small[0,:,:,0], vmin=0, vmax=1)
plt.colorbar()
```

The following is the output.

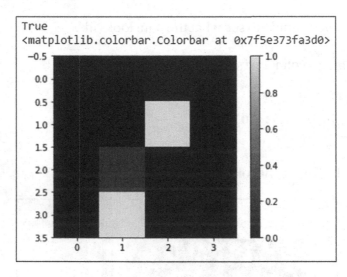

The next step is to encode the data as quantum circuits by representing each pixel with a qubit, where the qubit state depends on the pixel's value. The first step is to convert to a binary encoding.

Listing 8-4h. Binary Encoding qnn_mnist_tfq.ipynb

```
THRESHOLD = 0.5
x_train_bin = np.array(x_train_nocon > THRESHOLD, dtype=np.float32)
x_test_bin = np.array(x_test_small > THRESHOLD, dtype=np.float32)
```

Now, we rotate the qubits through an X gate at pixel indices with values that exceed a threshold to form a circuit.

Listing 8-4i. Rotate Qubits Through X Gate qnn_mnist_tfq.ipynb

```
def convert_to_circuit(image):
    """Encode truncated classical image into quantum datapoint."""
    values = np.ndarray.flatten(image)
    qubits = cirq.GridQubit.rect(4, 4)
    circuit = cirq.Circuit()
    for i, value in enumerate(values):
        if value:
            circuit.append(cirq.X(qubits[i]))
    return circuit
x_train_circ = [convert_to_circuit(x) for x in x_train_bin]
x_test_circ = [convert_to_circuit(x) for x in x_test_bin]
SVGCircuit(x_train_circ[0])
```

The following is the circuit created for the first example (circuit diagrams do not show qubits with zero gates).

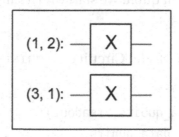

Now, we can compare this circuit to the indices where the image value exceeds the threshold.

Listing 8-4j. Comparison qnn_mnist_tfq.ipynb

```
bin_img = x_train_bin[0,:,:,0]
indices = np.array(np.where(bin_img)).T
indices
```

The following output.

```
array([[1, 2],
       [3, 1]])
```

The quantum data is loaded as a *tensor*, defined as a quantum circuit written in Cirq. TensorFlow executes this tensor on the quantum computer to generate a *quantum dataset*. So, we now convert these Cirq circuits to *tensors* for TensorFlow Quantum (tfq), as shown in Listing 8-4k.

Listing 8-4k. Comparison qnn_mnist_tfq.ipynb

```
x_train_tfcirc = tfq.convert_to_tensor(x_train_circ)
x_test_tfcirc = tfq.convert_to_tensor(x_test_circ)
```

Now we are ready to build the QNN. However, guidance for a quantum circuit structure that classifies images is sparse. Since the classification is based on the expectation of the readout qubit, Farhi et al. propose using two-qubit gates, with the readout qubit always acted upon. This is similar in some ways to running a small unitary RNN proposed by Arjovsky et al. [190] across the pixels.

To achieve this, we start with building the model circuit. The following code block shows a layered approach. Each layer uses *n* instances of the same gate, with each of the data qubits acting on the readout qubit. We start with a simple class that adds a layer of these gates to a circuit.

Listing 8-4l. Build the QNN Model Circuit qnn_mnist_tfq.ipynb

```
class CircuitLayerBuilder():
    def __init__(self, data_qubits, readout):
        self.data_qubits = data_qubits
        self.readout = readout

    def add_layer(self, circuit, gate, prefix):
        for i, qubit in enumerate(self.data_qubits):
            symbol = sympy.Symbol(prefix + '-' + str(i))
            circuit.append(gate(qubit, self.readout)**symbol)
demo_builder = CircuitLayerBuilder(data_qubits = cirq.GridQubit.rect(4,1),
                                   readout=cirq.GridQubit(-1,-1))
circuit = cirq.Circuit()
demo_builder.add_layer(circuit, gate = cirq.XX, prefix='xx')
SVGCircuit(circuit)
```

The output shows the circuit.

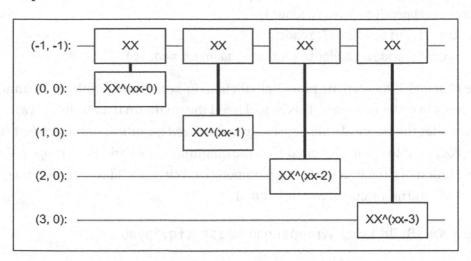

Next, we build a two-layered model, matching the data circuit's size, and include the preparation and readout operations.

Listing 8-4m. Build a Two-Layer Model `qnn_mnist_tfq.ipynb`

```python
def create_quantum_model():
    """Create a QNN model circuit and readout operation to go along with
        it."""
    data_qubits = cirq.GridQubit.rect(4, 4)  # a 4x4 grid
    readout = cirq.GridQubit(-1, -1)          # a single qubit at [-1,-1]
    circuit = cirq.Circuit()

    # Prepare the readout qubit.
    circuit.append(cirq.X(readout))
    circuit.append(cirq.H(readout))
    builder = CircuitLayerBuilder(
        data_qubits = data_qubits,
        readout=readout)
    # Then add layers (experiment by adding more).
    builder.add_layer(circuit, cirq.XX, "xx1")
    builder.add_layer(circuit, cirq.ZZ, "zz1")
```

```
    # Finally, prepare the readout qubit.
    circuit.append(cirq.H(readout))
    return circuit, cirq.Z(readout)
model_circuit, model_readout = create_quantum_model()
```

The next task is to wrap the model-circuit in a tfq-keras model with the quantum components in order to train it. This model is fed the quantum data from x_train_circ, which encodes the classical data. It uses a *parametrized quantum circuits* layer, tfq.layers.PQC, to train the model circuit, on the quantum data. Farhi et al. proposed taking the expectation of a readout qubit in a parameterized circuit to classify these images. The expectation returns a value between 1 and –1.

Listing 8-4n. Build Keras Wrapper qnn_mnist_tfq.ipynb

```
# Build the Keras model.
model = tf.keras.Sequential([
    # The input is the data-circuit, encoded as a tf.string
    tf.keras.layers.Input(shape=(), dtype=tf.string),
    # The PQC layer returns the expected value of the readout gate, range
        [-1,1].
    tfq.layers.PQC(model_circuit, model_readout),
])
```

Since the expected readout is in the range $[-1,1]$, optimizing the hinge loss is a natural fit.

Note   Another valid approach would be to shift the output range to [0,1] and treat it as the probability the model assigns to class 1. This could be used with a standard a tf.losses.BinaryCrossentropy loss.

To use the hinge loss here, we need two small adjustments. First, we convert the labels, y_train_nocon, from boolean to $[-1,1]$, as expected by the hinge loss.

```
y_train_hinge = 2.0*y_train_nocon-1.0
y_test_hinge = 2.0*y_test-1.0
```

We use a custom `hinge_accuracy` metric that correctly handles [-1, 1] as the y_true labels argument.

Listing 8-4o. Label Argument `qnn_mnist_tfq.ipynb`

```
def hinge_accuracy(y_true, y_pred):
    y_true = tf.squeeze(y_true) > 0.0
    y_pred = tf.squeeze(y_pred) > 0.0
    result = tf.cast(y_true == y_pred, tf.float32)
    return tf.reduce_mean(result)

model.compile(
    loss=tf.keras.losses.Hinge(),
    optimizer=tf.keras.optimizers.Adam(),
    metrics=[hinge_accuracy])

print(model.summary())
```

This prints out the model summary.

```
Model: "sequential_3"

Layer (type)                 Output Shape              Param #
=================================================================
pqc_1 (PQC)                  (None, 1)                 32
=================================================================
Total params: 32
Trainable params: 32
Non-trainable params: 0


None
```

We now train the QNN model, which takes about 30 to 45 minutes—depending on your system. If you don't want to wait that long, use a small subset of the data (set NUM_EXAMPLES=500, below). This does not affect the model's progress during training as it has only 32 parameters, and hence, data needed to constrain these are not substantial. Using fewer examples ends training earlier (~5 minutes) but runs long enough to show that it is making progress in the validation logs.

Listing 8-4p. Train the Quantum Model `qnn_mnist_tfq.ipynb`

```
EPOCHS = 3
BATCH_SIZE = 32
NUM_EXAMPLES = len(x_train_tfcirc)
x_train_tfcirc_sub = x_train_tfcirc[:NUM_EXAMPLES]
y_train_hinge_sub = y_train_hinge[:NUM_EXAMPLES]
qnn_history = model.fit(
        x_train_tfcirc_sub, y_train_hinge_sub,
        batch_size=32,
        epochs=EPOCHS,
        verbose=1,
        validation_data=(x_test_tfcirc, y_test_hinge))
qnn_results = model.evaluate(x_test_tfcirc, y_test)
```

This gives the following output for hinge accuracy, time to run and loss.

```
Epoch 1/3
233/233 [==============================] - 425s 2s/step - loss: 0.7841 - hinge_accuracy: 0.7421 - val_loss: 0.8572 - val_hinge_accuracy: 0.4990
Epoch 2/3
233/233 [==============================] - 426s 2s/step - loss: 0.5164 - hinge_accuracy: 0.7477 - val_loss: 0.6971 - val_hinge_accuracy: 0.4999
Epoch 3/3
233/233 [==============================] - 427s 2s/step - loss: 0.4978 - hinge_accuracy: 0.7819 - val_loss: 0.6460 - val_hinge_accuracy: 0.8733
68/68 [==============================] - 19s 278ms/step - loss: 0.6460 - hinge_accuracy: 0.8733
```

The training accuracy reports the average over the epoch. The validation accuracy is evaluated at the end of each epoch.

Now that we have had a look at the quantum side of things, we may wish to find out how a classical network would perform with the same dataset. Although a QNN works for this simplified MNIST problem, a basic classical neural network can outperform a QNN on this task. After a single epoch, a classical neural network can achieve >98% accuracy on the holdout set.

In the following example, a classical neural network is used for the 1–7 classification problem using the *entire* 28 × 28 image instead of subsampling the image. This easily converges to nearly 100% accuracy of the test set.

Listing 8-4q. Build the Classical Model `qnn_mnist_tfq.ipynb`

```
def create_classical_model():
    # A simple model based off LeNet from https://keras.io/examples/
      mnist_cnn/
    model = tf.keras.Sequential()
```

```
model.add(tf.keras.layers.Conv2D(32, [3, 3], activation='relu', input_
shape=(28,28,1)))
model.add(tf.keras.layers.Conv2D(64, [3, 3], activation='relu'))
model.add(tf.keras.layers.MaxPooling2D(pool_size=(2, 2)))
model.add(tf.keras.layers.Dropout(0.25))
model.add(tf.keras.layers.Flatten())
model.add(tf.keras.layers.Dense(128, activation='relu'))
model.add(tf.keras.layers.Dropout(0.5))
model.add(tf.keras.layers.Dense(1))
    return model
model = create_classical_model()
model.compile(loss=tf.keras.losses.BinaryCrossentropy(from_logits=True),
          optimizer=tf.keras.optimizers.Adam(),
          metrics=['accuracy'])
model.summary()
```

This produces the model summary as follows.

```
Model: "sequential_4"
_____
Layer (type)                 Output Shape              Param #
=================================================================
conv2d_2 (Conv2D)            (None, 26, 26, 32)        320
_____
conv2d_3 (Conv2D)            (None, 24, 24, 64)        18496
_____
max_pooling2d_1 (MaxPooling2 (None, 12, 12, 64)        0
_____
dropout_2 (Dropout)          (None, 12, 12, 64)        0
_____
flatten_2 (Flatten)          (None, 9216)              0
_____
dense_4 (Dense)              (None, 128)               1179776
_____
dropout_3 (Dropout)          (None, 128)               0
_____
dense_5 (Dense)              (None, 1)                 129
=================================================================
Total params: 1,198,721
Trainable params: 1,198,721
Non-trainable params: 0
_____
```

This model has nearly 1.2 million parameters. Next, we train the classical model as follows.

Listing 8-4r. Train the Classical Model `qnn_mnist_tfq.ipynb`

```
model.fit(x_train,
          y_train,
          batch_size=128,
          epochs=1,
          verbose=1,
          validation_data=(x_test, y_test))

cnn_results = model.evaluate(x_test, y_test)
```

This code block produces the output on accuracy, losses, and so on for the classical model with 1.2 million parameters. For a fairer comparison, we try a 37-parameter model on the subsampled images.

Listing 8-4s. 37-Parameter Model `qnn_mnist_tfq.ipynb`

```
def create_fair_classical_model():
    # A simple model based off LeNet from https://keras.io/examples/mnist_
      cnn/
    model = tf.keras.Sequential()
    model.add(tf.keras.layers.Flatten(input_shape=(4,4,1)))
    model.add(tf.keras.layers.Dense(2, activation='relu'))
    model.add(tf.keras.layers.Dense(1))
    return model
model = create_fair_classical_model()
model.compile(loss=tf.keras.losses.BinaryCrossentropy(from_logits=True),
              optimizer=tf.keras.optimizers.Adam(),
              metrics=['accuracy'])

model.summary()
```

This produces the 37-parameter model we want.

```
Model: "sequential_5"

_____
Layer (type)                 Output Shape              Param #
=================================================================
flatten_3 (Flatten)          (None, 16)                0
_____
dense_6 (Dense)              (None, 2)                 34
_____
dense_7 (Dense)              (None, 1)                 3
=================================================================
Total params: 37
Trainable params: 37
Non-trainable params: 0
_____
```

Now, we train this model with the new range of 37 parameters.

Listing 8-4t. Train the 37-Parameter Model `qnn_mnist_tfq.ipynb`

```
model.fit(x_train_bin,
          y_train_nocon,
          batch_size=128,
          epochs=20,
          verbose=2,
          validation_data=(x_test_bin, y_test))

fair_nn_results = model.evaluate(x_test_bin, y_test)
```

This gives the following (truncated) output.

```
59/59 - 0s - loss: 0.3374 - accuracy: 0.7742 - val_loss: 0.4067 - val_accuracy: 0.5146
Epoch 20/20
59/59 - 0s - loss: 0.3348 - accuracy: 0.7742 - val_loss: 0.4014 - val_accuracy: 0.5146
68/68 [==============================] - 0s 1ms/step - loss: 0.4014 - accuracy: 0.5146
```

In the last step, we compare the quantum and classical NNs.

Listing 8-4u. Comparison of Quantum and Classical `qnn_mnist_tfq.ipynb`

```
qnn_accuracy = qnn_results[1]
cnn_accuracy = cnn_results[1]
fair_nn_accuracy = fair_nn_results[1]
sns.barplot(["Quantum", "Classical, full", "Classical, fair"],
            [qnn_accuracy, cnn_accuracy, fair_nn_accuracy])
```

This code block gives the following output as comparison.

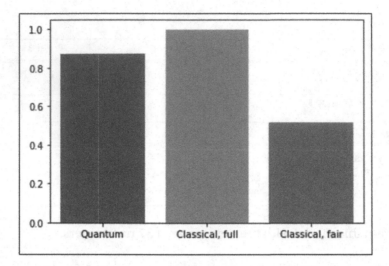

Higher-resolution input and a more powerful model make this problem an easy one for CNN. While a classical model of similar power (~32 parameters) trains to a similar accuracy in a fraction of the time, the classical neural network outperforms the quantum neural network.

In a *classical dataset*, it is difficult for a quantum system to beat a classical neural network, given the state of near-term quantum systems today. However, exercises on a quantum system with classical input data do promise leveraging the multi-dimensionality of Hilbert spaces and the power of superposition for certain specific problems. The insights gained with classical data also prepare us for algorithm implementation on the next-generation quantum computing systems, which, it is hoped, brings unprecedented computational power.

Quantum Convolutional Neural Networks

As seen in Chapter 3, classical convolutional neural network (CNN) is a popular model in image processing and computer vision-related fields. The structure of CNNs consists of applying alternating *convolutional layers* (plus an *activation function*) and *pooling* layers to an input array, typically followed by some fully connected layers before the output. Convolutional layers work by sweeping across the input array and applying different filters (often 2×2 or 3×3 matrices) block by block. They detect specific features of the image wherever they might appear. Pooling layers are then used to down-sample the results of these convolutions to extract the most relevant features and reduce the size of the data, making it easier to process in subsequent layers. Common pooling methods involve replacing blocks of the data with their maximum or average values.

CNN has the attribute of facilitating efficient use of the correlation information of data. However, if the given dimension of data or model becomes too large (which is likely in this age of big data), the computational expense associated with CNN can quickly become a challenge. A quantum convolutional neural network (QCNN) provides a new solution with CNN using a quantum computing environment to improve the performance of an existing learning model.

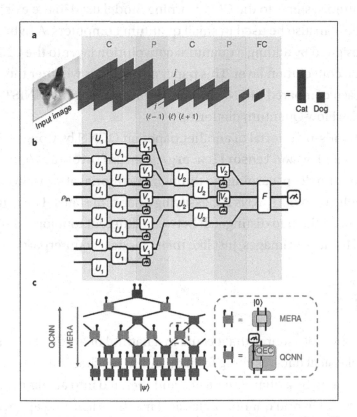

Figure 8-8. *Schematic of a QCNN (Source Cong et al. 2018 [193])*

QCNNs were first introduced by Cong et al. (2018) [193]. The structure of QCNNs is motivated by that of CNNs, as shown in Figure 8-8. In a QCNN, convolutions are operations (parameterized unitary rotations) performed on neighboring pairs of qubits, similar to a regular variational circuit. These convolutions are followed by pooling layers, which are affected by measuring a subset of the qubits and using the measurement

457

results to control subsequent operations. The analog of a fully connected layer is a multi-qubit operation on the remaining qubits before the final measurement. Parameters of all these operations are learned during training.

In studies done by Oh, Choi, and Kim in 2020 [192] proposes a model to effectively solve the classification problem in quantum physics and chemistry by applying the structure of CNN to the quantum computing environment. The research also proposes the model that can be calculated with $O(\log(n))$ depth using *multi-scale entanglement renormalization ansatz* (MERA).

The study introduces a method to improve the model's performance by adding a layer using quantum systems to the CNN learning model used in the existing computer vision. This model can also be used in small quantum computers. A hybrid learning model can be designed by adding a quantum convolution layer to the CNN model or replacing it with a convolution layer. This paper also verifies whether the QCNN model can efficiently learn compared to CNN through training using the MNIST dataset through the TensorFlow Quantum platform.

A thorough hands-on tutorial of another paper on QCNN by Cong et al. [193] is made available by Google AI at `www.tensorflow.org/quantum/tutorials/qcnn`. If you are interested in quantum deep learning, go through the tutorial at your own pace.

A natural application of QCNNs is classifying quantum states. The original work by Cong et al. [193] used them to distinguish between different topological phases. QCNNs can also be used to classify images, just like their classical counterparts.

Summary

Quantum data is any data source that occurs in a natural or artificial quantum system. This can be the classical data resulting from quantum mechanical experiments or data directly generated by a quantum device and then fed into an algorithm as input. There is evidence that hybrid quantum-classical machine learning applications on *quantum data* could provide a quantum advantage over classical-only machine learning due to quantum mechanical properties. Quantum data exhibits superposition and entanglement, leading to joint probability distributions that could require an exponential amount of classical computational resources to represent or store.

Perhaps the most immediate application of quantum machine learning is quantum data—the actual states generated by quantum systems and processes. Many quantum machine learning algorithms find patterns in classical data by mapping the data to quantum mechanical states and then manipulating those states using basic quantum linear algebra subroutines. These quantum machine learning algorithms can be applied directly to the quantum states of light and matter to reveal their underlying features and patterns. The resulting quantum modes of analysis are frequently much more efficient and illuminating than the classical analysis of data taken from quantum systems.

This chapter explored traveling salesperson problem and solutions on D-Wave, QNN with Xanadu's PennyLane, and Google's TensorFlow Quantum and explored quantum convolutional neural networks. We also explored recent research and advances in the fields of QCNN.

Chapter 9 explores advanced fields, such as quantum chemistry, quantum walks, and tomography, where quantum machine learning is becoming active and expected to grow.

CHAPTER 9

QML: Way Forward

The best way to predict your future is to create it.

—Abraham Lincoln

Quantum machine learning (QML) is a cross-disciplinary subject made up of two of the most exciting research areas: quantum computing and classical machine learning.

So far, we have explored several algorithms, concepts, and insights. We can solve a problem with a classical computer, whether a laptop or a GPU cluster. There is little value in solving it by a quantum computer that is at present very, very expensive (tens of millions of dollars). The interesting question in quantum machine learning is whether there are problems in machine learning and AI that fit quantum computers naturally but are challenging on classical hardware. This requires a good understanding of both machine learning and contemporary quantum computers.

Note   This chapter focuses on discussions about ongoing research areas into processes and applications related to quantum machine learning, such as quantum walks, Hamiltonian simulation, QBoost, quantum finance, quantum chemistry, and tensor networks, including two hands-on exercises.

Quantum computation and quantum information are disciplines that have appeared in various areas of computer science, such as information theory, cryptography, emotion representation, and image processing, because tasks that appear inefficient on classical computers can be achieved by exploiting the power of quantum computation [3] [233], [234], [235], [236]. Novel methods for speeding up certain tasks and interdisciplinary research between computer science and several other scientific fields have attracted deep scientific interest to this emerging field.

461

The growing confidence in the promises harbored by quantum computing and machine learning fields has given rise to many start-ups boosted by the quantum promise. Besides the well-known outfits, such as Google, IBM, Rigetti, Xanadu, D-Wave, to name a few. The following are examples of some young, promising members of the recent quantum tech start-up ecosystem.

- **Quantropi** (`www.quantropi.com`): Founded in Ottawa, Canada, in 2018, it is a rising player in quantum communication. It leverages the mathematical laws of quantum physics to encrypt data in transit without requiring the processing power of a quantum computer. Their solution offers true, secure quantum key distribution (QKD) over the current internet, allowing today's users to send quantum-encrypted data with their own computer. They boast three unique patented technologies forming a total end-to-end solution.

  - Quantum Entropy Expansion Protocol (QEEP)

  - Galois Public Key (GPK)

  - Quantum Public Key Envelope (QPKE) (patent pending)

  Quantropi's technology offers users the option to achieve unconditional security, along with being cost-effective, scalable, and efficient. Their published work includes research (and associated patents) addressing secrecy [286] and Quantum Public Key Distribution [287]. Recently, they were elected as one of the finalists for the Berlin Falling Walls Breakthrough of the Year 2020 in the category of Science Start-Ups and received the 2020 BOBs Best New (Ottawa) Business Award.

- **Quacoon** (`https://quacoon.com`): A quantum AI–based platform that provides collaborative supply chain intelligence for building and managing resilient and sustainable supply chains, as well as for solving product shortages, congestion, and other disruptions due to unexpected events. Quacoon was founded in 2020 by a team led by Tina Anne Sebastian.

  TQuT Inc. (`http://tqut.co/`): Started by Ritaban Chowdhury and registered in the United States, TQuT is an exciting quantum start-up. A new era of quantum technology unlocks new

possibilities in terms of material discoveries. There is a need for
a modern, cloud-based, API-driven, scalable material discovery
platform to efficiently explore those horizons. It is developing a
web-based material discovery platform, which leverages high-
performance computing and quantum computing hardware
in the cloud to make material discovery pipeline development
scalable, reusable, and easier.

Among all the applications of quantum computing that are attracting attention
at scale today are the ones that are related to the simulation of atoms, molecules, and
other biochemical systems which are uniquely suited to quantum computation. This
is because the ground state energy of large systems, the dynamical behavior of an
ensemble of molecules, or complex molecular behavior such as protein folding, are often
computationally hard or downright impossible to determine via classical computation or
experimentation [151][182].

Quantum Computing for Chemistry

Anticipation that a useful quantum computer will be realized soon has motivated
intense research into developing quantum algorithms that can potentially make
progress on classically intractable computational problems. While many research areas
expect to see transformative change with the development of such quantum devices,
computational chemistry is poised to be among the first domains to significantly benefit
from such new technologies. Due to the exponential growth in the size of the Hilbert
space with increasing orbitals, a quantum computer with tens of qubits could potentially
surpass classical algorithms [182], [183], [184]. Achieving such a capability depends not
only on the quality of the qubits but also on the algorithms' efficiency.

For the field of quantum chemistry, training nonlinear regression models—such as
Gaussian processes or neural networks—on accurate reference calculations allows direct
prediction of chemical properties [174], [175], [176]. ML predictions of atomic forces can
drive molecular dynamics simulations, while uncertainty estimates and active learning
guide the acquisition of more reference calculations to improve previously trained
models [179] [180]. Having immediate access to such accurate predictions of chemical
properties allows for extensive, large-scale studies that would not have been feasible
with conventional quantum chemistry simulations.

Rational design of molecules and materials with desired properties requires both the ability to calculate accurate microscopic properties, such as energies, forces, and electrostatic multipoles of specific configurations and efficient sampling of potential energy surfaces to obtain corresponding macroscopic properties. The tools that provide this are accurate first-principles calculations rooted in quantum mechanics and statistical mechanics, respectively. All of these come with a high computational cost that prohibits calculations for large systems or sampling intensive applications, such as long-timescale molecular dynamics simulations, thus presenting a severe bottleneck for searching the vast chemical compound space. To overcome this challenge, there have been increased efforts to accelerate quantum calculations with machine learning.

In a many-electron system, it is impossible to obtain the exact solution of the Schrödinger equation by using the present mathematical approach. The Hartree-Fock method, for which there have been widespread QML algorithms, was developed to solve approximately the time-independent Schrödinger equation. The total energy of a many-electron system can be represented by using one-electron and two-electron operators. The Schrödinger equation can be mathematically transformed to the one-electron Hartree-Fock equation by minimizing the total energy. The eigenvalue and wave function denotes orbital energy and molecular, atomic orbital.

As our understanding of quantum computers continues to mature, so too will the development of new methods and techniques which can benefit chemistry and molecular simulations. As it stands today, we are confident that quantum computers can aid those quantum chemistry computations that require an explicit representation of the wave function, either because of a high accuracy requirement of simulated properties or a high degree of entanglement in the system. In these cases, the exponential growth of the dimension of the wave function makes manipulation and storage very inefficient on a classical computer. Indeed, for even moderately large systems, it is intractable to explicitly maintain the full wave function.

As an example, a penicillin molecule, which comprises 42 atoms, would require 10^{86} classical bits. The simulation of the exponentially large parameter space of electron configurations would require more state than atoms in the Universe! As opposed to the number of bits required by the classical computer, a quantum computer would need about 286 qubits. A caffeine molecule would need about 10^{48} classical bits as opposed to about 160 qubits on a quantum computing platform.

As both the quantum hardware and software communities continue to make rapid progress, an immediate role of quantum computing for quantum chemistry has emerged. The advancement of quantum algorithms for quantum chemistry requires quantum information theory and classical quantum chemistry techniques.

OpenFermion

Quantum simulation of chemistry and materials is predicted to be an important application for both near-term and fault-tolerant quantum devices. However, developing and studying algorithms for these problems can be difficult due to the prohibitive amount of domain knowledge required in both the area of chemistry and quantum algorithms. To help bridge this gap and open the field to more researchers, Google has developed the OpenFermion software package (`www.openfermion.org`). OpenFermion is an open source software library written largely in Python under an Apache 2.0 license to enable the simulation of fermionic and bosonic models and quantum chemistry problems on quantum hardware. An interface to common electronic structure packages simplifies the translation between a molecular specification and a quantum circuit for solving or studying the electronic structure problem on a quantum computer, minimizing the amount of domain expertise required to enter the field.

OpenFermion can be installed in *user* mode as follows.

```
$ python -m pip install --user openfermion
```

Note   The Google AI research team has comprehensive online tutorials on OpenFermion at `https://quantumai.google/openfermion/tutorials/intro_to_openfermion`. You are urged to go through them for deeper insights.

Quantum simulation of chemical systems is one of the most promising near-term applications of quantum computers. One of the most promising quantum computing applications is the ability to efficiently model quantum systems in nature that are considered intractable for classical computers. Since the advent of electronic computers in the last century, computation has played a fundamental role in developing chemistry. Through the years, numerical methods for computing the static and dynamic properties of chemicals have revolutionized chemistry as a discipline. With the emergence of quantum computing, there is proven potential for similarly disruptive progress.

Quantum Walks

Quantum walks [223] are the quantum-mechanical analog of the well-known classical random walk. It has been established for a while in research that quantum walks have significant roles in various areas of quantum information processing [198], [199], [200]. They are central to quantum algorithms created to govern database search [201], graph isomorphism [202], [203], [204], network analysis and navigation [205], [206], quantum simulation [207], [208], [209], and modeling biological processes [210] [211]. Physical properties of quantum walks have been demonstrated in a variety of systems, such as nuclear magnetic resonance [212] [213], bulk [214] and fiber [215] [216] optics, trapped ions [216], [217], [218], trapped neutral atoms [219] and photonics [221], [222]. Quantum walks is one of the most important Hamiltonian simulation methods.

Classically, mathematical concepts of *random walks* have been around for a while and can be described as follows. This is a random process involving a "walker" (perhaps someone inebriated to the point where the person has no prior clue to where he/she will go with the next) who is placed in some n-dimensional medium, such as a grid or a graph. We then repeatedly query some random variable and based on the outcome of our measurement, the walker's position vector (i.e., position on the graph or grid, is updated).[1]

Let's look at a very basic example of a random walk. We start at some specific vertex v of a graph and repeat the following sequence several times: verify if v is marked, and if not, then choose one of its neighbors at random and set v to be that neighbor. This indicates that a random walk is a dynamic path with a randomly evolving time system. This process can also be explained in a one-dimensional graphical case, where we may consider a marker placed on the origin of a number line with markings at each of the integers.

A classical random walk is illustrated in Figure 9-1 where the walker at v at time $t = 0$ has the probability of ending up either at $(v + 1)$ or $(v - 1)$ at time $t = 1$.

---

[1]Analyzing graphs by looking at the eigenvalues of their adjacency matrix is called *algebraic graph theory* or *spectral graph theory*.

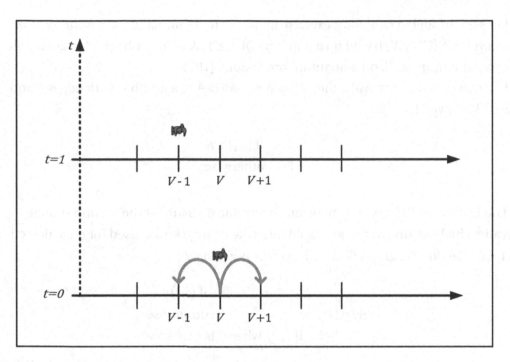

Figure 9-1. *Classical random walk example*

Looking at the classical random walk, it may appear at first glance that it is a superfluous algorithm. After all, it only needs a space complexity of $O(\log N)$, because we need to keep track of the current vertex v, and perhaps a counter to keep track of how many steps the walker has already taken. However, it is not at all superfluous! On the contrary, the algorithm can, for example, decide whether there is a path from a specific vertex v to a specific vertex u using $O(\log N)$ space, which becomes clearer and very useful when dealing in practical quantum physics at very low dimensions where statistical learnings and probabilities rule. The walk would start at v, and only u would be marked; it can be shown that if a path exists from v to u in G, then the walker reaches u in *polynomial*(N) steps. Random walks are similar to Grover search and the Ising model in the sense that wherever they appear, you can attempt replacing them with their quantum version and hope for an improvement.

Quantum walks are analogous to classical random walks, except the function determining the steps that the random walker takes and the position vector of the walker is encoded using qubits and unitary transformations. In other words, the ideas behind the construction of a regular random walk are translated to quantum analogs.

Let's go through a hands-on exercise in pyQuil to demonstrate a *continuous-time quantum walk* (CTQW) based on a paper by Qiang, Lok et al., published in 2016 called "Efficient quantum walk on a quantum processor." [197].

In a classical random walk, the adjacency matrix A of a graph G with edges E and vertices V is given by

$$A_{ij} = \begin{cases} 1 & \text{if } (i,j) \in E \\ 0 & \text{otherwise} \end{cases} \qquad 9.1$$

The Laplacian L is a two-dimensional isotropic measure of the second spatial derivative, highlighting regions of rapid intensity change and is used for edge detection. The Laplacian for the graph G of equation 9.1 is given by

$$A_{ij} = \begin{cases} 1 & \text{if } (i,j) \in E \\ 0 & \text{otherwise} \\ -i & \text{if } i = j, \text{where } i \text{ is in } degree \end{cases} \qquad 9.2$$

The Laplacian is sometimes defined differently than in equation 9.2. This definition is used here because it allows L to be a discrete approximation of the Laplacian operator ∇^2 in the continuum.

The behavior of a classical continuous random walk is determined by the Laplacian operator and described by the length vector $|V|$ of probabilities $p(t)$, where $p(t)$ is the probability of being at the vertex i at time t and is given by a differential equation [200], [201].

$$\frac{dp_i(t)}{dt} = \sum_{(i,j) \in E} L_{i,j} p_j(t) \qquad 9.3$$

Equation 9.3 has the following solution: $p(t) = e^{Lt} p(0)$. This can be viewed as a discrete analog of the diffusion equation, which is instrumental in quantum walk algorithms being important in physical systems such as trapped ions and photonics.

To bring things to a full circle: The Schrödinger equation, given by

$i\hbar \dfrac{d|\psi(t)\rangle}{dt} = \mathcal{H}|\psi(t)\rangle$ can be used to obtain the Laplacian.[2] We have encountered the Laplacian previously, especially when solving the NP-hard traveling salesperson

---

[2]For example, https://en.wikipedia.org/wiki/Hamiltonian_(quantum_mechanics)

problem in Chapter 8. The exponential operator on the left side of the Schrödinger equation is usually defined by the corresponding power series in $\mathcal{H}$. By the properties of functional calculus, the following operator describes our well-known unitary operator widely used: $U = e^{-i\mathcal{H}t}$. From a mathematical and physics point of view, this implies that the Laplacian preserves the normalization of the state of the system.

Therefore, the equation 9.3 can be expressed in Laplacian terms by the following.

$$i\frac{d\psi_i(t)}{dt} = \sum_{(i,j)\in E} L_{i,j}\psi_j(t)$$

9.4

The differential equation 9.4 determines the behavior of the quantum analog of the continuous random walk. This solution is given by

$$|\psi(t)\rangle = e^{-iLt}|\psi(0)\rangle$$

We do not necessarily require a Laplacian to define a quantum walk. It can be defined with the help of any operator that respects the structure of the graph G (i.e., allows transitions only to neighboring vertices in the graph or remains stationary [222, 223]).

To grasp the behavior of a quantum walk, we first look at an example of a continuous-time quantum walk on a line. In this case, the eigenstates of the Laplacian operator for the graph representing the infinite line are the momentum states with eigenvalues $2(\cos p - 1)$ for momentum $p \in [-\pi, \pi]$. The momentum states are given in terms of the position states $|x\rangle$ after applying, the Laplacian operator is given by

$$|p\rangle = \sum_x e^{ixp}|x\rangle$$

And the Laplacian is given by

$$L|p\rangle = \sum_x e^{ixp}|(x+1)\rangle + e^{ixp}|(x-1)\rangle - 2e^{ixp}|x\rangle$$

$$= \sum_x \left(e^{ip} + e^{-ip} - 2\right)e^{ixp}|x\rangle$$

$$\Rightarrow L|p\rangle = 2(\cos p - 1)|p\rangle$$

9.5

From equation 9.5, we find the probability distribution at time t.

$$p(x,t) = \left| \langle x | e^{-iLt} | \psi(0) \rangle \right|^2$$

9.6

where the initial position $|\psi(0)\rangle = |0\rangle$ is given by

$$\left| \langle x | e^{-iLt} | 0 \rangle \right|^2 = \left| J_x(2t) \right|^2$$

9.7

The classical quantum behavior in this context resembles the illustration in Figure 9-2.

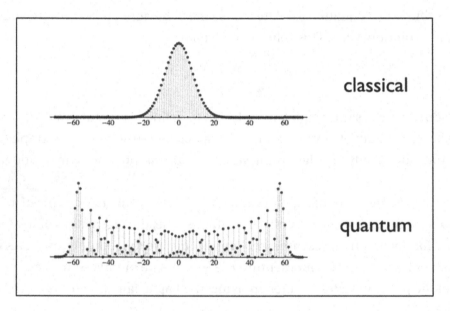

Figure 9-2. *Quantum to classical random walk*

While the probability distribution for the classical continuous-time random walk approaches its peak at $x = 0$ with a Gaussian width of $2\sqrt{t}$, the quantum walk has its largest peaks at the extrema, with oscillations between that decreasing in amplitude as one approaches the starting position at $x = 0$.

Note   Random walks can be of several types. The binomial distribution of the classical random walk is illustrated in the Cirq tutorial at `https://colab.research.google.com/github/quantumlib/Cirq/blob/master/docs/tutorials/quantum_walks.ipynb#scrollTo=yaKGvbmieSIL`. You are urged to visit the site and go through the Cirq version of a quantum walk for greater exposure to the versatility of the problem.

The behavior is different for the quantum walk due to the destructive interference between states of different phases that *does not occur* in the classical case. On the contrary, the probability distribution of the classical walk has no oscillations but instead a single peak centered at $x = 0$, which widens and flattens as time t increases. Hence, by replacing probabilities with quantum amplitudes, we see that interference in the quantum domain can produce radically different results.

A numerical or matrix representation of a similar graph is shown in Figure 9-3.

Graph G:

$$A = \begin{pmatrix} 0 & 1 & 1 & 0 & 0 \\ 1 & 0 & 0 & 1 & 1 \\ 1 & 0 & 0 & 1 & 0 \\ 0 & 1 & 1 & 0 & 1 \\ 0 & 1 & 0 & 1 & 0 \end{pmatrix}$$

Adjacency Matrix

$$L = \begin{pmatrix} 2 & -1 & -1 & 0 & 0 \\ -1 & 3 & 0 & -1 & -1 \\ -1 & 0 & 2 & -1 & 0 \\ 0 & -1 & -1 & 3 & -1 \\ 0 & -1 & 0 & -1 & 2 \end{pmatrix}$$

Laplacian

Figure 9-3. Classical random to quantum walk

Coding Quantum Walk

Equipped with the understanding of the basic behavior of the CTQW, we start our hands-on exercise with Rigetti's QVM and pyQuil based on the paper by Qiang, Lok et al. [197].

Open the Jupyter notebook file titled `quantum_walk_pyquil.ipynb` from the book's website and start three different terminals: start the QVM with the `qvm -S` command, another to start Quilc with the `quilc -S` command, and a third to start the Jupyter notebook.

Listing 9-1a. Libraries and QVM Connection quantum_walk_pyquil.ipynb

```python
import numpy as np
import networkx as nx
import matplotlib.pyplot as plt
from scipy.linalg import expm
%matplotlib inline
from pyquil import Program
from pyquil.gates import H, X, CPHASE00
from pyquil.api import WavefunctionSimulator
from pyquil.api import ForestConnection
from pyquil.api import QVMConnection

qvm = QVMConnection()
wfn_sim = WavefunctionSimulator()
```

In the next step, we start by looking at coding the graph *G* for a continuous-time classical walk with four nodes leveraging the NetworkX graphical library.

Listing 9-1b. Classical Walk with Four Nodes quantum_walk_pyquil.ipynb

```python
G = nx.complete_graph(4)
nx.draw_networkx(G)
```

This code block produces a four-node K_4 graph *G* in Figure 9-4 for the classical walk.

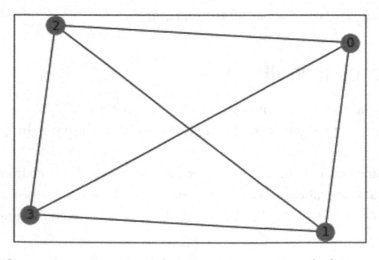

Figure 9-4. *Classical random walk graph with four nodes (K_4)*

The spectrum of a complete graph requires one eigenvalue equal to $(N - 1)$, where N is the number of nodes, and the remaining eigenvalues equal to -1. In the next step, we define the adjacency matrix A and corresponding eigenvalues.

Listing 9-1c. Adjacency Matrices `quantum_walk_pyquil.ipynb`

```
A = nx.adjacency_matrix(G).toarray()
eigvals, _ = np.linalg.eigh(A)
print(eigvals)
```

This gives the following matrix as output.

$$[-1. \quad -1. \quad -1. \quad 3.]$$

For the CTQW the usual Hamiltonian is the adjacency matrix A. This is slightly modified here by adding the identity matrix; that is, we take $\mathcal{H} = A + I$ (please note that we have changed the font of the Hamiltonian here to distinguish between the "$\mathcal{H}$" of the Hamiltonian and "H" of the Hadamard gate). This modification reduces the number of gates we need to apply since the eigenvectors with 0 eigenvalues do not acquire a phase. Hence, next, we use the matrix A to calculate the Hamiltonian, for which we use the NumPy kron function to generate the Kronecker product of two arrays.

Listing 9-1d. Hamiltonian and Adjacency Matrices `quantum_walk_pyquil. ipynb`

```
hamil = A + np.eye(4)
had = np.sqrt(1/2) * np.array([[1, 1], [1, -1]])
Q = np.kron(had, had)
Q.conj().T.dot(hamil).dot(Q)
```

This code block produces the following matrix as the product.

```
array([[ 4.00000000e+00, -4.93038066e-32, -4.93038066e-32,
         4.93038066e-32],
       [ 0.00000000e+00,  0.00000000e+00,  0.00000000e+00,
         0.00000000e+00],
       [ 0.00000000e+00,  0.00000000e+00,  0.00000000e+00,
         0.00000000e+00],
       [ 0.00000000e+00,  0.00000000e+00,  0.00000000e+00,
         0.00000000e+00]])
```

This output shows that K_n graphs are Hadamard diagonalizable. This allows us to write $\mathcal{H} = Q \wedge Q^\dagger$, where $Q = H \otimes H$, and H denotes the Hadamard gate. We also see that the time evolution operator $e^{-i\mathcal{H}t}$ is s also diagonalized by the same transformation.

$$Qe^{-i\mathcal{H}t}Q = \begin{pmatrix} e^{-i4t} & 0 & 0 & 0 \\ 0 & 1 & 0 & 0 \\ 0 & 0 & 1 & 0 \\ 0 & 0 & 0 & 1 \end{pmatrix}$$

Now, it is time to compute the continuous-time quantum walk. To achieve this, we first define a function that changes the matrix to a diagonal basis, then calculates its time evolution and eventually changes it back to the computational basis. From the reference paper [197], $\gamma = 1$ in our case. Hence, the circuit to simulate these is shown in Listing 9-1e.

Listing 9-1e. Functions for Matrices and Time Evolution `quantum_walk_pyquil.ipynb`

```
def k_4_ctqw(t):
    #   Change to diagonal basis
    p = Program(H(0), H(1), X(0), X(1))

    #   Time evolve
    p += CPHASE00(-4*t, 0, 1)

    #   Change back to computational basis
    p += Program(X(0), X(1), H(0), H(1))
        return p
```

We now compare the quantum walk with a classical random walk. The classical time evolution operator is $e^{-(T - I)t}$, where T is the transition matrix of the graph. As an initial condition, we choose $|\psi(0)\rangle = |0\rangle$ same as in equations 8.6 and 8.7. This means the walker starts at the first node. Afterward, due to properties of symmetry, the probability of occupation of all nodes besides $|0\rangle$ is the same.

Listing 9-1f. Time Evolution `quantum_walk_pyquil.ipynb`

```
T = A / np.sum(A, axis=0)
time = np.linspace(0, 4, 40)
```

```
quantum_probs = np.zeros((len(time), 4))
classical_probs = np.zeros((len(time), 4))
for i, t in enumerate(time):
    p = k_4_ctqw(t)
    wvf = wfn_sim.wavefunction(p)
    vec = wvf.amplitudes
    quantum_probs[i] = np.abs(vec)**2
    classical_ev = expm((T-np.eye(4))*t)
    classical_probs[i] = classical_ev[:, 0]
f, (ax1, ax2) = plt.subplots(2, sharex=True, sharey=True)
ax1.set_title("Quantum evolution")
ax1.set_ylabel('p')
ax1.plot(time, quantum_probs[:, 0], label='Initial node')
ax1.plot(time, quantum_probs[:, 1], label='Remaining nodes')
ax1.legend(loc='center left', bbox_to_anchor=(1, 0.5))
ax2.set_title("Classical evolution")
ax2.set_xlabel('t')
ax2.set_ylabel('p')
ax2.plot(time, classical_probs[:, 0], label='Initial node')
ax2.plot(time, classical_probs[:, 1], label='Remaining nodes')
```

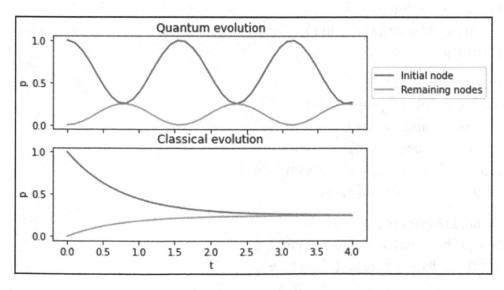

Figure 9-5. *Quantum vs. classical random walk*

Listing 9-1f produces the plots in Figure 9-5 to compare the time-dependent probabilistic evolution of the CTQW and the classical walk.

As expected, the quantum walk exhibits coherent oscillations while the classical walk converges to the stationary distribution of $p_i = \dfrac{d_n}{\sum_n d_n} \cong 0.25$.

This comparison between CTQW and classical walks for a K_4 graph can be easily generalized to any K_{2^n} graphs as follows.

Listing 9-1g. Time Evolution quantum_walk_pyquil.ipynb

```python
def k_2n_ctqw(n, t):
    p = Program()
    #    Change to diagonal basis
    for i in range(n):
        p += Program(H(i), X(i))
    #    Create and apply CPHASE00
    big_cphase00 = np.diag(np.ones(2**n)) + 0j
    big_cphase00[0, 0] = np.exp(-1j*4*t)
    p.defgate("BIG-CPHASE00", big_cphase00)
    args = tuple(["BIG-CPHASE00"] + list(range(n)))
    p.inst(args)
    #    Change back to computational basis
    for i in range(n):
        p += Program(X(i), H(i))
    return p

def k_2n_crw(n, t):
    G = nx.complete_graph(2**n)
    A = nx.adjacency_matrix(G)
    T = A / A.sum(axis=0)
    classical_ev = expm((T-np.eye(2**n))*t)
    return classical_ev[:, 0]

time = np.linspace(0, 4, 40)
quantum_probs = np.zeros((len(time), 8))
classical_probs = np.zeros((len(time), 8))

for i, t in enumerate(time):
```

```
p = k_2n_ctqw(3, t)
wvf = wfn_sim.wavefunction(p)
vec = wvf.amplitudes
quantum_probs[i] = np.abs(vec)**2
classical_probs[i] = k_2n_crw(3, t)
f, (ax1, ax2) = plt.subplots(2, sharex=True, sharey=True)
ax1.set_title("Quantum evolution")
ax1.set_ylabel('p')
ax1.plot(time, quantum_probs[:, 0], label='Initial node')
ax1.plot(time, quantum_probs[:, 1], label='Remaining nodes')
ax1.legend(loc='center left', bbox_to_anchor=(1, 0.5))
ax2.set_title("Classical evolution")
ax2.set_xlabel('t')
ax2.set_ylabel('p')
ax2.plot(time, classical_probs[:, 0], label='Initial node')
ax2.plot(time, classical_probs[:, 1], label='Remaining nodes')
```

The last code block produces a comparison plot as shown next, which is very similar to Figure 9-5, where a larger distribution has caused higher smoothening for the classical part, as expected.

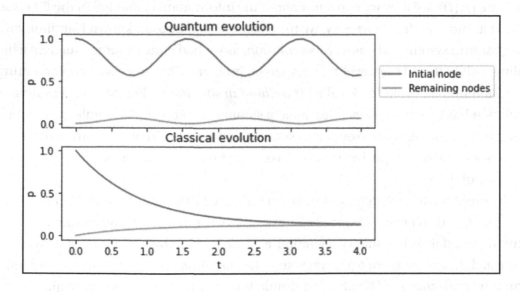

477

Polynomial Time Hamiltonian Simulation

Throughout our journey in this book, we have made healthy use of unitary transformations when dealing with the dynamics of quantum systems. Apart from measurement, a quantum state described by a vector of amplitudes can change by multiplication with a unitary matrix: for instance, a two-qubit gate tensor tied with identities on the other qubits. However, this mathematical paradigm does not clarify which unitary occurs in a physical system. The answer to this is determined by the Hamiltonian system, which is the observable $\mathcal{H}$ corresponding to the total energy E in the system. Here $\langle \psi | \mathcal{H} | \psi \rangle$ is the expectation value of energy E of time-dependent state $| \psi(t) \rangle$.

We can interpret the Hamiltonian $\mathcal{H}$ as describing the physical characteristics of the system. These characteristics do not determine the initial state $| \psi_i(t) \rangle$ of the system, but they do determine the evolution of the state in time (i.e., the state $| \psi_f(t) \rangle$) as a function of the time-parameter t, given initial state, for example, $| \psi_i(t) \rangle = | \psi(0) \rangle$. This is governed by the most important equation in quantum mechanics: the Schrödinger equation.

One of the main Hamiltonian simulation methods is quantum walks. We have explored Hamiltonians and referenced their properties in various coding exercises and calculations.

As a quantum state evolves through time from an initial n-qubit state of $| \psi_i(t) \rangle$ to a final state $| \psi_f(t) \rangle$, we may wish to know about the information contained in the final state of the Hamiltonian $\mathcal{H}$. Conversely, we may also wish to impose a known Hamiltonian on a quantum system and observe its evolution, as we had done in some exercises when dealing with VQE and the traveling salesperson problem. This process of implementing a given Hamiltonian evolution is called *Hamiltonian simulation*. Research in Hamiltonian simulation has two broad categories: analog simulations and digital simulations, with analog simulation relevant to quantum systems that simulate Hamiltonians in their natural state and decomposition of time evolution into quantum gates being dealt with digital simulation.

The problem of *Hamiltonian simulation* is defined as follows. Given a Hamiltonian $\mathcal{H}$, a time t, and an error tolerance ϵ (for example, trace distance), find a quantum circuit that performs the unitary operation $e^{-i\mathcal{H}t}$ on an unknown quantum state with an error which is at most ϵ. For a general result, the Hamiltonian can be time-dependent. Common applications of Hamiltonian simulations include continuous-time quantum algorithms, such as quantum walk, adiabatic optimization, and linear equations.

If we have a time-independent Hamiltonian consisting of a sum of several other Hamiltonians, such that $\mathcal{H} = \sum_n \mathcal{H}_n$, then each of those "member" Hamiltonians should be relatively easier to simulate. As for non-commuting $\mathcal{H}_n$ the factorization rules cannot be applied. Seth Lloyd, in his paper in 1996 [207] used the following formula to achieve this.

$$e^{-i\sum_n \mathcal{H}_n t} = \prod_n e^{-i\mathcal{H}_n t} + O\left(t^2\right)$$

9.8

Now, if we subject this equation to Trotterization (see Chapter 7, equation 7.24), which states that for small t the factorization rule is approximately valid, we can write the evolution of $\mathcal{H}$ for time t as a sequence of small time-steps of length Δt as

$$e^{-i\mathcal{H}_n t} = \left(e^{-i\mathcal{H}_n \Delta t}\right)^{t/\Delta t}$$

9.9

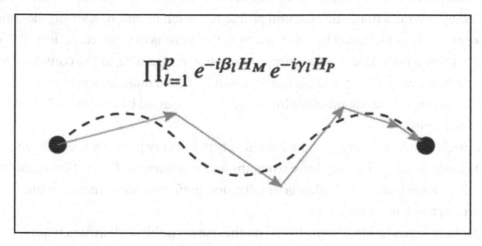

$$\prod_{l=1}^{p} e^{-i\beta_l H_M} e^{-i\gamma_l H_P}$$

Figure 7-9. *Trotterization. Source [122]*

Figure 7-9 illustrating Trotterization is reproduced for ease of reading. The expression on the left side of equation 9.9 shows a non-trivial error for each time step Δt. However, in this case, the error in equation 9.8 becomes negligible. Note that the *number of times* the sequence has to be repeated increases as the length of time-step Δt decreases. Given the trade-off between the quantity of sequence and size of time-steps, this method allows us to simulate Hamiltonians by manipulating the terms of $\mathcal{H}_n$. This approach of *polynomial-time Hamiltonian simulation* shows that if the method to decompose Hamiltonians into sums of elementary Hamiltonians is known, and if we know how to simulate those elementary Hamiltonians, then we can approximately evolve a quantum system in our desired way.

In the discrete-variable qubit model, efficient methods of Hamiltonian simulation have been discussed in literature extensively, providing several implementations that depend on the properties of the Hamiltonians and result in a linear simulation time [224, 225]. Efficient implementations of Hamiltonian simulation also exist in the continuous-variable (CV) formulation [226], with specific application to Bose-Hubbard Hamiltonians, which describe a system of interacting bosonic particles on a lattice of orthogonal position states [227]. fittingly, this method is suited to photonic quantum computation and communications, both of which are the subject of active research.

Ensembles and QBoost

QBoost was proposed by researchers at Google and D-Wave in 2008 and 2009 [228], [229], [230] as an iterative training algorithm in which a subset of weak classifiers is selected by solving a hard optimization problem in each iteration. A strong classifier is incrementally constructed by concatenating the subsets of weak classifiers. It was introduced as a novel discrete optimization method for training in the context of a boosting framework for large-scale binary classifiers. The main goal is to cast the machine learning training problem into the format required by existing adiabatic quantum hardware.

A single machine learning model is unlikely to fit every possible scenario since any learning algorithm will always have strengths and weaknesses. Ensembles combine multiple models to achieve higher generalization performance than any of the constituent models is capable of.

Machine learning algorithms based on the use of multiples learners belong to a broader class of ensemble learning algorithms [231], [232]. Ensemble methods construct better classifiers by considering a group of trained models. An ensemble of trained models (or classifiers) decides for a new prediction together, which helps combine the individual advantages of its member models.

A *constituent* model in an ensemble is also called a *base classifier* or *weak learner*, and the composite model a *strong learner*. Ensembles typically produce better results when there is considerable diversity among the base classifiers. The base classifiers generate various types of errors in the presence of sufficient diversity when a strategic combination may reduce the total error, thereby ideally improving performance.

The procedure of ensemble methods has two main steps.

1. Develop a set of base classifiers from the training data.

2.   Combine them to form the ensemble.

In the simplest combination, the base learners vote, and the label prediction is based on the majority. More involved methods weigh the votes of the base learners.

However, this leads us to the challenge of assembling the weak learners to produce the maximum possible efficiency. There are several ways to achieve this. For example, given the current collection of models, we can add one more based on where that model performs well. Alternatively, we can look at all the correlations of the predictions between all the models and optimize for the most uncorrelated predictors. Since the latter approach is global, it maps to a quantum computer by default.

Intuitively, we realize that before getting our hands dirty with ensembles, we need insight into *loss functions* and *regularization*—two key concepts in machine learning—before we can be confident enough to make a judgment call on the collective performance of error functions of various models.

This is where some hands-on experience is beneficial. Hence, we look at ensemble methods and QBoost via hands-on coding.

Ensembles

To start, we download and open the Jupyter notebook file called qboost_ensemble. ipynb from the book's website. Now that we have discussed the fundamentals of ensemble, let us import some packages and define our figure of merit as *accuracy* in a balanced dataset.

Listing 9-2a. Libraries qboost_ensemble.ipynb

```
import matplotlib.pyplot as plt
import numpy as np
import sklearn
import sklearn.datasets
import sklearn.metrics
%matplotlib inline
metric = sklearn.metrics.accuracy_score
```

Next, we generate a random dataset of two classes that form concentric circles.

Listing 9-2b. Generate Dataset qboost_ensemble.ipynb

```
np.random.seed(0)
data, labels = sklearn.datasets.make_circles()
idx = np.arange(len(labels))
np.random.shuffle(idx)
# train on a random 2/3 and test on the remaining 1/3
idx_train = idx[:2*len(idx)//3]
idx_test = idx[2*len(idx)//3:]
X_train = data[idx_train]
X_test = data[idx_test]
y_train = 2 * labels[idx_train] - 1  # binary -> spin
y_test = 2 * labels[idx_test] - 1

scaler = sklearn.preprocessing.StandardScaler()
normalizer = sklearn.preprocessing.Normalizer()

X_train = scaler.fit_transform(X_train)
X_train = normalizer.fit_transform(X_train)

X_test = scaler.fit_transform(X_test)
X_test = normalizer.fit_transform(X_test)
plt.figure(figsize=(6, 6))
plt.subplot(111, xticks=[], yticks=[])
plt.scatter(data[labels == 0, 0], data[labels == 0, 1], color='navy')
plt.scatter(data[labels == 1, 0], data[labels == 1, 1], color='red');
```

This code block produces the following circular dataset as output.

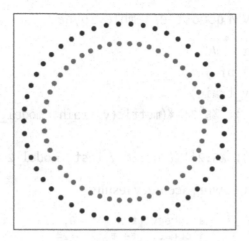

It may be recalled that we dealt with classical perceptron in Chapter 3.

Note   If you are interested in *quantum perceptrons,* go through the tutorial at `https://qml.entropicalabs.io/`.

In this lab, let us train a perceptron again.

Listing 9-2c. Train a Perceptron qboost_ensemble.ipynb

```
from sklearn.linear_model import Perceptron
model_1 = Perceptron()
model_1.fit(X_train, y_train)
print('accuracy (train): %5.2f'%(metric(y_train, model_1.predict(X_
train))))
print('accuracy (test): %5.2f'%(metric(y_test, model_1.predict(X_test))))
```

This gives the following as accuracy.

```
accuracy (train):  0.44
accuracy (test):  0.65
```

The decision surface of the perceptron is linear, and as such, we get a poor accuracy. It may be worth trying a support vector machine (SVM) with a nonlinear kernel and see what happens.

Listing 9-2d. Apply SVM qboost_ensemble.ipynb

```
from sklearn.svm import SVC
model_2 = SVC(kernel='rbf')
model_2.fit(X_train, y_train)
print('accuracy (train): %5.2f'%(metric(y_train, model_2.predict(X_
train))))
print('accuracy (test): %5.2f'%(metric(y_test, model_2.predict(X_test))))
```

This time we get the following accuracy results.

```
accuracy (train):  0.62
accuracy (test):   0.24
```

The output indicates that the SVM performs better on the training set but poorly on the test data, hence, unsuitable for generalization.

Boosting is an ensemble method that explicitly seeks models that complement one another. The variation between boosting algorithms is how they combine weak learners.

Adaptive boosting (AdaBoost) is a popular method that sequentially combines the weak learners based on their individual accuracies. It has a convex objective function that does not penalize for complexity. AdaBoost is likely to include all available weak learners in the final ensemble. Listing 9-2e is a training exercise with AdaBoost with a few weak learners.

Listing 9-2e. AdaBoost qboost_ensemble.ipynb

```
from sklearn.ensemble import AdaBoostClassifier
model_3 = AdaBoostClassifier(n_estimators=3)
model_3.fit(X_train, y_train)
print('accuracy (train): %5.2f'%(metric(y_train, model_3.predict(X_
train))))
print('accuracy (test): %5.2f'%(metric(y_test, model_3.predict(X_test))))
```

AdaBoost produces the following output which is also about the same in accuracy (only very marginal improvement over SVM).

```
accuracy (train):  0.65
accuracy (test):   0.29
```

QBoost

The idea of QBoost [230] is that optimization on a quantum computer is not constrained to convex objective functions. This allows us to add arbitrary penalty terms and rephrase our objective [230]. The QBoost problem can be defined as follows.

$$\text{argmin}_w \left[\frac{1}{N} \sum_{i=1}^{N} \left(\sum_{k=1}^{K} w_k h_k(x_i) - y_i \right)^2 + \lambda \|w\|_0 \right]$$

where $h_k(x_i)$ is the prediction for weak learner k for training instance k. The weights in this formulation are binary. Hence, this objective function maps to an Ising model by default. The regularization ensures sparsity. It is not a classical regularization; it is hard to optimize this term on a digital computer.

Expanding the quadratic part of the objective, we have

$$\text{argmin}_w \left[\frac{1}{N} \sum_{i=1}^{N} \left(\sum_{k=1}^{K} w_k h_k(x_i) \right)^2 - 2 \sum_{k=1}^{K} w_k h_k(x_i) y_i + y_i^2 + \lambda \|w\|_0 \right]$$

where, y_i^2 is a constant. Therefore, the optimization reduces to

$$\text{argmin}_w \left[\frac{1}{N} \sum_{k=1}^{K} \sum_{l=1}^{K} w_k w_l \left(\sum_{i=1}^{N} h_k(x_i) h_l(x_i) \right) - \frac{2}{N} \sum_{k=1}^{K} w_k \sum_{i=1}^{N} h_k(x_i) y_i + y_i^2 + \lambda \|w\|_0 \right]$$

This form tells us that we consider all correlations between the predictions of the weak learners as there is a summation of $h_k(x_i)h_l(x_i)$. Since this term has a positive sign, we penalize for correlations. On the other hand, the correlation with the true label, $h_k(x_i)y_i$, has a negative sign, which means the regularization term remains unchanged.

Let us consider all three models from the previous section as weak learners, calculate their predictions, and set it to 1. The predictions are scaled to reflecting the averaging in the objective.

Listing 9-2f. Model Definition `qboost_ensemble.ipynb`

```
models = [model_1, model_2, model_3]
n_models = len(models)
predictions = np.array([h.predict(X_train) for h in models], dtype=np.float64)
```

```
# scale hij to [-1/N, 1/N]
predictions *= 1/n_models
λ = 1
```

We create the quadratic binary optimization (QUBO) of the objective function.

Listing 9-2g. Define QUBO qboost_ensemble.ipynb

```
w = np.dot(predictions, predictions.T)
wii = len(X_train) / (n_models ** 2) + λ - 2 * np.dot(predictions, y_train)
w[np.diag_indices_from(w)] = wii
W = {}
for i in range(n_models):
    for j in range(i, n_models):
        W[(i, j)] = w[i, j]
```

We solve the quadratic binary optimization with simulated annealing and read out the optimal weights.

Listing 9-2h. Solve QUBO qboost_ensemble.ipynb

```
import dimod
sampler = dimod.SimulatedAnnealingSampler()
response = sampler.sample_qubo(W, num_reads=10)
weights = list(response.first.sample.values())
```

We define a prediction function to help with measuring accuracy.

Listing 9-2i. Prediction Function qboost_ensemble.ipynb

```
def predict(models, weights, X):
    n_data = len(X)
    T = 0
    y = np.zeros(n_data)
    for i, h in enumerate(models):
        y0 = weights[i] * h.predict(X)  # prediction of weak classifier
        y += y0
        T += np.sum(y0)
```

```
    y = np.sign(y - T / (n_data*len(models)))
    return y
print('accuracy (train): %5.2f'%(metric(y_train, predict(models, weights,
X_train))))
print('accuracy (test): %5.2f'%(metric(y_test, predict(models, weights,
X_test))))
```

This gives the following accuracy as output.

accuracy (train): 0.65
accuracy (test): 0.29

The accuracy coincides with our strongest weak learners: the AdaBoost model. Looking at the optimal weights, this is apparent as the output of the weights command gives.

[0, 0, 1]

Only AdaBoost made it to the final ensemble. The first two models show poor performances, and their predictions are correlated. Yet, if we remove regularization by setting $\lambda = 0$, the second model also enters the ensemble, decreasing overall performance. This is a demonstration that regularization is important.

Quantum Image Processing (QIMP)

Quantum image processing, which utilizes the characteristic of quantum parallelism to speed up many processing tasks, is a subfield of quantum information processing [3], [237]. The first steps taken in quantum image processing have involved proposals on representations to capture and store the image on quantum computers. Various representations for images on quantum computers have been proposed, such as qubit lattice, wherein the images are two-dimensional arrays of qubits [239]; Real Ket, wherein the images are quantum states having gray levels as coefficients of the states [239]; grid qubit; in which geometric shapes are encoded in quantum states [240]; quantum lattice, wherein color pixels are stored in quantum systems qubit by qubit [241]; flexible representation of the quantum image (FRQI), wherein the images are normalized states that capture the essential information about every point in an image (i.e., it's color and position [237], [242], [243]); multi-channel quantum image (MCQI) [244], which is an extension of FRQI representation that contains the R, G and B channels for processing color information and novel enhanced quantum representation (NEQR) of digital images, which works with an internal representation of an image in a QPU [250].

Quantum computation and quantum information are disciplines that have penetrated every aspect of human activities and have appeared in various areas of computer science, such as information theory, cryptography, emotion representation, and image processing because inefficient tasks on classical computers can be overcome by exploiting the power of quantum computation [3], [233], [234], [235], [236]. Novel methods for speeding up certain tasks and interdisciplinary research between computer science and several other scientific fields have attracted deep scientific interest.

Technically, efforts in quantum-related image processing can be classified into three main groups [63].

- **Quantum-assisted digital image processing (QDIP)**: These applications aim to improve some well-known digital or classical image processing tasks and applications by exploiting some useful properties of quantum computing algorithms [245].

- **Optics-based quantum imaging (OQI)**: These applications focus on techniques for optical imaging and parallel information processing at the quantum level by exploiting the quantum nature of light and the intrinsic parallelism of optical signals [247], [248].

- **Quantum image processing (QIMP)**: Inspired by the impending realization of quantum computing hardware, these applications focus on extending classical image processing tasks and applications to a quantum computing framework [245], [249].

Research has shown that quantum computers can utilize maximally entangled qubits to reconstruct images without additional information [240], improving both storage and retrieval. Storage of images in qubit arrays has been researched and established, but retrieval and reconstruction are another important aspect that quantum computing improves upon. Using a quantum measurement to probe the entanglement shared between the vertex qubits can determine their location. Furthermore, the parallelism inherent in quantum systems has been used in fast image searching [42] and image reconstruction. FRQI and the real *ket* image model can process all pixels of the image simultaneously, demonstrating their inherent parallelism.

Image segmentation is also an important aspect of image processing in quantum computation. Image segmentation is the process of dividing an image into separate regions, for example, finding faces in a picture. This concept is especially important for machine learning when the detection of this region is automated.

In traditional image segmentation, complexities associated with the intensity and spatial location of objects in an image make this task challenging. In other words, image segmentation fails in traditional computer image recognition often because of a lack of previous information to form an expectation of an image. Venegas-Andraca et al. [240] expand further on their techniques and improve image segmentation relative to traditional systems. They continue to use maximally entangled states to store points that correspond to objects within the image. They then can detect quantum entanglement to determine which pixels belong to which objects by knowing which vertices correspond to which objects. Detected corners are then used to build geographic shapes that correspond to the object shapes and then using mathematical morphology operators (techniques used to analyze geometric structures) to fill out space bordered by those shapes. The result is knowledge on which pixels belong to which objects.

Of all the areas, some of the most researched is [243] and NEQR [250].

FRQI, a *flexible representation of the quantum image* FRQI [243], attempts to encode the colors of a classical image through angles and is similar to the pixel representation of images on traditional computers. It captures the essential information about the colors and the position of every point in an image and integrates them into a quantum state with the formula.

$$|I\rangle = \frac{1}{2^n}\sum_{j=0}^{2^{2n}-1}|q_j\rangle \otimes |j\rangle$$

where, $|q_j\rangle = \cos\theta_j|0\rangle + \sin\theta_j|1\rangle$, $|0\rangle$ and $|1\rangle$ are two-dimensional computational basis states, $|j\rangle$ are 2^{2n}-dimensional computational basis states and $\theta_j \in \left[0, \frac{\pi}{2}\right]$ is the vector of angles encoding the colors. The two parts in the FRQI representation of an image, $|q_j\rangle$ and $|j\rangle$ encode information about the colors and corresponding positions respectively in the image.

FRQI is mainly implemented via the Identity and Hadamard's gates. In quantum computation, computers are usually initialized in well-prepared states. Hence, a preparation process is necessary to transform quantum computers from the initialized state to the desired quantum image state.

A *multi-channel representation for quantum images* (MCQI) [243], which uses R, G, and B channels to represent color information about the image while retaining its normalized state, was proposed in [244]. The MCQI proposal was created by extending the grayscale information encoded in an FRQI image to a color representation allowing the design of some low-complexity color information operators. The MCQI proposal implies that the computational complexities of associated operators are independent of the image size. MCQI and related operations also appear to provide efficient tools for quantum watermarking algorithms based on color images by hiding secret information in the design of quantum circuits [252].

Novel enhanced quantum representation (NEQR) [250] for digital images improve on FRQI and MCQI. In the FRQI and MCQI representations, the color information is encoded by the superposition of one and three qubits separately. Hence, these quantum images may require multiple measurements to retrieve information. The new model or NEQR uses the basis state of a qubit sequence to store the grayscale value of every pixel. Therefore, two-qubit sequences, representing the grayscale and positional information of all the pixels, are used in NEQR representation to store the whole image.

The computational complexity of preparing an NEQR image shows a quadratic decrease (i.e., $O(qn. \ 2^n)$), compared to FRQI and MCQI images [250]. However, it is to be noted that NEQR representation uses *more qubits* to encode a quantum image. From its representation, $(q + 2n)$ qubits are needed to construct the quantum image model for a $2^n \times 2^n$ image with the gray range 2^q. However, the $2n$ qubits for position information are the same as for FRQI and MCQI representation. NEQR uses q qubits for color information, while FRQI and MCQI use one qubit and three qubits, respectively. Researchers have devised *improved* NEQR (INEQR) and generalized QIR (GQIR) by resizing the quantum image to an arbitrary size for wider applications [253].

Note   At the time of authoring this book a collaborative research team had just updated the Qiskit documentation site covering some excellent tutorials for FRQI and NEQR [254]. You can gain more in-depth insight at `https://qiskit.org/ textbook/ch-applications/image-processing-frqi-neqr.html?utm_ source=Medium&utm_medium=Social&utm_campaign=Community&utm_ content=Textbook-CTA`.

Tensor Networks

Tensor networks were originally designed for simulating quantum physics, especially in low-dimension. They are now increasingly applied for solving machine learning tasks such as image recognition. Some of the hardest scientific problems, such as understanding the true nature of space and time or developing high-temperature superconductors, involve dealing with the complexity of quantum systems. The number of quantum states in these systems is exponentially large, negating brute-force computation as a possible path to a solution. To deal with this, tensor networks, which are a form of data structures, are used. Tensor networks are becoming increasingly attractive for researchers in machine learning (ML) due to their flexibility.

Despite their obvious attraction, traditionally, some difficulties prohibited tensor networks from widespread use in the ML community.

- A production-level tensor network library for accelerated hardware has not been available to run tensor network algorithms at scale.

- Most of the tensor network literature is geared toward physics applications and creates the false impression that expertise in quantum mechanics is required to understand the algorithms.

Keeping this in mind, Google, the Perimeter Institute for Theoretical Physics (www. perimeterinstitute.ca), and X (https://x.company) created the TensorNetwork library, which was open sourced in 2019 on the back of two publications [257][258].

For the interested, it is recommended that they visit the GitHub site for TensorNetwork at https://github.com/google/TensorNetwork. If you are interested in tensor network simulation, a tool called TensorTrace (www.tensortrace.com) [259] allows users to easily design and simulate tensor networks. This is illustrated in Figure 9-6.

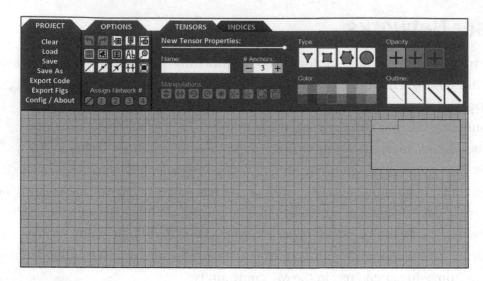

Figure 9-6. *TensorTrace (Source www.tensortrace.com)*

Quantum Finance

Quantum finance is an upcoming cross-discipline of financial engineering with quantum field theory, classical finance theory, computer science, and AI technology [260]. With recent advances in quantum computing technologies and the promised quantum advantage in QML, researchers have considered how to utilize them in industries. A major focus area is aspects of finance (see review at [260]).

Since financial institutions are performing enormous tasks of numerical calculation in their daily works, the promise of quantum computers speeding up tasks is too enticing to ignore. One of such tasks is the pricing of financial derivatives.[3] Large banks typically have a huge number of derivatives written on various types of assets such as stock price, foreign exchange rate, interest rate, commodity, and so on. Therefore, the pricing of derivatives is an important issue for them.

Another important aspect of quantum finance is option pricing, closely related to financial derivatives [261]. Efforts of marrying quantum machine learning to stock market dynamics by leveraging the probabilistic nature of quantum computers and algorithms have been ongoing for forecasting and risk-analysis; websites have been created toward this end; for example, see QFFC at http://qffc.uic.edu.hk/.

---

[3]Financial derivatives, or simply derivatives, are contracts in which payoffs are determined in reference to the prices of underlying assets at some fixed times.

Collaborations such as Chicago Quantum Exchange (`https://quantum.uchicago.edu/about`) have been formed to investigate and develop this exciting cross-discipline subject. The research effort includes the following universities.

- University of Chicago

- Argonne National Laboratory

- Fermi National Accelerator Laboratory

- The University of Illinois at Urbana Champaign

- University of Wisconsin-Madison

- Northwestern University

Several startups have published research papers using QML algorithms on stock market portfolio modeling and financial forecasting, including Chicago Quantum (`www.chicagoquantum.com`) [262]. Mizuho–DL Financial Technology (`www.mizuhobank.co.jp`) in Japan proposed enhanced applications quantum Monte Carlo simulation of pricing to financial derivatives [263]. Quantum promises to be an exciting field that opens endless possibilities in financial industries, be that risk analysis, portfolio optimization, stock price prediction, or supply chain, just to name a few. The tools for use are in place in the forms of software libraries (for example, D-Wave, Qiskit, TensorFlow Quantum, PennyLane, pyQuil, etc.), quantum simulators, and hardware as well if needed.

Quantum Communication

There is currently a germinating but steady effort to build up a worldwide Quantum Internetwork that can generate, distribute, and process quantum information in addition to classical data. The US, UK, China, the EU, among others, are actively investing in relevant technologies and associated research. For example, in February 2020, the US Department of Energy (DOE), in conjunction with Brookhaven National Laboratories, held a workshop on building a worldwide quantum Internet and released their report (`www.osti.gov/servlets/purl/1638794`) [264] in July 2020.

Quantum communication researchers are trying to take advantage of the laws of quantum physics to connect users and protect data at scale. These laws allow particles—typically photons of light—to transmit data along optical cables utilizing superposition. Technologies that are currently being researched address some of the most important aspects of quantum communication, such as secure private-bid auctions by Guha et al. [271], generation of multiparty shared secrets [34], [266], distributed quantum computing [267], improved sensing [268], [269], and quantum computing on encrypted data (blind quantum computing) [270].

One of the fundamental differences for quantum networks (as opposed to existing classical ones) arises from the fact that entanglement, whose long-distance generation is an essential network function, is inherently present at the network's physical layer. This differs from classical networking, where shared states typically are established only at higher layers. In this context, solutions must be found to guarantee network device fidelity levels capable of supporting entanglement distribution and deterministic teleportation, as well as quantum repeater schemes that can compensate for loss and allow for operation error correction.

The last step in the evolution of quantum communication networks is reached when devices capable of error correction and concatenated quantum repeaters are available. Quantum error correction (QEC) has been a major challenge in areas of quantum computing communications. Widespread efforts are underway by worldwide collaborations between research and industry outfits to tackle QEC challenges. One area of focus is using machine learning to optimize error in near-term quantum systems; for example, see Nautrup et al. [284] and Colomer et al. [285].

One of the longstanding challenges is networking superconducting quantum computers. So far, the typical approach has been to cascade transducers by converting to optical frequencies at the transmitter and microwave frequencies at the receiver. However, the small microwave-optical coupling and added noise have proven formidable obstacles. One way to bypass this challenge has been proposed by Krastanov et al. in 2020 [276] where they employed optical networking via heralding end-to-end entanglement with one detected photon and teleportation. This scheme absorbs the low optical-microwave coupling efficiency into the heralding step, breaking the rate-fidelity trade-off, and simplifying entanglement generation between superconducting and other physical devices in quantum networks.

Some of the most important aspects of a quantum communication network are resource management, visibility, simulation of test networks, and analysis. Related to this, a great review [277] on state-of-the-art tools for evaluating the performance of quantum networks, focusing on information-theoretic benchmarks, analytical tools, and simulation, was published in 2020.

No communication network is complete without security protocols. Certain security solutions such as QKD (protocol, entanglement, and teleportation-based security have been proposed and are being researched currently. However, implementing these protocols on optical platforms for communication challenges such as physical qubit noise, distance-dependent coherence, fidelity, memory, and regeneration of keys at (distance challenge mitigating) repeaters [282] name a few. One of the ways some members of the US Department of Energy (DOE) Quantum Internet Blueprint Workshop [264] held at Brookhaven National Laboratory in July 2020 recommended bypassing memory requirements in quantum networks of the near future by utilizing *software-defined networking* (SDN) [281].

Among several different ways that the quantum systems research community approaches feasible solutions for challenges in qubit noise, memory, and resource allocation, machine learning is one of the more popular ones. As time, research, funding, and energy are invested in ongoing efforts. The future holds the promise of unprecedented technological supremacy via this alliance between machine learning and physics.

Summary

Machine learning methods have been effectively used in various quantum information processing problems, including quantum metrology, Hamiltonian estimation, quantum signal processing, and qubit mapping using quantum walk algorithms, problems of quantum control, and many others. The construction of advanced quantum devices, including quantum computers, uses quantum machine learning and artificial intelligence techniques. Automated machines can control complex processes, for example, the execution of a sequence of simple gates, as used in quantum computation.

The amount of data needed to reliably train a classical computation model is ever-growing and reaching the limits which normal computing devices can handle. In such a scenario, the quantum computation can aid in continuing training with huge data. Quantum machine learning looks to devise learning algorithms more efficiently than their classical counterparts. Classical machine learning is about finding patterns in data and using those patterns to predict further events. Quantum systems, on the other hand, produce atypical patterns which are not producible by classical systems, thereby postulating that quantum computers may overtake classical computers on machine learning tasks.

As we venture deeper into the twenty-first century, an exciting future awaits us with unprecedented promises of advancement in finance, science, biology, chemistry, physics, and very likely, into applications that we cannot even think of today.

APPENDIX A

Mathematical Review

This appendix is a brief refresher on some of the mathematical theories used in this book.

Preliminaries

Real coordinate space: A *real coordinate space* of dimension n, often represented as $\mathbb{R}^n$, is a coordinate space over the set of the n-tuples of real numbers (sequences of n real numbers) with component-wise addition and scalar multiplication. It forms a real vector space.

Subspace: A *subspace* of a vector (or coordinate) space **V** is a subset **H** of **V**, such that **H** is itself a vector space under the same operations of vector addition and scalar multiplication that are already defined on **V**.

Basis of a subspace: If **V** is a subspace of $\mathbb{R}^n$ *basis* of **V** is a set of vectors, $\mathbb{V} \equiv \{v_1, v_2, \ldots v_k\}$, such that $\mathbb{V}$ is the minimum set of vectors that span **V** and $\mathbb{V}$ is linearly independent.

Completeness: Let's say that (X, d) is a metric space, and (x_n) is a sequence of points of X. $\{x_n\}$ is a *Cauchy sequence* if for every $\varepsilon > 0$ there exists $N \in \mathbb{N}$, such that $d(x_n, x_m) < \varepsilon$ for all $n, m \geq N$. A *metric space* (X, d) is called *complete* if every Cauchy sequence (x_n) in X converges to some point of X. A subset A of X is called *complete* if A as a metric subspace of (X, d) is *complete*; that is, if every Cauchy sequence (x_n) in A converges to a point in A.

© Santanu Ganguly 2021
S. Ganguly, *Quantum Machine Learning: An Applied Approach*, https://doi.org/10.1007/978-1-4842-7098-1

Orthogonal vectors: Two vectors are said to be *orthogonal* if they are perpendicular to each other (i.e., the *dot* or *inner* product of the two vectors is zero). A set of vectors $\{v_1, v_2, \ldots v_n\}$ are said to be mutually orthogonal if every pair of vectors is orthogonal.

$$v_i.v_j = 0, \textit{for all } i \neq j$$

For example, the following set of vectors is mutually orthogonal.

$$v_1 = \begin{pmatrix} 1 \\ 0 \\ -1 \end{pmatrix}, v_2 = \begin{pmatrix} 1 \\ \sqrt{2} \\ 1 \end{pmatrix}, v_3 = \begin{pmatrix} 1 \\ -\sqrt{2} \\ 1 \end{pmatrix}$$

because

$$(1,0,-1).\left(1,\sqrt{2},1\right) = 0$$

$$(1,0,-1).\left(1,-\sqrt{2},1\right) = 0$$

$$\left(1,\sqrt{2},1\right).\left(1,-\sqrt{2},1\right) = 0$$

Orthonormal vectors: A set of vectors S is *orthonormal* if every vector in S has a unit magnitude of 1 and the set of vectors are mutually orthogonal. The vectors v_1, v_2, v_3 are mutually orthogonal but not orthonormal because they are not normalized. The word *normalized* often means that the vectors are not of magnitude 1. Now, let's define the following.

$$u_1 = \frac{v_1}{|v_1|} = \frac{1}{\sqrt{2}}\begin{pmatrix} 1 \\ 0 \\ -1 \end{pmatrix} = \begin{pmatrix} 1/\sqrt{2} \\ 0 \\ -1/\sqrt{2} \end{pmatrix}$$

$$u_2 = \frac{v_2}{|v_2|} = \frac{1}{2}\begin{pmatrix} 1 \\ \sqrt{2} \\ 1 \end{pmatrix} = \begin{pmatrix} 1/2 \\ 1/\sqrt{2} \\ 1/2 \end{pmatrix}$$

$$u_3 = \frac{v_3}{|v_3|} = \frac{1}{2}\begin{pmatrix} 1 \\ -\sqrt{2} \\ 1 \end{pmatrix} = \begin{pmatrix} 1/2 \\ -1/\sqrt{2} \\ 1/2 \end{pmatrix}$$

The set of vectors $\{u_1, u_2, u_3\}$ is orthonormal.

Orthonormal basis: An *orthonormal basis* for an inner product space V with finite dimension is a basis for V whose vectors are orthonormal; that is, they are all unit vectors and orthogonal to each other. A set of basis vectors $|\varphi_1\rangle, |\varphi_2\rangle \ldots\ldots |\varphi_n\rangle$ are said to form an orthonormal basis if

$$\langle \varphi_i | \varphi_j \rangle = \delta_{ij} \qquad\qquad\qquad\qquad \text{A.1}$$

The term δ_{ij} is called the *Kronecker delta*. The following describes its properties.

$$\delta_{ij} = \begin{cases} 0 \; if \; i \neq j \\ 1 \; if \; i = j \end{cases} \qquad\qquad\qquad\qquad \text{A.2}$$

Linear operators: A *linear operator* A maps a vector $|v\rangle$ in vector space V to a vector $|w\rangle$ in the vector space W and is linear in its inputs. If $|v\rangle = \sum_i a_i |v_i\rangle$, then for a linear operator

$$A \sum_i a_i |v_i\rangle = \sum_i a_i A |v_i\rangle \qquad\qquad\qquad\qquad \text{A.3}$$

Any vector can be expressed as a linear sum of its basis vectors. Hence, it is enough to know how the linear operator transforms its basic vectors to understand how a linear operator transforms any given vector in a vector space.

A vector can be represented in terms of an underlying basis, as is usual in quantum computing. If the state $|\psi\rangle = \alpha|0\rangle + \beta|1\rangle$ of a qubit is represented as $\begin{bmatrix} \alpha \\ \beta \end{bmatrix}$, then $|0\rangle$ and $|1\rangle$ are the basis vectors of this column vector representation of the qubit.

Let's suppose the matrix A denotes the transformation with respect to the input and output bases B_1 and B_2, respectively, where B_1 consists of the basis vectors $|\phi_0\rangle, |\phi_1\rangle.... |\phi_{n-1}\rangle$, and B_2 consists of the basis vectors $|\omega_0\rangle, |\omega_1\rangle...|\omega_{m-1}\rangle$. Then, any basis vector $|\phi_k\rangle$ is represented as a column vector of all 0s but one 1 in the k-th row when represented in its own set of basis vectors, as shown next.

$$|\phi_k\rangle_{B_1} = \begin{bmatrix} 0 \\ .. \\ 1 \\ .. \\ 0 \end{bmatrix} \leftarrow index\ k \qquad\qquad \text{A.4}$$

In equation A.2, $|\phi_k\rangle_{B_1}$ represents $|\phi_k\rangle$ with respect to basis B_1. Hence, a transformation of $|\phi_k\rangle$ with respect to basis B_1 would form the vector $|w\rangle$ with respect to basis B_2, as shown next.

$$|w\rangle_{B_2} = \hat{A}|\phi_k\rangle = A|\phi_k\rangle_{B_1} = A\begin{bmatrix} 0 \\ .. \\ 1 \\ .. \\ 0 \end{bmatrix} = \begin{bmatrix} a_{0k} \\ a_{1k} \\ .. \\ a_{(m-1)k} \end{bmatrix} \qquad\qquad \text{A.5}$$

If we want to expand the vector to the usual basis representation $|w\rangle$, then

$$|w\rangle_{B_2}[|w_1\rangle|w_2\rangle\cdot|w_m\rangle] = \begin{bmatrix} a_{0k} \\ a_{1k} \\ .. \\ a_{(m-1)k} \end{bmatrix}[|w_1\rangle|w_2\rangle\cdot|w_m\rangle] = \sum_{i=0}^{m-1}a_{ik}|w_i\rangle \qquad \text{A.6}$$

From A.3 and A.6, we get

$$\hat{A}|\phi_k\rangle = A|\phi_k\rangle_{B_1} = \sum_{i=0}^{m-1}a_{ik}|w_i\rangle \qquad \text{A.7}$$

Schmidt Decomposition: Examples

The following is an example of Schmidt decomposition (other solutions are possible).

$$|\psi_1\rangle = \frac{|00\rangle+|11\rangle+|22\rangle}{\sqrt{3}} = \sum_{i=0,1,2}\frac{1}{\sqrt{3}}|i\rangle|i\rangle$$

The following is another example of Schmidt decomposition (other solutions are possible).

$$|\psi_2\rangle = \frac{|00\rangle+|01\rangle+|10\rangle+|11\rangle}{2} = |+\rangle|+\rangle$$

where $|+\rangle = \frac{1}{\sqrt{2}}[|0\rangle+|1\rangle]$

Tensor Product

If we have two vector spaces, V that is n-dimensional, and W that is m-dimensional, the *tensor product* of these two vector spaces is (nm)-dimensional. It is depicted by $V \otimes W$, where the symbol $\otimes$ denotes the tensor product[1] operation. If A and B are linear

---

[1] A good reference for vectors, tensors, and so forth is at http://hitoshi.berkeley.edu/221a/ tensorproduct.pdf.

operators on vectors $|v\rangle$ and $|w\rangle$ in vector spaces V and W, respectively, then the linear operator on vector $|v\rangle \otimes |w\rangle$ is given by $A \otimes B$. The linear operator $A \otimes B$ works on $|v\rangle \otimes |w\rangle$ as follows.

$$A \otimes B|v\rangle \otimes |w\rangle = A|v\rangle \otimes B|w\rangle \qquad \text{A.8}$$

For any two vectors, $|v\rangle \in \mathbb{R}^m$ and $|w\rangle \in \mathbb{R}^n$, the tensor product is given by

$$|v\rangle \otimes |w\rangle = \begin{bmatrix} v_1 \\ v_2 \\ . \\ . \\ v_m \end{bmatrix} \otimes \begin{bmatrix} w_1 \\ w_2 \\ . \\ . \\ w_n \end{bmatrix} = \begin{bmatrix} v_1|w\rangle \\ v_2|w\rangle \\ \cdots \\ v_m|w\rangle \end{bmatrix} \qquad \text{A.9}$$

For instance, if $|v\rangle = \begin{bmatrix} a \\ b \end{bmatrix}$ and $|w\rangle = \begin{bmatrix} c \\ d \\ e \end{bmatrix}$, then, their tensor product is given by

$$|v\rangle \otimes |w\rangle = \begin{bmatrix} a \\ b \end{bmatrix} \otimes \begin{bmatrix} c \\ d \\ e \end{bmatrix} = \begin{bmatrix} a\begin{bmatrix} c \\ d \\ e \end{bmatrix} \\ b\begin{bmatrix} c \\ d \\ e \end{bmatrix} \end{bmatrix} = \begin{bmatrix} ac \\ ad \\ ae \\ bc \\ bd \\ be \end{bmatrix}$$

Similarly, the tensor product of two matrices—A of dimension $m \times n$ and B of dimension $p \times q$—can be written as follows.

$$A \otimes B = \begin{bmatrix} a_{11}B & \cdots & a_{1n}B \\ \vdots & \ddots & \vdots \\ a_{m1}B & \cdots & a_{mn}B \end{bmatrix} \qquad \text{A.10}$$

For example, let's take the Pauli matrices X and Y and work through their tensor product based on equation A.10.

$$X \otimes Y = \begin{bmatrix} 0 & 1 \\ 1 & 0 \end{bmatrix} \otimes \begin{bmatrix} 0 & -i \\ i & 0 \end{bmatrix}$$

$$= \begin{bmatrix} 0 \begin{bmatrix} 0 & -i \\ i & 0 \end{bmatrix} & 1 \begin{bmatrix} 0 & -i \\ i & 0 \end{bmatrix} \\ 1 \begin{bmatrix} 0 & -i \\ i & 0 \end{bmatrix} & 0 \begin{bmatrix} 0 & -i \\ i & 0 \end{bmatrix} \end{bmatrix}$$

$$= \begin{bmatrix} 0 & 0 & 0 & -i \\ 0 & 0 & i & 0 \\ 0 & -i & 0 & 0 \\ i & 0 & 0 & 0 \end{bmatrix}$$

The following are some of the properties of the tensor product of linear operators A and B.

- Complex conjugation of $A \otimes B$ is given by $(A \otimes B)^* = A^* \otimes B^*$
- Transpose of $A \otimes B$ is given by $(A \otimes B)^T = A^T \otimes B^T$
- Complex conjugation or adjoint of $A \otimes B$ is given by $(A \otimes B)^\dagger = A^\dagger \otimes B^\dagger$

For a one-qubit system, the basis vectors are $|0\rangle = [1\ 0]^T$ and $|1\rangle = [0\ 1]^T$. For a two-qubit system, the basis vectors are the tensor product of the individual qubit's basis states (i.e., $|i\rangle \otimes |j\rangle$, where i denotes the basis state of qubit 1, and j represents the basis state for qubit 2). This produces four combinations: $|0\rangle \otimes |0\rangle$, $|0\rangle \otimes |1\rangle$, $|1\rangle \otimes |0\rangle$, and $|1\rangle \otimes |1\rangle$. The column vector representation can be derived by expanding the tensor product: $|1\rangle \otimes |0\rangle = [0\ 1]^T \otimes [1\ 0]^T = [0\ 0\ 1\ 0]^T$. In general, for a n-qubit system, there would 2^n basis state vectors of the form $|k_o\rangle \otimes |k_1\rangle \otimes |k_2\rangle ... \otimes |k_{n-1}\rangle$, where k_i stands for the basis vector of the i-th qubit.

Eigenvalues and Eigenvectors

Observables, represented by linear Hermitian operators, have *real* expectation values since every *measured* value of an observable A *must be real*. As a rule, the allowed values of observable A are found by solving an eigenvalue equation for operator A. For example, the eigenvalue equation for energy is the time-independent Schrödinger equation.

$$\mathcal{H}\psi_i(x) = E_i\psi_i(x)$$

where, $\mathcal{H}$ is the Hamiltonian operator, and eigenvalues E_i are the possible energies of the system. More generally, we can write the *eigenvalue equation* for any observable A in the form of

$$A\psi_i(x) = \lambda_i\psi_i(x)$$

or, in the language of *ket*-vectors in the form of

$$A|\psi\rangle_i = \lambda_i|\psi\rangle_i$$

The functions $\psi_i(x)$ are called the *eigenfunctions* of operator A, and the corresponding vectors $|\psi_i\rangle$ are called *eigenvectors*. The numbers λ_i are the eigenvalues, which are interpreted as the possible values of observable A. Since A is an observable, operator A is Hermitian.

The *eigenvector* of a linear operator A on a vector space V is a vector $|v\rangle$ such that $A|v\rangle = \lambda|v\rangle$. Here λ is the eigenvalue, and $|v\rangle$ is the corresponding eigenvector.

The eigenvalues of an operator can be found by solving for its characteristic equation, $\det|A - \lambda I| = 0$. The characteristic function corresponding to the characteristic equation of the linear operator is defined as $c(\lambda) = \det|A - \lambda I|$.

The Fourier Transform (also known as Discrete Fourier Transform)

Any general periodic signal has the automatic property $f(t) = f\left(\dfrac{2\pi t}{T}\right)$, where T is the period of the signal. The 2π is used because the trigonometric functions are good examples of periodicity. The complexity of $f(t)$ is irrelevant if it repeats itself consistently.

Note that t is an arbitrary variable (and not necessarily time). What follows is applicable provided the variable has *functional repetition* in some way with a period T. Thus, spatial repetition is another important variable to which we may apply the theory.

Fourier discovered that such a complex signal could be decomposed into an infinite series made up of cosine and sine terms and many coefficients that can be readily determined.

$$f(t) = \frac{1}{2}a_0 + \sum_{n=1}^{\infty} a_n \cos\left(\frac{2\pi nt}{T}\right) + \sum_{n=1}^{\infty} b_n \sin\left(\frac{2\pi nt}{T}\right)$$

We decomposed the original function $f(t)$ into a series of basis states. This immediately begs a question. Is this the only decomposition possible? The answer is no. Many more are possible.

The coefficients can be determined by integration.

$$a_n = \frac{2}{T} \int_{-T/2}^{T/2} f(t)\cos\left(\frac{2\pi nt}{T}\right) dt$$

$$b_n = \frac{2}{T} \int_{-T/2}^{T/2} f(t)\sin\left(\frac{2\pi nt}{T}\right) dt$$

By introducing complex notations, we can simplify the preceding.

$$f(t) = \sum_{n=-\infty}^{\infty} c_n \exp\left(-i\frac{2\pi nt}{T}\right)$$

$$c_n = \frac{1}{T} \int_{-T/2}^{T/2} f(t)\exp\left(i\frac{2\pi nt}{T}\right) dt$$

Here $c_0 = \frac{1}{2}a_0$, $c_n = \frac{1}{2}(a_n + ib_n)$, and $c_{-n} = \frac{1}{2}(a_n - ib_n)$.

There is an excellent pictorial example of how addition of cosine time function terms are Fourier transformed into coefficients on page 21 of the book *"The fast Fourier Transform and its Applications"* by *E. Oran Brigham* [289]. In this case, the author have showcased only $c_n = a_n/2$. Note that the centre line with the big arrow is to *mark the axis* only – it is **not** part of the coefficient display. Notice also that **two** coefficient lines appear for every frequency. The latter is related to the *Nyquist* sampling theorem and is also why the coefficient magnitudes are halved. Notice also the **spacing** of the coefficients to be an integral multiple of $f_0 = 1/T$ with the sign consistent with the input waveform.

It is important to stress that it is an *intrinsic* property that the coefficients c_n are discrete. This is one area that is sometimes confusing in some textbooks because they are drawn as continuous functions. It *is* possible to make them a continuous function by doing a simple trick and imagining that T is enormously large or tends to infinity.

Hence, with pure mathematics and substitutions, $s = \dfrac{2\pi n}{T}$, leading smoothly to $ds = \dfrac{2\pi}{T} dn$ as $T \to \infty$, and introducing the continuous function $F(s)$ to replace the discrete c_n, we get a *functional symmetry*.

$$f(t) = \frac{1}{2\pi} \int_{-\infty}^{+\infty} F(s) \exp(-ist)\, ds$$

$$F(s) = \int_{-\infty}^{+\infty} f(t) \exp(ist)\, dt$$

(Reminder: t and s are *arbitrary variables!*)

It is to be noted that 's' has units of frequency and so we talk of transforming the repetitive time function into the frequency domain. What has happened is an analysis of time behaviour into its corresponding frequency components.

APPENDIX B

Buzzwords in Quantum Tech

This glossary covers some common terms in quantum machine learning (adapted from [290]).

Barren plateaus: Areas in the cost landscape where the gradient of a parameterized circuit disappears. The gradient variance at these points is also close to zero in all directions. Hence, in the future, depending on developments, this may cause many variational algorithms to become redundant.

Circuit ansatz: An ansatz is a basic architecture of a circuit (i.e., a set of gates that act on specific subsystems). The architecture defines which algorithms a variational circuit can implement by fixing the trainable parameters. A circuit ansatz is analogous to the architecture of a neural network.

Hybrid computation: A computation that includes classical and quantum subroutines executed on different devices.

Parameter-shift rule: A recipe for estimating the gradients of quantum circuits.

Quantum Approximate Optimization Algorithm (QAOA): A hybrid variational algorithm that finds approximate solutions for combinatorial optimization problems and characterized by a circuit ansatz featuring two alternating parameterized components.

© Santanu Ganguly 2021
S. Ganguly, *Quantum Machine Learning: An Applied Approach*, https://doi.org/10.1007/978-1-4842-7098-1

Quantum convolutional neural network: A quantum neural network that mirrors a classical convolutional neural network structure. Characterized by alternating convolutional layers and pooling layers affected by performing quantum measurements.

Quantum differentiable programming: The paradigm of making quantum algorithms differentiable and thereby trainable. See also quantum gradient.

Quantum embedding: A representation of classical data as a quantum state.

Quantum feature map: The mathematical map that embeds classical data into a quantum state. Usually executed by a variational quantum circuit whose parameters depend on the input data.

Quantum gradient: The derivative of a quantum computation concerning the parameters of a circuit.

Quantum machine learning: A research area that explores ideas at the intersection of machine learning and quantum computing.

Quantum neural network: A term with many different meanings, usually referring to a generalization of artificial neural networks to quantum information processing. Also, increasingly refers to variational circuits in the context of quantum machine learning.

Quantum node: A quantum computation executed as part of a larger hybrid computation.

Quanvolutional neural network: A hybrid classical-quantum model in which layers of variational quantum circuits augment classical CNNs.

Variational circuit: Quantum algorithms that depend on tunable parameters and can therefore be optimized.

Variational Quantum Classifier (VQC): A supervised learning algorithm in which variational circuits (QNNs) are trained to perform classification tasks.

Variational Quantum Eigensolver (VQE): A variational algorithm for finding the ground-state energy of a quantum system. This hybrid algorithm involves incorporating measurement results obtained from a quantum computer running a series of variational circuits into a classical optimization routine to find optimal variational parameters.

Variational Quantum Linear Solver (VQLS): An algorithm for solving systems of linear equations on quantum computers. Based on short variational circuits, it is amenable to running on near-term quantum hardware.

References

[1] John Preskill's lectures: http://www.theory.caltech.edu/people/preskill/ph229/

[2] Ramamurti Shankar. *Principles of Quantum Mechanics.* Springer, 2013.

[3] Michael A. Nielsen and Isaac L. Chuang. *Quantum Computation and Quantum Information* (10th Anniversary edition). Cambridge University Press, 2011.

[4] W. K. Wootters and W. H. Zurek. "A single quantum cannot be cloned." *Nature,* vol. 299, 802–803, 1982.

[5] Maria Schuld and Francesco Petruccione. *Supervised Learning with Quantum Computers.* Springer, 2018.

[6] Quirk quantum circuits simulator: https://algassert.com/2016/05/22/quirk.html

[7] N. Zettili. *Quantum Mechanics: Concepts and Applications.* Wiley, 2009.

[8] A. Y. Khrennikov. *Ubiquitous Quantum Structure: From Psychology to Finance.* Springer, 2010.

[9] Sundaraja Sitharama Iyengar, Latesh K. J. Kumar, and Mario Mastriani. "Analysis of five techniques for the internal representation of a digital image inside a quantum processor." *Quantum Physics,* 2020.

[10] Li, Y., Zhou, R., Xu, R., Luo, J., and Hu, W. "A quantum deep convolutional neural network for image recognition." *Quantum Science and Technology,* October 2020.

© Santanu Ganguly 2021
S. Ganguly, *Quantum Machine Learning: An Applied Approach,* https://doi.org/10.1007/978-1-4842-7098-1

[11] Kaustav Basu, Chenyang Zhou, and Arunabha Sen, Victoria Horan Goliber. "A Novel Graph Analytic Approach to Monitor Terrorist Networks." 2018 IEEE Intl Conf on Parallel and Distributed Processing with Applications, Ubiquitous Computing and Communications, Big Data and Cloud Computing, Social Computing and Networking, Sustainable Computing and Communications (ISPA/IUCC/BDCloud/SocialCom/SustainCom), 1159–1166, March 20, 2019.

[12] Wisconsin Breast Cancer Dataset: `https://archive.ics.uci.edu/ml/datasets/Breast+Cancer+Wisconsin+(Diagnostic`

[13] Iris datset: `https://archive.ics.uci.edu/ml/datasets/iris`

[14] scikit-learn cross-val-score: `https://scikit-learn.org/stable/modules/generated/sklearn.model_selection.cross_val_score.html`

[15] Aurélien Géron. *Hands-on Machine Learning with Scikit-Learn and TensorFlow*. O'Reilly Media, 2017.

[16] Scikit-Learn Iris SVM: `https://scikit-learn.org/stable/auto_examples/svm/plot_iris_svc.html`

[17] Kevin P. Murphy. *Machine Learning: A Probabilistic Perspective*. Cambridge, MA: MIT Press, 2012.

[18] J. J. Hopfield (1982). "Neural networks and physical systems with emergent collective computational abilities." Proceedings of the National Academy of Sciences of the United States of America, vol. 79, 2554–8, April 1, 1982.

[19] Hongzhou Lin and Stefanie Jegelka. "ResNet with one-neuron hidden layers is a Universal Approximator." NeurIPS, 2018.

[20] D. Hubel and T. Wiesel. "Receptive fields and functional architecture of monkey striate cortex." *The Journal of Physiology*, vol. 195, 1968.

[21] `https://medium.com/@cdabakoglu/what-is-convolutional-neural-network-cnn-with-keras-cab447ad204c`

[22] Wikipedia for CNN: `https://en.wikipedia.org/wiki/Convolutional_neural_network`

[23] TesnorBoard: `https://www.tensorflow.org/tensorboard`

[24] Färber, Stephan Günnemann, Hans-Peter Kriegel, Peer Kröger,Emmanuel Müller, Erich Schubert, Thomas Seidl, and Arthur Zimek. "On Using Class-Labels in Evaluation of Clusterings." Proc. 1st International Workshop on Discovering, Summarizing and Using Multiple Clusterings, 2010.

[25] J. C. Dunn. "Well-Separated Clusters and Optimal Fuzzy Partitions." *Journal of Cybernetics*, vol. 4, 1974.

[26] ReLU: `https://en.wikipedia.org/wiki/Rectifier_(neural_networks)`

[27] TensorFlow Playground: `http://playground.tensorflow.org`

[28] Entropy of an Ideal Gas: `http://hyperphysics.phy-astr.gsu.edu/hbase/Therm/entropgas.html`

[29] Daniel M. Greenberger, Michael A. Horne, and Anton Zeilinge." Going Beyond Bell's Theorem," 2007.

[30] Cirq documentation: `https://quantumai.google/cirq`

[31] Anton Frisk Kockum. "Quantum optics with artificial atoms." Dissertation. Chalmers University of Technology, 2014. `http://publications.lib.chalmers.se/records/fulltext/206197/206197.pdf`

[32] A. Einstein, B. Podolsky, and N. Rosen. "Can Quantum-Mechanical Description of Physical Reality Be Considered Complete?". *Physical Review Journals Archive*. 47, 777–780, May 15, 1935.

[33] Yong Yu, Fei Ma, Xi-Yu Luo, et al. "Entanglement of two quantum memories via fibres over dozens of kilometres." *Nature*, vol. 578, 240–245, February 2020.

[34] Arthur K. Ekert. "Quantum cryptography based on Bell's theorem." *Physical Review Letters*, vol. 67, 661–663, August 5, 1991.

[35] Joe Lykken lecture notes: `https://home.fnal.gov/~lykken/`

[36] Aram W. Harrow, Avinatan Hassidim, and Seth Lloyd. "Quantum Algorithm for Linear Systems of Equations." *Physical Review Letters*, 2009.

[37] David Deutsch. "Quantum theory, the Church- Turing principle and the universal quantum computer." Proceedings of the Royal Society A. vol. 400, July 1985.

[38] Mark Fingerhuth, Tomáš Babej, and Peter Wittek. "Open source software in quantum computing." PLOS ONE, December 20, 2018. `https://doi.org/10.1371/journal.pone.0208561`, `https://journals.plos.org/plosone/article?id=10.1371/journal.pone.0208561`

[39] N. Cody Jones, Rodney Van Meter, Austin G. Fowler, Peter L. McMahon, Jungsang Kim, Thaddeus D. Ladd, and Yoshihisa Yamamoto. "Layered Architecture for Quantum Computing." *Physical Review X*, 2010.

[40] Edward Farhi, Jeffrey Goldstone, and Sam Gutmann. "A Quantum Approximate Optimization Algorithm." November 2014.

[41] Richard P. Feynman, Tony Hey (ed.), and Robin W. Allen (ed.). *Feynman Lectures on Computation*, CRC Press, 2000.

[42] Lov K. Grover. "A Quantum Approximate Optimization Algorithm." *Physical Review Letters*, vol. 79, 325, July 1997.

[43] John Watrous. "The Theory of Quantum Information." Cambridge University Press, 2018.

[44] Ethan Bernstein and Umesh Vazirani. "Quantum Complexity Theory." *SIAM Journal on Computing*, vol. 26, 1411–1473, 1997.

[45] John Watrous. "Quantum Computational Complexity." *Encyclopedia of Complexity and Systems Science*, pp. 7174–7201. Springer, 2009.

[46] Maria Schuld and Nathan Killoran. "Quantum machine learning in feature Hilbert spaces." *Physical Review Letters*, 2019.

[47] Vojtěch Havlíček, Antonio D. Córcoles, Kristan Temme, Aram W Harrow, Abhinav Kandala, Jerry M. Chow, and Jay M. Gambetta. "Supervised learning with quantum-enhanced feature spaces." *Nature*, vol. 567, 209–212, 2019.

[48] Frank Leymann, Johanna Barzen, Michael Falkenthal, Daniel Vietz, Benjamin Weder, and Karoline Wild. "Quantum in the cloud: application potentials and research opportunities." Proc. Closer 2020, 10th Int. Conf. on Cloud Computing and Services Science (Prague), pp. 7–9, 2020.

[49] Quantum Neural Network: `https://quantumcomputing.stackexchange.com/questions/11370/how-is-data-encoded-in-a-quantum-neural-network`

[50] Vivek V. Shende and Igor L. Markov. "Quantum circuits for incompletely specified two-qubit operators." *Quantum Information and Computation*, vol. 5, 49–57, 2004.

[51] Vivek V. Shende, S. S. Bullock, and I. L. Markov. "Synthesis of quantum-logic circuits." *IEEE Transactions on Computer-Aided Design of Integrated Circuits and Systems*, vol. 25, 1000–10, 2006.

[52] Schuld M and Killoran N (2019). Quantum machine learning in feature Hilbert spaces Phys. Rev. Lett. 122 040504

[53] John A. Cortese and Timothy M. Braje. "Loading classical data into a quantum computer." March 5, 2018.

[54] Kosuke Mitarai, Masahiro Kitagawa, and Keisuke Fujii. "Quantum analog-digital conversion." *Physical Review A*, vol. 99, January 2, 2019.

[55] Frank Leymann and Johanna Barzen. "The bitter truth about gate-based quantum algorithms in the NISQ era." *Quantum Science Technology*, vol. 5, September 1, 2020.

[56] K. M. Svore, A. V. Aho, A. W. Cross, I. Chuang, and I. L. Markov. "A layered software architecture for quantum computing design tools." *Computer*, vol. 39, 74–83, 2006.

[57] Vittorio Giovannetti, Seth Lloyd, and Lorenzo Maccone. "Quantum random access memory." *Physical Review Letters*, July 2008.

[58] Andrei N. Soklakov and Ruediger Schack. "Efficient state preparation for a register of quantum bits." *Physical Review A*, January 9, 2006.

[59] Maria Schuld, Alex Bocharov, Krysta M. Svore, and Nathan Wiebe. "Circuit-centric quantum classifiers." *Physical Review A*, March 6, 2020.

[60] Hartmut Neven, Vasil S. Denchev, Geordie Rose, William G. Macready. "Training a large-scale classifier with the quantum adiabatic algorithm." December 2009.

[61] Guang Hao Low, Theodore J. Yoder, and Isaac L. Chuang. "Quantum inference on Bayesian networks." *Physical Review A*. 89, June 13, 2014.

[62] Maris Ozols, Martin Roetteler, and Jérémie Roland. "Quantum rejection sampling." *ACM Transactions on Computation Theory* (TOCT), vol. 5, 2011.

[61] Rigetti Forest: `https://pyquil-docs.rigetti.com/en/stable/`

[62] Angel Rivas and Susana F. Huelga. *Open Quantum Systems: An Introduction*. Springer, 2012.

[63] MIT Open Courseware on Open Quantum Systems: `https://ocw.mit.edu/courses/nuclear-engineering/22-51-quantum-theory-of-radiation-interactions-fall-2012/lecture-notes/MIT22_51F12_Ch8.pdf`

[64] Boltzmann Distribution: `https://en.wikipedia.org/wiki/Boltzmann_distribution`

[65] Karl Pearson. (1901)."On Lines and Planes of Closest Fit to Systems of Points in Space." https://doi.org/ 10.1080/14786440109462720.

[66] Seth Lloyd, Masoud Mohseni, and Patrick Rebentrost. "Quantum principal component analysis." *Nature Physics*, vol. 10, 631–633, 2014.

[67] Ryan LaRose, Arkin Tikku, Étude O'Neel-Judy, Lukasz Cincio, and Patrick J. Coles. "Variational quantum state diagonalization." *npj Quantum Information*, vol. 5, 57, 2019.

[68] John Preskill. "Quantum Computing in the NISQ era and beyond." *Quantum*, 2018.

[69] Edward Farhi, Jeffrey Goldstone, and Sam Gutmann, (2014) "A quantum approximate optimization algorithm," arXiv preprint arXiv:1411.4028, accessed on on 15th of March 2021.

[70] G. G. Guerreschi and M. Smelyanskiy. "Practical optimization for hybrid quantum-classical algorithms." 2017.

[71] Robert S. Smith, Michael J. Curtis, William J. Zeng. "A Practical Quantum Instruction Set Architecture." 2016.

[72] Rigetti pyQuil: github: http://github.com/rigetti/pyquil

[73] Rigetti QVM: https://github.com/rigetti/qvm

[74] Charles H. Bennett, Herbert J. Bernstein, Sandu Popescu, and Benjamin Schumacher. "Concentrating partial entanglement by local operations." *Physical Review A*, vol. 53, 2046–2052. April 1, 1996.

[75] Phillip Kaye and Michele Mosca. Quantum networks for concentrating entanglement. *Journal of Physics A*, vol. 34, 6939–6948, August 24, 2001.

[76] J. Bell. "On the Einstein-Podolsky-Rosen paradox." *Physics*, vol. 1, 195–200, 1964.

[77] Fabian Hassler. Majorana Qubits, "Quantum Information Processing. Lecture Notes of the 44th IFF Spring School 2013."

[78] Jian Pan, Yudong Cao, Xiwei Yao, Zhaokai Li, Chenyong Ju, Xinhua Peng, Sabre Kais, Jiangfeng Du. "Experimental realization of quantum algorithm for solving linear systems of equations." *Physical Review A*, vol. 89, February 2013.

[79] Maria Schuld, Ilya Sinayskiy, and Francesco Petruccione. "Prediction by linear regression on a quantum computer." *Physical Review A*, vol. 94, August 30, 2016.

[80] P. McCullagh and John A. Nelder. *Generalized Linear Models*. London: Chapman and Hall/CRC, 1989.

[81] Frank Nielsen and Vincent Garcia. "Statistical exponential families: A digest with flash cards." November 2009.

[82] Colleen M. Farrelly. "KNN Ensembles for Tweedie Regression: The Power of Multiscale Neighborhoods." July 2017.

[83] Colleen M. Farrelly. "Topology and Geometry in Machine Learning for Logistic Regression." 2017.

[84] Nathan Killoran, Thomas R. Bromley, Juan Miguel Arrazola, Maria Schuld, Nicolás Quesada, Seth Lloyd. "Continuous-variable quantum neural networks." June 2018.

[85] Seth Lloyd, Masoud Mohseni, and Patrick Rebentrost. "Quantum algorithms for supervised and unsupervised machine learning." 2013.

[86] Seth Lloyd, Masoud Mohseni, and Patrick Rebentrost. "Quantum principal component analysis." *Nature Physics*, vol. 10, 631–633, 2014.

[87] Harry Buhrman, Richard Cleve, John Watrous, and Ronald de Wolf. "Quantum fingerprinting." *Physical Review Letters*, vol. 87, 2001.

[88] Steph Foulds, Viv Kendon, and Tim Spiller. "The controlled SWAP test for determining quantum entanglement." 2020.

[89] Vittorio Giovannetti, Seth Lloyd, and Lorenzo Maccone. "Quantum-Enhanced Measurements: Beating the Standard Quantum Limit." *Science*, vol. 306, 1330–1336, 2004.

[90] Mach-Zehnder Interferometer: https://en.wikipedia.org/wiki/Mach%E2%80%93Zehnder_interferometer

[91] Central Limit Theorem: https://en.wikipedia.org/wiki/Central_limit_theorem

[92] Patrick Rebentrost, Masoud Mohseni, and Seth Lloyd. "Quantum support vector machine for big feature and big data classification." *Physical Review Letters*, vol. 113, 2014.

[93] Bernhard Schölkopf and Alexander J. Smola. *Learning with Kernels: Support Vector Machines, Regularization, Optimization, and Beyond*. MIT Press, 2001.

[94] Bernhard Schölkopf, Ralf Herbrich, and Alexander J. Smola. *A Generalized Representer Theorem*. COLT/EuroCOLT, 2001.

[95] Maria Schuld, Mark Fingerhuth, Francesco Petruccione. "Quantum machine learning with small-scale devices: Implementing a distance-based classifier with a quantum interference circuit." 2017.

[96] E. Miles Stoudenmire and David J. Schwab. "Supervised Learning with Tensor Networks." NIPS, 2016.

[97] Havlícek, V., Córcoles, A., Temme, K., Harrow, A., Kandala, A., Chow, J., and Gambetta, J.M. (2019). Supervised learning with quantum-enhanced feature spaces. Nature, 567, 209-212.

[98] S.P. Lloyd. "Least squares quantization in PCM." *IEEE Transactions on Information Theory*, vol. 28, 129–136, March 1982. https://ieeexplore.ieee.org/document/1056489.

[99] Christopher D. Manning, Prabhakar Raghavan and Hinrich Schütze. *Introduction to Information Retrieval*. Cambridge University Press, 2008

[100] Lloyd, S., Mohseni, M., and Rebentrost, P. (2013). Quantum algorithms for supervised and unsupervised machine learning. arXiv: Quantum Physics.

[101] Philipp Niemann, Rhitam Datta, and Robert Wille. "Logic Synthesis for Quantum State Generation." 2016 IEEE 46th International Symposium on Multiple-Valued Logic (ISMVL), 247–252, May 2016.

[102] Esma Aïmeur, Gilles Brassard, and Sébastien Gambs. "Quantum clustering algorithms." ICML, 2007.

[103] Wikipedia for QUBO: https://en.wikipedia.org/wiki/ Quadratic_unconstrained_binary_optimization

[104] Martin Anthony, Endre Boros, Yves Crama, and Aritanan Gruber. "Quadratic reformulations of nonlinear binary optimization problems." *Mathematical Programming*, vol. 162, 115–144, 2017.

[105] Gary Kochenberger, Jin-Kao Hao, Fred Glover, Mark Lewis, Zhipeng Lü, Haibo Wang, and Yang Wang. "The unconstrained binary quadratic programming problem: a survey." *Journal of Combinatorial Optimization*, vol. 28, 58–81, 2014.

[109] Gary Kochenberger and Fred Glover. *A Unified Framework for Modeling and Solving Combinatorial Optimization Problems: A Tutorial*. Springer, 2016.

[110] Andrew Lucas. "Ising formulations of many NP problems." February 2014. https://doi.org/10.3389/fphy.2014.00005

[111] Wikipedia for Ising Model: https://en.wikipedia.org/ wiki/Ising_model

[112] Michael Booth, Steven P. Reinhardt, and Aidan Roy. "Partitioning Optimization Problems for Hybrid Classical / Quantum Execution." D-Wave Technical Report Series." October 18, 2017.

[113] Zhikuan Zhao, Alejandro Pozas-Kerstjens, Patrick Rebentrost, and Peter Wittek. "Bayesian deep learning on a quantum computer." *Quantum Machine Intelligence*, vol. 1, 41–51, 2018.

[114] Pan, J., Cao, Y., Yao, X., Li, Z., Ju, C., Peng, X., Kais, S., and Du, J. (2014). Experimental realization of quantum algorithm for solving linear systems of equations. Physical Review A, 89, 022313.

[115] A. Harrow. "Quantum algorithm for solving linear systems of equations." *Bulletin of the American Physical Society*, 2010.

[116] Peter W. Shor. "Polynomial-Time Algorithms for Prime Factorization and Discrete Logarithms on a Quantum Computer." SIAM Journal on Computing, October 1997. `https://doi.org/10.1137/S0097539795293172`.

[117] Nathan Wiebe, Daniel Braun, and Seth Lloyd. "Quantum algorithm for data fitting." *Physical Review Letters*, vol. 109, August 2012.

[118] Danial Dervovic, Mark Herbster, Peter Mountney, Simone Severini, Naïri Usher, and Leonard Wossnig. "Quantum linear systems algorithms: a primer." February 2018.

[119] HHL with Cirq: `https://github.com/quantumlib/Cirq/blob/master/examples/hhl.py`

[120] D. Wecker, M. Hastings, and M. Troyer. "Progress towards practical quantum variational algorithms." *Physical Review A*, vol. 92, 2015.

[121] Alberto Peruzzo, Jarrod McClean, Peter Shadbolt, Man-Hong Yung, Xiao-Qi Zhou, Peter J. Love, Alán Aspuru-Guzik, and Jeremy L. O'Brien. "A variational eigenvalue solver on a photonic quantum processor." *Nature Communications*, vol. 5, 2014.

[122] M. Cerezo, Andrew Arrasmith, Ryan Babbush, Simon C. Benjamin, Suguru Endo, Keisuke Fujii, Jarrod R. McClean, Kosuke Mitarai, Xiao Yuan, Lukasz Cincio, and Patrick J. Coles. "Variational Quantum Algorithms." 2020.

[123] Hubbard Model: `https://en.wikipedia.org/wiki/Hubbard_model`

[124] Stuart Hadfield, Zhihui Wang, Bryan O'Gorman, Eleanor G. Rieffel, Davide Venturelli, and Rupak Biswas. "From the Quantum Approximate Optimization Algorithm to a Quantum Alternating Operator Ansatz." *Algorithms*, 12, vol. 34, 2019.

[125] Seth Lloyd. "Quantum approximate optimization is computationally universal." 2018.

[126] Lie Product Formula: `https://en.wikipedia.org/wiki/Lie_product_formula`

[127] Zhihui Wang, Nicholas C. Rubin, Jason M. Dominy, and Eleanor G. Rieffel. "X Y mixers: Analytical and numerical results for the quantum alternating operator ansatz." *Physical Review A*, vol. 101, January 2020.

[128] Nikolaj Moll, Panagiotis Barkoutsos, Lev S. Bishop, Jerry M. Chow, Andrew Cross, Daniel J. Egger, Stefan Filipp, Andreas Fuhrer, Jay M. Gambetta, Marc Ganzhorn, Abhinav Kandala, Antonio Mezzacapo, Peter Müller, Walter Riess, Gian Salis, John Smolin, Ivano Tavernelli, and Kristan Temme. "Quantum optimization using variational algorithms on near-term quantum devices." 2017.

[129] Zhaokai Li, Xiaomei Liu, Nanyang Xu, and Jiangfeng Du. "Experimental realization of a quantum support vector machine." *Physical Review Letters*, vol. 114, April 8, 2015.

[130] Iordanis Kerenidis, Anupam Prakash, and Dániel Szilágyi. "Quantum Algorithms for Second-Order Cone Programming and Support Vector Machines." 2019.

[131] A.K. Bishwas, Ashish Mani, and V. Palade. "Big Data Quantum Support Vector Clustering." 2018.

[132 Seyran Saeedi and Tom Arodz. "Quantum Sparse Support Vector Machines." 2019.

[133] Davide Anguita, Sandro Ridella, Fabio Rivieccio, and Rodolfo Zunino. "Quantum optimization for training support vector machines." *Neural networks: the official journal of the International Neural Network Society*, vol. 16, 763–70.

[134] Qiskit Community. Qiskit aqua-tutorials. `https://github.com/Qiskit/aqua-tutorials/blob/master/artificial_intelligence/svm_qkernel.ipynb`.

[135] Havlíček, V., Córcoles, A., Temme, K., Harrow, A., Kandala, A., Chow, J., and Gambetta, J.M. (2019). Supervised learning with quantum-enhanced feature spaces. Nature, 567, 209–212.

[136] Qiskit Community. QSVVM. `https://qiskit.org/documentation/tutorials/machine_learning/01_qsvm_classification.html`

[137] Jacob Biamonte, Peter Wittek, Nicola Pancotti, Patrick Rebentrost, Nathan Wiebe, Seth Lloyd. "Quantum machine learning." *Nature*, vol. 549, 195–202, 2016.

[138] Sau Lan Wu, Jay Chan, Wen Guan, Shaojun Sun, Alex Wang, Chen Zhou, Miron Livny, Federico Carminati, Alberto Di Meglio, Andy C. Y. Li, Joseph Lykken, Panagiotis Spentzouris, Samuel Yen-Chi Chen, Shinjae Yoo, and Tzu-Chieh Wei. "Application of Quantum Machine Learning to High Energy Physics Analysis at LHC using IBM Quantum Computer Simulators and IBM Quantum Computer Hardware."

[139] Mohammad H. Amin, Evgeny Andriyash, Jason Rolfe, Bohdan Kulchytskyy, and Roger Melko. "Quantum Boltzmann machine." 2016.

[140] Jacob D. Biamonte and Peter J. Love. "Realizable Hamiltonians for Universal Adiabatic Quantum Computers." *Physical Review A*, vol. 78, 2008.

[141] D-Wave Systems GitHub: `https://github.com/dwavesystems`

[142] D-Wave Leap resources: `https://cloud.dwavesys.com/leap/resources`

[143] D-Wave Website: `https://www.dwavesys.com/quantum-computing`

[144] D-Wave Quantum Computing Primer: `https://www.dwavesys.com/tutorials/background-reading-series/quantum-computing-primer#h2-8`

[145] PennyLane documentation: `https://pennylane.readthedocs.io/en/stable/introduction/pennylane.html`

[146] Ville Bergholm, Josh A. Izaac, Maria Schuld, Christian Gogolin, and Nathan Killoran. "PennyLane: Automatic differentiation of hybrid quantum-classical computations." 2018.

[147] Wecker, D., Hastings, M., and Troyer, M. (2015). Progress towards practical quantum variational algorithms. Physical Review A, 92, 042303.

[148] Google Cirq VQE: `https://quantumai.google/cirq/tutorials/variational_algorithm`

[149] Nelder-Mead: `https://en.wikipedia.org/wiki/Nelder%E2%80%93Mead_method`

[150] Rigetti grove documentation: Variational-Quantum-Eigensolver (VQE). Retrieved from `https://grove-docs.readthedocs.io/en/latest/vqe.html`

[151] Phillip W. K. Jensen and Sumner Alperin-Lea. "Quantum computing for quantum chemistry." 2018.

[152] Teng Bian, Daniel Murphy, Rongxin Xia, Ammar Daskin, and Sabre Kais. "Quantum computing methods for electronic states of the water molecule." *Molecular Physics*, vol. 117, 2069–2082, 2019.

[153] Sanjeev Arora and Boaz Barak. *Computational Complexity: A Modern Approach*. Cambridge University Press, 2009.

[154] D-Wave Systems' dimod: `https://github.com/dwavesystems/dimod`

[155] D-Wave unit cell Chimera Graph: `https://docs.dwavesys.com/docs/latest/c_gs_4.html`

[156] Richard M. Karp. *Reducibility Among Combinatorial Problems. 50 Years of Integer Programming*. Springer, 2009.

[157] Francisco Barahona, Martin Grötschel, Michael Jünger, and Gerhard Reinelt. "An Application of Combinatorial Optimization to Statistical Physics and Circuit Layout Design." *Operations Research*, vol. 36, 493–513, 1988.

[158] G. G. Guerreschi and A. Y. Matsuura. "QAOA for Max-Cut requires hundreds of qubits for quantum speed-up." *Scientific Reports*, vol. 9, 2019.

[159] J. S. Otterbach, R. Manenti, N. Alidoust, A. Bestwick, M. Block, B. Bloom, S. Caldwell, N. Didier, E. Schuyler Fried, S. Hong, P. Karalekas, C. B. Osborn, A. Papageorge, E. C. Peterson, G. Prawiroatmodjo, N. Rubin, Colm A. Ryan, D. Scarabelli, M. Scheer, E. A. Sete, P. Sivarajah, Robert S. Smith, A. Staley, N. Tezak, W. J. Zeng, A. Hudson, Blake R. Johnson, M. Reagor, M. P. da Silva, and C. Rigetti. "Unsupervised Machine Learning on a Hybrid Quantum Computer." 2017.

[160] Lagrange Multiplier: https://en.wikipedia.org/wiki/Lagrange_multiplier#:~:text=In%20mathematical%20optimization%2C%20the%20method,chosen%20values%20of%20the%20variables).

[161] D-Wave Qbsolve: https://github.com/dwavesystems/qbsolv

[162] The Travelling Salesman Problem: https://en.wikipedia.org/wiki/Travelling_salesman_problem#cite_note-6

[163] Yang Wang, Zhipeng Lü, Fred Glover, and Jin-Kao Hao. "A Multilevel Algorithm for Large Unconstrained Binary Quadratic Optimization." CPAIOR, 2012.

[164] Steve Adachi. "Application of Quantum Annealing to Training of Deep Neural Networks." Lockheed Martin Corp., 2016.

[165] Misha Denil and Nando de Freitas. "Toward the Implementation of a Quantum RBM." Neural Information Processing Systems (NIPS) Conf. on Deep Learning and Unsupervised Feature Learning Workshop, vol. 5, 2011.

[166] Vincent Dumoulin, Ian J. Goodfellow, Aaron Courville, and Yoshua Bengio. "On the Challenges of Physical Implementations of RBMs." AAAI, 2014.

[167] Classical Boltzmann Machine: https://en.wikipedia.org/wiki/Boltzmann_machine

[168] David H. Ackley, Geoffrey E. Hinton, and Terrence J. Sejnowski. "A Learning Algorithm for Boltzmann Machines." *Cognitive Science*, vol. 9, 147–169, January 1985.

[169] Ke-Lin Du and M. N. S. Swamy. *Neural Networks and Statistical Learning.* Springer London, 2019.

[170] Ludwig Boltzmann. *Ueber die Natur der Gasmoleküle.* 1877. https://doi.org/10.1002/andp.18772360120

[171] J. W. Gibbs. *Elementary Principles in Statistical Mechanics.* Cambridge University Press, 1902.

[172] PennyLane Function fitting for a QNN: https://pennylane.readthedocs.io/en/user-docs-refactor/tutorials/pennylane_quantum_neural_net.html

[173] PennyLane AdamOptimizer: https://pennylane.readthedocs.io/en/stable/code/api/pennylane.AdamOptimizer.html

[174] Fock state: https://en.wikipedia.org/wiki/Fock_state#:~:text=In%20quantum%20mechanics%2C%20a%20Fock,the%20Soviet%20physicist%20Vladimir%20Fock.

[175] PennyLane documentation for StronglyEntanglingLayers: https://pennylane.readthedocs.io/en/stable/code/api/pennylane.templates.layers.StronglyEntanglingLayers.html

[176] Jörg Behler and Michele Parrinello. "Generalized neural-network representation of high-dimensional potential-energy surfaces." *Physical Review Letters*, vol. 98, 2007.

[177] Albert P. Bartók, Mike C. Payne, Risi Kondor, and Gábor Csányi. "Gaussian approximation potentials: the accuracy of quantum mechanics, without the electrons." *Physical Review Letters*, vol. 104, April 2010.

[178] Matthias Rupp, Alexandre Tkatchenko, Klaus-Robert Müller, and O. Anatole von Lilienfeld. "Fast and accurate modeling of molecular atomization energies with machine learning." *Physical Review Letters*, vol. 108, January 31, 2012.

[179] Zhenwei Li, James R. Kermode, and Alessandro De Vita. "Molecular dynamics with on-the-fly machine learning of quantum-mechanical forces." *Physical Review Letters*, vol. 114, March 6, 2015.

[180] Jörg Behler. "Constructing high-dimensional neural network potentials: A tutorial review." *International Journal of Quantum Chemistry*, vol. 115, 1032–1050, March 6, 2015.

[181] Harper R. Grimsley, Sophia E. Economou, Edwin Barnesm and Nicholas J. Mayhall. "An adaptive variational algorithm for exact molecular simulations on a quantum computer." *Nature Communications*, vol. 10, 2019.

[182] Alán Aspuru-Guzik, Anthony D. Dutoi, Peter J. Love, and Martin Head-Gordon. "Simulated Quantum Computation of Molecular Energies." *Science*, vol. 309, 1704 – 1707, 2005.

[183] Sam McArdle, Suguru Endo, Alan Aspuru-Guzik, Simon Benjamin, and Xiao Yuan. "Quantum computational chemistry." *Reviews of Modern Physics*, vol. 92, 2020.

[184] Yudong Cao, Jonathan Romero, Jonathan P. Olson, Matthias Degroote, Peter D. Johnson, Mária Kieferová, Ian D. Kivlichan, Tim Menke, Borja Peropadre, Nicolas P. D. Sawaya, Sukin Sim, Libor Veis, and Alán Aspuru-Guzik. "Quantum Chemistry in the Age of Quantum Computing." *Chemical Reviews*, August 30, 2019.

[185] Edward Farhi and Hartmut Neven. "Classification with Quantum Neural Networks on Near Term Processors." 2018.

[186] Marcello Benedetti, Erika Lloyd, Stefan Sack, Mattia Fiorentini. "Parameterized quantum circuits as machine learning models." 2019.

[187] Ahmed Younes and Julian F. Miller. "Representation of Boolean quantum circuits as reed–Muller expansions." *International Journal of Electronics*, vol. 91, 431–444, 2006.

[188] TensorFlow Quantum documentation on QNN: `https://www.tensorflow.org/quantum/tutorials/mnist`

[189] TensorFlow Quantum: `https://www.tensorflow.org/quantum`

[190] Martin Arjovsky, Amar Shah, and Yoshua Bengio. "Unitary Evolution Recurrent Neural Networks." 2016.

[191] Michael Broughton, Guillaume Verdon, Trevor McCourt, Antonio J. Martinez, Jae Hyeon Yoo, Sergei V. Isakov, Philip Massey, Murphy Yuezhen Niu, Ramin Halavati, Evan Peters, Martin Leib, Andrea Skolik, Michael Streif, David Von Dollen, Jarrod R. McClean, Sergio Boixo, Dave Bacon, Alan K. Ho, Hartmut Neven, Masoud Mohseni. "TensorFlow Quantum: A Software Framework for Quantum Machine Learning." 2020.

[192] Seunghyeok Oh, Jaeho Choi, and Joongheon Kim. "A Tutorial on Quantum Convolutional Neural Networks (QCNN)." 2020 International Conference on Information and Communication Technology Convergence (ICTC), 236–239, 2020.

[193] Iris Cong, Soonwon Choi, and Mikhail D. Lukin. "Quantum convolutional neural networks." *Nature Physics*, 1-6, 2018.

[194] Hans J. Briegel and Gemma De las Cuevas. "Projective simulation for artificial intelligence." *Scientific Reports*, vol. 2, 2012.

[195] B. L. Douglas and J. B. Wang. "Efficient quantum circuit implementation of quantum walks." *Physical Review A*, vol. 79, May 27, 2009.

[196] Andrew M. Childs, Richard Cleve, Enrico Deotto, Edward Farhi, Sam Gutmann, Daniel A. Spielman. "Exponential algorithmic speedup by a quantum walk." STOC, 2003.

[197] Xiaogang Qiang, Thomas Loke, Ashley Montanaro, Kanin Aungskunsiri, Xiao-Qi Zhou, Jeremy O'Brien, Jingbo Wang, and Jonathan Matthews. "Efficient quantum walk on a quantum processor." *Nature Communications*, vol. 7, 2016.

[198] Edward Farhi and Sam Gutmann. "Quantum computation and decision trees." *Physical Review A*, vol. 58, 915–928, August 1998.

[199] Julia Kempe. "Quantum random walks: An introductory overview." *Contemporary Physics*, 44, 307–327, 2003.

[200] Andrew M. Childs, David Gosset, and Zak Webb. "Universal Computation by Multiparticle Quantum Walk." *Science*, vol. 339, 791–794, 2013.

[201] Andrew M. Childs and Jeffrey Goldstone. "Spatial search by quantum walk." *Physical Review A*, vol. 70, August 23, 2004.

[202] B. L. Douglas and J. B. Wang. "Classical approach to the graph isomorphism problem using quantum walks." *Journal of Physics A*, vol. 41, 2008.

[203] John King Gamble, Mark Friesen, Dong Zhou, Robert Joynt, and S. N. Coppersmith. "Two-particle quantum walks applied to the graph isomorphism problem." *Physical Review A*, vol. 81, May 13, 2010.

[204] Scott D. Berry and Jingbo B. Wang. "Two-particle quantum walks: Entanglement and graph isomorphism testing." *Physical Review A*, vol. 83, April 2011.

[205] Scott D. Berry and Jingbo B. Wang. "Quantum-walk-based search and centrality." *Physical Review A*, vol. 82, 1–12, October 2010.

[206] Eduardo Sánchez-Burillo, Jordi Duch, Jesús Gómez-Gardeñes, and David Zueco. "Quantum Navigation and Ranking in Complex Networks." *Scientific Reports*, vol. 2, 2012.

[207] Seth Lloyd. "Universal Quantum Simulators." *Science*, 273, 1073– 1078, August 23, 1996.

[208] Dominic Berry and Andrew M. Childs. "Black-box Hamiltonian simulation and unitary implementation." *Quantum Information and Computation*, vol. 12, 29–62, January 2012.

[209] Andreas Schreiber, Aurel Gabris, Peter P. Rohde, Kaisa Laiho, Martin Stefanak, Vaclav Potocek, Craig Hamilton, Igor Jex, and Christine Silberhorn. "A 2D Quantum Walk Simulation of Two-Particle Dynamics." *Science*, vol. 336, 55–58, 2012.

[210] Gregory S. Engel, Tessa R. Calhoun, Elizabeth L. Read, Tae-Kyu Ahn, Tomáš Mančal, Yuan-Chung Cheng, Robert E. Blankenship, and Graham R. Fleming. "Evidence for wavelike energy transfer through quantum coherence in photosynthetic systems." *Nature*, vol. 446, 782–786, 2007.

[211] Patrick Rebentrost, Masoud Mohseni, Ivan Kassal, Seth Lloyd, Alán Aspuru-Guzik. "Environment-assisted quantum transport." *New Journal of Physics*, vol. 11, 2009.

[212] Jiangfeng Du, Hui Li, Xiaodong Xu, Mingjun Shi, Jihui Wu, Xianyi Zhou, and Rongdian Han. "Experimental implementation of the quantum random-walk algorithm." *Physical Review A*, vol. 67, April 22, 2003.

[213] C. A. Ryan, M. Laforest, J. C. Boileau, and R. Laflamme. "Experimental implementation of a discrete-time quantum random walk on an NMR quantum-information processor." *Physical Review A*, vol. 72, December 9, 2005.

[214] Binh Do, Michael L. Stohler, Sunder Balasubramanian, Daniel S. Elliott, Christopher Eash, Ephraim Fischbach, Michael A. Fischbach, Arthur Mills, and Benjamin Zwickl. "Experimental

realization of a quantum quincunx by use of linear optical elements." *Journal of The Optical Society of America B-optical Physics*, vol. 22, 499–504, 2005.

[215] Qin-Qin Wang, Xiao-Ye Xu, Wei-Wei Pan, Kai Sun, Jin-Shi Xu, Geng Chen, Yong-Jian Han, Chuan-Feng Li, and Guang-Can Guo. "Photons walking the line: a quantum walk with adjustable coin operations." *Physical Review Letters*, vol. 104, 2018.

[216] Peng Xue, Barry C. Sanders, and Dietrich Leibfried. "Quantum walk on a line for a trapped ion." *Physical Review Letters*, vol. 103, October 28, 2009.

[217] H. Schmitz, R. Matjeschk, Ch. Schneider, J. Glueckert, M. Enderlein, T. Huber, and T. Schaetz. "Quantum walk of a trapped ion in phase space." *Physical Review Letters*, vol. 103, August 28, 2009.

[218] F. Zähringer, G. Kirchmair, R. Gerritsma, E. Solano, R. Blatt, and C. F. Roos. "Realization of a quantum walk with one and two trapped ions." *Physical Review Letters*, vol. 104, March 9, 2010.

[219] Michal Karski, Leonid Förster, Jai-Min Choi, Andreas Steffen, Wolfgang Alt, Dieter Meschede, and Artur Widera. "Quantum Walk in Position Space with Single Optically Trapped Atoms." *Science*, vol. 325, 174– 177, 2009.

[220] Hagai B. Perets, Yoav Lahini, Francesca Pozzi, Marc Sorel, Roberto Morandotti, and Yaron Silberberg. Realization of quantum walks with negligible decoherence in waveguide lattices. *Physical Review Letters*, vol. 100, May 2, 2008.

[221] Jacques Carolan, Jasmin D. A. Meinecke, Peter J S Shadbolt, Nick J Russell, Nur Ismail, Kerstin Wörhoff, Terry Rudolph, Mark G. Thompson, Jeremy L. O'Brien, Jonathan C. F. Matthews, Anthony Laing. "On the experimental verification of quantum complexity in linear optics." *Nature Photonics*, vol. 8, 621–626, 2014.

[222] M. Szegedy. "Quantum speed-up of Markov chain based algorithms." 45th Annual IEEE Symposium on Foundations of Computer *Science*, 32–41, 2004.

[223] Renato Portugal. *Quantum Walks and Search Algorithms.* Springer, 2013.

[224] Andrew M. Childs and Nathan Wiebe. "Hamiltonian simulation using linear combinations of unitary operations." *Quantum Information and Computation*, vol. 12, 901–924, November 2012.

[225] Dominic W. Berry, Graeme Ahokas, Richard Cleve, and Barry C. Sanders. "Efficient Quantum Algorithms for Simulating Sparse Hamiltonians." *Communications in Mathematical Physics*, vol. 270, 359-371, 2007.

[226] Timjan Kalajdzievski, Christian Weedbrook, and Patrick Rebentrost. "Continuous-variable gate decomposition for the Bose-Hubbard model." *Physical Review A*, vol. 97, June 7, 2018.

[227] Tomasz Sowiński, Omjyoti Dutta, Philipp Hauke, Luca Tagliacozzo, and Maciej Lewenstein. "Dipolar molecules in optical lattices." *Physical Review Letters*, vol. 108, March 13, 2012.

[228] Hartmut Neven, Vasil S. Denchev, Geordie Rose, and William G. Macready. "Training a Large-Scale Classifier with the Quantum Adiabatic Algorithm." 2009.

[229] Hartmut Neven, Vasil S. Denchev, Geordie Rose, and William G. Macready. "QBoost: Large-Scale Classifier Training with Adiabatic Quantum Optimization." ACML, 2012.

[230] Hartmut Neven, Vasil S. Denchev, Geordie Rose, and William G. Macready. "Training a Binary Classifier with the Quantum Adiabatic Algorithm." 2008.

[231] Lior Rokach. *Ensemble Learning: Pattern Classification Using Ensemble Methods.* WSPC, 2019.

[232] Martin Sewell. "Ensemble Learning." January 20, 2011.
http://www.cs.ucl.ac.uk/fileadmin/UCL-CS/research/
Research_Notes/RN_11_02.pdf

[233] D. G. Cory, R. Laflamme, E. Knill, L. Viola, T.F. Havel,
N. Boulant, G. Boutis, E. Fortunato, S. Lloyd, R. Martinez,
C. Negrevergne, M. Pravia, Y. Sharf, G. Teklemariam,
Y.S. Weinstein, and W. H. Zurek. "NMR Based Quantum
Information Processing: Achievements and Prospects." *Protein
Science*, vol. 48, 875–907, 2000.

[234] Fei Yan, Abdullah M. Iliyasu, Z. Liu, A. S. Salama, F. Dong,
and K. Hirota. "Bloch Sphere-Based Representation for Quantum
Emotion Space." *Journal of Advanced Computational Intelligence
and Intelligent Informatics*, vol. 19, 134–142, 2015.

[235] Yang, Y., Xia, J., Jia, X., and Zhang, H. "Novel image
encryption/decryption based on quantum Fourier transform and
double phase encoding." *Quantum Information Processing*,
vol. 12, 3477–3493, 2013.

[236] Ri-Gui Zhou, Qian Wu, Man-Qun Zhang, and Chen-Yi Shen.
"Quantum Image Encryption and Decryption Algorithms Based
on Quantum Image Geometric Transformations." *International
Journal of Theoretical Physics*, vol. 52, 1802–1817, 213, 2013.

[237] Phuc Q. Le, Fangyan Dong, and Kaoru Hirota. "A flexible
representation of quantum images for polynomial preparation,
image compression, and processing operations." *Quantum
Information Processing*, vol. 10, 63–84, 2011.

[238] Salvador E. Venegas-Andraca and Sougato Bose. "Storing,
processing, and retrieving an image using quantum mechanics."
SPIE Defense + Commercial Sensing, 2003.

[239] Jose I. Latorre. "Image compression and
entanglement." 2005.

[240] Salvador E. Venegas-Andraca and J. L. Ball. "Processing
images in entangled quantum systems." *Quantum Information
Processing*, vol. 9, 1–11, 2010.

[241] Ulrich Mutze. "Quantum Image Dynamics—an entertainment application of separated quantum dynamics." October 2008.

[242] Phuc Q. Le, Fangyan Dong, A. Yoshinori, and Kaoru Hirota. "Flexible Representation of Quantum Images and Its Computational Complexity Analysis." ISIS 2009, August 2009.

[243] Phuc Q. Le, Abdullah M. Iliyasu, Fangyan Dong, and Kaoru Hirota. *A Flexible Representation and Invertible Transformations for Images on Quantum Computers.* Springer, 2011.

[244] Bo Sun, Abdullah M. Iliyasu, Fei Yan, Fangyan Dong, and Kaoru Hirota. "An RGB Multi-Channel Representation for Images on Quantum Computers." *Journal of Advanced Computational Intelligence and Intelligent Informatics*, vol. 17, 404–417, 2013.

[245] Fei Yan, Abdullah M. Iliyasu, and Salvador E. Venegas-Andraca. "A survey of quantum image representations." *Quantum Information Processing*, vol. 15, 1–35, 2016.

[246] Fei Yan, Abdullah M. Iliyasu, and Phuc Q. Le. "Quantum image processing: A review of advances in its security technologies." *International Journal of Quantum Information*, vol. 15, 2017.

[247] Robert W. Boyd and Peter J. Reynolds. "Introduction to the special issue on quantum imaging." *Quantum Information Processing*, vol. 11, 887–889, 2012.

[248] Lugiato, L., Gatti, A., Brambilla, E., Caspani, L., Trapani, P., Jedrkiewicz, O., Ferri, F., and Magatti, D. "Some topics in Quantum Imaging." *Journal of Physics: Conference Series*, February 2010.

[249] Abdullah M. Iliyasu. "Towards Realising Secure and Efficient Image and Video Processing Applications on Quantum Computers." *Entropy*, vol. 15, 2874–2974, 2013.

[250] Zhang, Y., Lu, K., Gao, Y., and Wang, M. "NEQR: a novel enhanced quantum representation of digital images." *Quantum Information Processing*, vol. 12, August 2013.

[251] Ruan, Y., Chen, H., Tan, J., and Li, X. "Quantum computation for large-scale image classification." *Quantum Information Processing*, vol. 15, 4049–4069, 2016.

[252] Bo Sun, Abdullah M. Iliyasu, Fei Yan, Fangyan Dong, and Kaoru Hirota. "A Multi-Channel Representation for images on quantum computers using the RGBa color space." 2011 IEEE 7th International Symposium on Intelligent Signal Processing.

[253] Jiang, N., Lu, X., Hu, H., Dang, Y., and Cai, Y. "A Novel Quantum Image Compression Method Based on JPEG." *International Journal of Theoretical Physics*, vol. 57, 611-636, 2017.

[254] Quantum Image Processing with Qiskit: `https://qiskit.org/textbook/ch-applications/image-processing-frqi-neqr.html?utm_source=Medium&utm_medium=Social&utm_campaign=Community&utm_content=Textbook-CTA`

[256] TensorNetwork: `https://github.com/google/TensorNetwork`

[257] Ashley Milsted, Martin Ganahl, Stefan Leichenauer, Jack Hidary, and Guifre Vidal. "TensorNetwork on TensorFlow: A Spin Chain Application Using Tree Tensor Networks." 2019.

[258] Roberts, Chase and Milsted, Ashley and Ganahl, Martin and Zalcman, Adam and Fontaine, Bruce and Zou, Yijian and Hidary, Jack and Vidal, Guifre and Leichenauer, Stefan. "TensorNetwork: A Library for Physics and Machine Learning." 2019.

[259] TensorTrace: `https://www.tensortrace.com/`

[260] Roman Orus, Samuel Mugel, and Enrique Lizaso. "Quantum computing for finance: Overview and prospects." *Reviews in Physics*, vol. 4, 2019.

[261] Nikitas Stamatopoulos, Daniel J. Egger, Yue Sun, Christa Zoufal, Raban Iten, Ning Shen, Stefan Woerner. "Option Pricing using Quantum Computers." *Quantum* 4, vol. 291, 2020.

[262] Jeffrey Cohen, Alex Khan, and Clark Alexander. "Portfolio Optimization of 40 Stocks Using the DWave Quantum Annealer." 2020.

[263] Kaneko, K., Miyamoto, K., Takeda, N., and Yoshino, K. "Quantum Pricing with a Smile: Implementation of Local Volatility Model on Quantum Computer." 2020.

[264] DOE: https://www.osti.gov/servlets/purl/1638794

[264] H. J. Kimble. "The quantum internet." *Nature*, vol. 453, 1023–1030, 2008.

[265] Stephanie Wehner, David Elkouss, and Ronald Hanson. "Quantum internet: A vision for the road ahead." *Science*, vol. 362, October 19, 2018.

[266] Charles H. Bennett and Gilles Brassard. "Quantum cryptography: Public key distribution and coin tossing." *Theoretical Computer Science*, vol. 560, 7–11, 2014.

[267] J. I. Cirac, A. K. Ekert, S. F. Huelga, and C. Macchiavello. (1999). "Distributed quantum computation over noisy channels." *Physical Review A*, vol. 59, 4249–4254, June 1, 1999.

[268] Daniel Gottesman, Thomas Jennewein, and Sarah Croke. "Longer-baseline telescopes using quantum repeaters." *Physical Review Letters*, vol. 109, August 16, 2012.

[269] Peter Kómár, Eric M. Kessler, Michael Bishof, Liang Jiang, Anders S. Sørensen, Jun Ye, and Mikhail D. Lukin. "A quantum network of clocks." *Nature Physics*, vol. 10, 582–587, 2013.

[270] Anne Broadbent, Joseph Fitzsimons, and Elham Kashefi. "Universal Blind Quantum Computation." 2009 50th Annual IEEE Symposium on Foundations of Computer Science, 517–526.

[271] Saikat Guha, Tad Hogg, David Fattal, Timothy Spiller, Raymond G. Beausoleil. "Quantum Auctions Using Adiabatic Evolution: The Corrupt Auctioneer and Circuit Implementations." *International Journal of Quantum Information*, vol. 06, 815–839, 2007.

[272] C. Angulo, F. J. Ruiz, L. G. Abril, and J. Ortega. "Multi-Classification by Using Tri-Class SVM." Neural Processing Letters, vol. 23, 89–101, 2005.

[273] Muhammad Hamdan. VHDL auto-generation tool for optimized hardware acceleration of convolutional neural networks on FPGA (VGT). 2018.

[274] Ian Goodfellow, Yoshua Bengio, and Aaron Courville. *Deep Learning*. MIT Press, 2016. https://www.deeplearningbook.org/

[275] Qiskit API documentation: https://qiskit.org/documentation/

[276] Stefan Krastanov, Hamza Raniwala, Jeffrey Holzgrafe, Kurt Jacobs, Marko Lončar, Matthew J. Reagor, and Dirk R. Englund. "Optically-Heralded Entanglement of Superconducting Systems in Quantum Networks." 2020.

[277] Koji Azuma, Stefan Bäuml, Tim Coopmans, David Elkouss, and Boxi Li. "Tools for quantum network design." 2020.

[278] Yin, J., Cao, Y., Li, Y., Liao, S., Zhang, L., Ren, J., Cai, W., Liu, W., Li, B., Dai, H., Li, G., Lu, Q., Gong, Y., Xu, Y., Li, S., Li, F., Yin, Y., Jiang, Z., Li, M., Jia, J., Ren, G., He, D., Zhou, Y., Zhang, X., Wang, N., Chang, X., Zhu, Z., Liu, N., Chen, Y., Lu, C., Shu, R., Peng, C., Wang, J., and Pan, J. "Satellite-based entanglement distribution over 1200 kilometers." *Science*, vol. 356, 1140–1144, 2017.

[279] Ren, J., Xu, P., Yong, H., Zhang, L., Liao, S., Yin, J., Liu, W., Cai, W., Yang, M., Li, L., Yang, K., Han, X., Yao, Y., Li, J., Wu, H., Wan, S., Liu, L., Liu, D., Kuang, Y., He, Z., Shang, P., Guo, C., Zheng, R., Tian, K., Zhu, Z., Liu, N., Lu, C., Shu, R., Chen, Y., Peng, C., Wang, J., and Pan, J. "Ground-to-satellite quantum teleportation." *Nature*, vol. 549, 70–73, 2017.

[280] R. Courtland. "China's 2,000-km quantum link is almost complete [News]." *IEEE Spectrum*, vol. 53, 11–12, 2016.

[281] Santanu Ganguly. "System and Architecture of a Quantum Key Distribution (QKD) Service over SDN," Technical Disclosure Commons, `https://www.tdcommons.org/dpubs_series/2885`

[282] Zhen-Sheng Yuan, Yu-Ao Chen, Bo Zhao, Shuai Chen, Jörg Schmiedmayer, and Jian-Wei Pan. "Experimental demonstration of a BDCZ quantum repeater node." *Nature*, vol. 454, 1098–1101, 2018.

[283] IEEE SDN `https://sdn.ieee.org/outreach/resources`

[284] Hendrik Poulsen Nautrup, Nicolas Delfosse, Vedran Dunjko, Hans J. Briegel, and Nicolai Friis. "Optimizing Quantum Error Correction Codes with Reinforcement Learning." 2018.

[285] Laia Domingo Colomer, Michalis Skotiniotis, and Ramon Muñoz-Tapia. "Reinforcement learning for optimal error correction of toric codes." 2019.

[286] R. Kuang and N. Bettenburg. "Shannon Perfect Secrecy in a Discrete Hilbert Space." 2020 IEEE International Conference on Quantum Computing and Engineering (QCE), 249–255.

[287] R. Kuang and N. Bettenburg. "Quantum Public Key Distribution using Randomized Glauber States." 2020 IEEE International Conference on Quantum Computing and Engineering (QCE), 191–196.

[288] Fock States and Quadrature operators: `https://courses.cit.cornell.edu/ece531/Lectures/handout7.pdf`

[289] E. Brigham. *The Fast Fourier Transform and Its Applications.* Pearson, 1988.

[290] PennyLane Key Concepts: `https://pennylane.ai/qml/glossary.html`

Index

© Santanu Ganguly 2021
S. Ganguly, *Quantum Machine Learning: An Applied Approach*, https://doi.org/10.1007/978-1-4842-7098-1

Printed in the United States
by Baker & Taylor Publisher Services